Teaching Engineering: A Beginner's Guide

OTHER IEEE PRESS BOOKS

Robust Control, *Edited by P. Dorato*
Writing Reports to Get Results: Guidelines for the Computer Age, *By R. S. Blicq*
Multi-Microprocessors, *Edited by A. Gupta*
Advanced Microprocessors, II, *Edited by A. Gupta*
Adaptive Signal Processing, *Edited by L. H. Sibul*
System Design for Human Interaction, *Edited by A. P. Sage*
Microcomputer Control of Power Electronics and Drives, *Edited by B. K. Bose*
Advances in Local Area Networks, *Edited by K. Kümmerle, J. O. Limb, and F. A. Tobagi*
Load Management, *Edited by S. Talukdar and C. W. Gellings*
Computers and Manufacturing Productivity, *Edited by R. K. Jurgen*
Being the Boss, *By L. K. Lineback*
Effective Meetings for Busy People, *By W. T. Carnes*
VLSI Signal Processing, II, *Edited by S. Y. Kung, R. E. Owen, and J. G. Nash*
Modern Acoustical Imaging, *Edited by H. Lee and G. Wade*
Low-Temperature Electronics, *Edited by R. K. Kirschman*
Undersea Lightwave Communications, *Edited by P. K. Runge and P. R. Trischitta*
Multidimensional Digital Signal Processing, *Edited by the IEEE Multidimensional Signal Processing Committee*
Adaptive Methods for Control System Design, *Edited by M. M. Gupta*
Residue Number System Arithmetic, *Edited by M. A. Soderstrand, W. K. Jenkins, G. A. Jullien, and F. J. Taylor*
Singular Perturbations in Systems and Control, *Edited by P. V. Kokotovic and H. K. Khalil*
Getting the Picture, *By S. B. Weinstein*
Space Science and Applications, *Edited by J. H. McElroy*
Medical Applications of Microwave Imaging, *Edited by L. Larsen and J. H. Jacobi*
Modern Spectrum Analysis, *Edited by S. B. Kesler*
The Calculus Tutoring Book, *By C. Ash and R. Ash*
Imaging Technology, *Edited by H. Lee and G. Wade*
Phase-Locked Loops, *Edited by W. C. Lindsey and C. M. Chie*
VLSI Circuit Layout: Theory and Design, *Edited by T. C. Hu and E. S. Kuh*
Monolithic Microwave Integrated Circuits, *Edited by R. A. Pucel*
Next-Generation Computers, *Edited by E. A. Torrero*
Kalman Filtering: Theory and Application, *Edited by H. W. Sorenson*
Spectrum Management and Engineering, *Edited by F. Matos*
Digital VLSI Systems, *Edited by M. I. Elmasry*
Introduction to Magnetic Recording, *Edited by R. M. White*
Insights into Personal Computers, *Edited by A. Gupta and H. D. Toong*
Television Technology Today, *Edited by T. S. Rzeszewski*
The Space Station: An Idea Whose Time Has Come, *Edited by T. R. Simpson*
Marketing Technical Ideas and Products Successfully! *Edited by L. K. Moore and D. L. Plung*
The Making of a Profession: A Century of Electrical Engineering in America, *By A. M. McMahon*
Power Transistors: Device Design and Applications, *Edited by B. J. Baliga and D. Y. Chen*
VLSI: Technology and Design, *Edited by O. G. Folberth and W. D. Grobman*
General and Industrial Management, *By H. Fayol; revised by I. Gray*
A Century of Honors, *An IEEE Centennial Directory*
MOS Switched-Capacitor Filters: Analysis and Design, *Edited by G. S. Moschytz*
Distributed Computing: Concepts and Implementations, *Edited by P. L. McEntire, J. G. O'Reilly, and R. E. Larson*
Engineers and Electrons, *By J. D. Ryder and D. G. Fink*
Land-Mobile Communications Engineering, *Edited by D. Bodson, G. F. McClure, and S. R. McConoughey*
Frequency Stability: Fundamentals and Measurement, *Edited by V. F. Kroupa*
Electronic Displays, *Edited by H. I. Refioglu*
Spread-Spectrum Communications, *Edited by C. E. Cook, F. W. Ellersick, L. B. Milstein, and D. L. Schilling*
Color Television, *Edited by T. Rzeszewski*
Advanced Microprocessors, *Edited by A. Gupta and H. D. Toong*
Biological Effects of Electromagnetic Radiation, *Edited by J. M. Osepchuk*
Engineering Contributions to Biophysical Electrocardiography, *Edited by T. C. Pilkington and R. Plonsey*
The World of Large Scale Systems, *Edited by J. D. Palmer and R. Saeks*
Electronic Switching: Central Office Systems of the World, *Edited by A. E. Joel, Jr.*
A Guide for Writing Better Technical Papers, *Edited by C. Harkins and D. L. Plung*

Teaching Engineering: A Beginner's Guide

Edited by
Madhu S. Gupta
Professor of Electrical Engineering
University of Illinois at Chicago

A volume in the IEEE PRESS Selected Reprint Series, published under the sponsorship of the IEEE Education Society

Published with the cooperation of the American Society for Engineering Education

IEEE PRESS

The Institute of Electrical and Electronics Engineers, Inc., New York

PRINTED IN THE UNITED STATES OF AMERICA

IEEE Order Number: PC02238

Library of Congress Cataloging-in-Publication Data

Teaching engineering.

(IEEE Press selected reprint series)
Includes bibliographies and indexes.
1. Engineering—Study and teaching. I. Gupta, Madhu, S. II. Series.
T65.T36 1987 620'.007'11 87-22543

ISBN 0-87942-234-3

Contents

Preface

I. Purpose of This Volume

THIS volume is intended for the engineering faculty members, and the teaching fellows and teaching assistants in the engineering departments of a university, who are new to the task of teaching. Its primary purpose is to make available to the new teachers of engineering a source of guidance and advice on the teaching of engineering, and to stimulate them to reflect on issues related to their teaching activities. Others, such as practicing engineers who may be involved in some teaching, and engineering graduates or students who are considering a teaching career, may also find this book helpful. It should also be useful as a source of background reading for short courses or seminars that are sometimes organized in an engineering college to help initiate new faculty and teaching assistants in the task of teaching.

This is not a "manual" on teaching, which provides rules or step-by-step procedures like a cookbook; that would be denegrating, depressing, and ineffective. Teaching engineering at the college level does not appear to be amenable to recipes and style manuals. Instead, this volume is based on the philosophy that all teaching can be improved, that all teachers are desirous of improving the effectiveness of their teaching, that beginning teachers especially need help in this task, and that one possible form of that help is a sourcebook such as this.

This volume is a compilation of approximately forty papers by various authors, published in a number of sources during the past few decades. This allows the reader to benefit from the collective wisdom of a number of authors accumulated over the years, and from the resulting variety of viewpoints. It has also allowed the scope of this volume to be considerably broader than the interest and expertise limitations of a single author would permit. The papers reprinted here are devoted to a variety of topics relevant to engineering education, and they share some qualities, to be mentioned shortly, which make them particularly valuable to those who are beginning a career of teaching engineering.

II. The Need for This Compilation

There are four reasons why a need for this compilation arises:

(1) Lack of organized training in teaching. University and college teaching has the distinction of being almost the only profession in which the practitioners are not required to undergo any specific preparation. The credentials required for teaching (usually a doctorate degree) are the same as those for non-teaching positions, and attest only to the subject matter expertise of the candidate, since training in teaching is seldom a necessary part of the doctoral education in engineering. Any training that takes place is on-the-job, incidental, and unplanned. A beginning engineering teacher is therefore a content specialist who may or may not have had any training in teaching. Such a teacher often finds himself in his first job with little preparation for the process of teaching, and little appreciation for the intricacies of the task. The consequences are often unsatisfactory, and sometimes disastrous. A beginning teacher usually has much to learn about matters such as effective oral presentation, alternative modes of delivery, selection and use of teaching aids and media, preparation of instructional objectives, examinations, and grades, and a host of other skills required in a teaching career. A background in instructional matters, to which this volume is devoted, should be an excellent complement to the subject matter expertise of the new teacher [1].

(2) The versatile role of a book. A new teacher may be able to draw on a number of resources in a university, such as workshops and seminars on teaching, teaching improvement clinics, classroom visitations by senior faculty or by expert counselors, teaching specialists who serve as consultants, and the services of "instructional development" offices on their campus. Each of these forms of help provided to beginning teachers has a different role, costs, and limitations. A book for the guidance of beginning teachers complements these roles in many ways: it can be consulted in the privacy of one's room; it is non-threatening and non-judgmental; it stays with the teacher long after the seminar is over; it can be used at one's convenience, or as and when the need arises; it can be more comprehensive in coverage; and it is particularly suitable in the role of a device for raising awareness of ideas. Admittedly, that is also the limitation of the book: it can do little more than raise the awareness. It will not supply the beginning teacher with feedback about his teaching, such as a videotape replay of his classroom would, nor the encouragement that a senior colleague could. The various ways of learning the art of teaching are, however, complementary, and a book in conjunction with other methods is better than either method alone. For example, a book such as this one can be a useful companion in a seminar on learning and teaching; it can serve as a source of background reading so that all seminar participants are sufficiently and equally informed in advance, and the seminar time is used more efficiently, rather than in verbally supplying the basic information to the lesser-informed members of the seminar group.

(3) The different needs of beginning teachers. The professional literature on the subject of engineering education is extensive, and extends back about a century [2]. There are journals conference proceedings, and books devoted exclusively to engineering education. In addition, there are books, and occasional articles and lectures in many other engineering journals and conferences. Finally, at least a small part of the

sizable literature of the field of higher education is relevant to engineering education. Why then should it be necessary to have another volume on engineering education? The answer is that the needs of the beginning teachers are different from those of their senior colleagues. A vast majority of this literature is of little immediate practical utility to new teachers. (This observation may be related to the fact that junior faculty are under-represented in the membership of professional societies devoted to engineering education [3]). While the senior educators may ponder such weighty matters as enrollments, co-op arrangements, engineering manpower projections, minority retention, and extension programs, with which a majority of the literature of engineering education deals, the beginning teacher is just beginning to grapple with the bread-and-butter issues of lectures, homeworks, tests, grading, and tenure requirements. This is not meant in a pejorative sense, as if the beginning teachers are any less capable, concerned, or competent. It is just that the there is a difference in what is most important at the time. A volume suitable for their needs should therefore deal with the nuts-and-bolts issues of common interest to beginning teachers of engineering, particularly those issues related to the improvement in the quality of teaching.

(4) Time pressures on engineering teachers. There are many demands on the time of an educator, and especially a new faculty member in a university. His or her activities will include research, publication, grant acquisition, consulting, administrative duties, committee meetings, professional activities, service activities, and, finally, teaching. As a result, there is little time available to reflect on teaching, to experiment with teaching, and to attempt to improve it. Given the time pressures, a beginning teacher can soon learn to get by with a minimal commitment of time to the preparation for teaching, a habit which can last a lifetime. Furthermore, it is often helpful to have someone to discuss teaching related issues with, particularly someone with an extensive experience of teaching. However, the same time pressures prevent the senior teachers from serving as mentors for their junior colleagues, and counseling or assisting them in this professional growth. Even when such a mentor is available, there may be a hesitation on the part of the new teacher in seeking such help, due to the perception that the mentor may be judgmental, busy, or critical. The literature on teaching can serve at least some of the functions of the mentor, such as introducing sound principles and efficient procedures, and providing ideas and suggestions. However, the large volume of literature is an inconvenience to a beginning teacher because finding a discussion of a specific topic in a large volume of literature can be very time consuming and is therefore discouraging. It is clear that an aid to learning about teaching and improving instruction will be more useful if it does not require a large initial investment of time. Short articles in the literature on the subject of teaching seem to be one such form of aid. Moreover, it would be helpful to have available in one place a compilation of such helpful printed material on topics of interest to beginning engineering teachers. This book is intended to fill this proverbial ''gap.''

III. Scope and Content of This Volume

It appears (to the editor) that there are four major areas in which a beginning teacher will find it helpful to have some knowledge and orientation:

(1) Content Management. This involves establishing the goals of teaching, developing course objectives and content, and ensuring that the teaching activities are valid, i.e., they serve a desirable purpose. Clearly, this is an area in which the subject matter expertise of the teacher is also important, since the objectives reflect the nature of the discipline to be taught.

(2) Learning Principles. Instruction must be based on knowledge of how people learn, so that the instructor can establish the conditions conducive to learning. This requires being aware of the learners' characteristics, motivations, expectations, needs, learning styles, and viewpoint, all of which influence the success of learning.

(3) Teaching Skills. Some of these skills, such as laboratory operation, test construction, and grade assignment, are procedural, while others, such as effective oral presentation, are sometimes called the ''art'' of teaching.

(4) Professional Development. Many aspects of the academic environment of the teachers, such as career prospects, other concurrent teaching-related or non-teaching responsibilities, and expectations, are important because they interact with his teaching, influence his effectiveness, and sometimes create a ''hidden agenda'' to which the teaching activities must adapt.

The above considerations led to the choice of topics like learning theory, syllabus construction, teaching methods, laboratory instruction, examinations, grade assignment, and teaching evaluation for this volume. In addition, a historical introduction is included to provide a perspective for understanding the profession and its role. This book is organized in ten parts, dealing with the following topics: (I) History of Engineering Education, (II) Overall Goals of Engineering Education, (III) Instructional Objectives and Syllabus, (IV) The Learning Process and Learner Characteristics, (V) Teaching Methods and Strategies, (VI) Laboratory Instruction, (VII) Examinations and Grading, (VIII) Evaluation of Teaching, (IX) Other Academic Activities, and (X) The Teaching Profession and Career Development.

To become familiar with a new territory, one should know the principal features of the terrain, have a road map, and actually visit some of the prominent landmarks. Each part in this volume consists of three components—introductory editorial remarks, an annotated bibliography, and a selection of reprinted papers—which serve somewhat analogous functions. The editorial remarks state the scope and purpose of the part, supply some definitions, point out the significant issues of interest, and put the subject matter of the reprinted papers in perspective. The annotated bibliography is intended to help the reader further explore some of those issues himself. Then there are some reprinted papers as a sample of the useful writings available in the literature.

IV. The Selection of Reprinted Papers

Given the size of the literature on engineering education, one would expect a plethora of articles that might be of interest to beginning teachers. What kind of articles should be included in this volume? In view of the goals of the volume, the selection of papers was based on the following three criteria:

(1) Practical Applicability. The primary requirement was that the papers should have a relevance to actual teaching or a long-term value to the teacher. This criterion excludes the numerous philosophical discussions, opinion papers, and discussions of the societal or administrative matters, such as student recruitment, enrollments, and cooperative and continuing educational programs, which account for a bulk of the literature on engineering education. Reports of individual educational experiments were also excluded. Despite their importance, such reports are of limited utility to beginning teachers, because they are difficult to critically evaluate for validity, transferability, and effectiveness, except by experts [4]. They are not as useful to the new teacher as a survey of many such experiments, along with conclusions and recommendations stated in practical, operational terms.

(2) Discipline Independence. A second requirement was that the papers should be useful to a wide cross-section of engineering teachers, regardless of their fields of speciality or subject matter of interest. This requirement excludes the many papers on subject-specific teaching materials, such as a new approach, a new proof, a new laboratory instrument, a new illustrative example or problem for introducing a topic, and tutorial material on a newly emerging topic, that a teacher may find useful for teaching a particular topic.

(3) Simplicity. Papers that are directly useful in themselves, without a great deal of effort in implementation or supplementary reading, were considered the most desirable. For this reason, papers on major undertakings, such as in educational technology or computer-aided instruction, were not included. Similarly, short, succinct articles were favored, because the teachers have only a limited amount of time to read about teaching, and can be easily discouraged by lengthy treatises. Moreover, the articles addressed to beginners were preferred over those written for experts, because they tend to assume less background preparation and use less technical jargon. As a result, for nearly any topic included in this volume, it is possible to find a more complete, advanced, or careful treatment elsewhere. Indeed, entire books are devoted to many of the topics represented here in a few pages.

Essentially similar criteria were used to select the papers included in the annotated bibliographies appended with each part of the volume, with the one additional requirement that the articles should be easily accessible. Given human nature, difficulty of access is a significant discouragement in the actual use of any form of information. For this reason, papers in major journals were favored over those in limited circulation newsletters and magazines, or conference papers "available from the author."

V. Acknowledgments

I would like to acknowledge two kinds of assistance I received that led to my involvement in, and preparation of, this book. The first is the "indirect" form of assistance over the years that helped me develop my thoughts and clarify my thinking about engineering education, which in turn led me to compile this book. The second is the "direct" form of assistance received during the compilation of this volume.

The "indirect" assistance came in the form of my own participation in teaching-related programs from the following organizations and individuals: (1) The Center for Learning and Teaching at the University of Michigan, which organized workshops for faculty and teaching assistants, and served to greatly enhance my awareness while I was a Teaching Fellow at the University of Michigan; (2) The Lilly Foundation of Indianapolis, IN, which funded a program of fellowships for young faculty at the Division for Study and Research in Education at M.I.T., and as a holder of the fellowship, allowed me to think about teaching-related issues and engage in intense weekly interaction with the other Lilly Fellows who, like myself, were beginning to grapple with educational issues and had an interest in them; Professor Roy Kaplow, and late Professor Thomas F. Jones, who organized and guided the Lilly Fellows program; and two colleagues and mentors, Professor Richard B. Adler and Professor Stephen D. Senturia, who helped shape my ideas about the teaching process and profession during the early years of my career as a faculty member at M.I.T.

The direct assistance in the compilation of this volume was provided by an advisory committee, set up by the then-President of the IEEE Education Society, Professor Jim Rowland, to review the compiled material and make recommendations. The members of this committee were:

(1) Professor Bruce A. Eisenstein
Department of Electrical & Computer Engineering
Drexel University, Philadelphia, PA.

(2) Professor Edward W. Ernst
Department of Electrical Engineering and Computer Science, University of Illinois at Urbana-Champaign, Urbana, IL.

(3) Professor Edwin C. Jones, Jr.
Department of Electrical Engineering, Iowa State University, Ames, IA.

I, however, remain responsible for the opinions expressed, and for the biases and omissions in the selection of topics and papers appearing in this volume.

References

[1] O. Lancaster, "Do We Believe in Teaching," *Engineering Education,* vol. 68, no. 3, p. 226, Dec. 1977.

[2] *Proceedings of the Society for the Promotion of Engineering Education,* vol. 1, 1894-present. (This society later became the American Society for Engineering Education.)

[3] D. G. Ullman, "Maintaining a First-rate Faculty: The Base Data," *Engineering Education,* vol. 71, no. 7, pp. 700-708, Apr. 1981.

[4] M. S. Gupta, "Evaluation of New Educational Programs," *IEEE Trans. Education,* vol. E-17, no. 3, pp. 140-142, Aug. 1974.

Part I
History of Engineering Education

Scope and Purpose

THE purpose of this part is to help the reader develop an awareness of the history of engineering education. This history is a part of the cultural aspects of the educator's profession, and is helpful in understanding both the reasons for the present and the directions for the future.

The Issues

The following are some of the issues that a historical study of engineering education brings out, and that the reader may wish to explore while pursuing the subject.

- What was the origin of the engineering colleges and engineering curricula?
- How did engineering curricula evolve to their present state?
- How have the economic climate and industrial innovation affected the development of engineering curricula?
- How did the traditional nature of the universities affect the development of engineering education?
- How did engineering programs within large universities differ from those in colleges started primarily for teaching engineering?
- What factors allowed some educational institutions to develop outstanding teaching programs, or perform superior service to industry?
- Why were some educational innovations dependent on single individuals or events at one institution, and others were ''inevitable'' in that they arose at many institutions almost simultaneously and independently?
- What role did engineering societies play in the growth of engineering programs on campus?
- How do curricula and colleges cope with the growth of technological knowledge, or increasing specialization, or the high cost of initiating new disciplines on campus?
- How does the cyclic nature of the demand for trained engineers in industry influence academia?

The Reprinted Papers

Two papers are reprinted in this part:

1. ''A Brief History of Engineering Education in the United States,'' L. P. Grayson, *Engineering Education*, vol. 68, no. 3, pp. 246–264, Dec. 1977; Reprinted in *IEEE Trans. Aerospace and Electronic Systems*, vol. AES-16, no. 3, pp. 373–391, May 1980.
2. ''A Brief History of Electrical Engineering Education,'' F. E. Terman, *Proc. IEEE*, vol. 64, no. 9, pp. 1399–1407, Sept. 1976.

Grayson's article is an extensive account of the major events in the development of engineering programs in U.S. universities. It relates the development of engineering education to the contemporaneous growth of the population and the needs of the people. It also identifies the role of some of the individuals who played a major part in the course of development.

Terman's history of engineering education is confined to the field of electrical engineering. It begins with a listing of the significant electrical discoveries in the later half of the nineteenth century, and gives some details of the early curricula. Terman's article examines the data on the graduate degrees granted at a few of the larger institutions over a period of several decades, and attempts to relate this to factors such as government sponsored research and the need for graduate degrees in the career paths of the engineers.

The Reading Resources

(The following list is restricted to the history of engineering education, although this subject requires at least some knowledge of the history of the corresponding engineering disciplines, a topic on which vast amounts of literature is more readily available.)

[1] D. F. Berth, ''The Story of Engineering Education,'' *IEEE Student Jour.,* vol. 3, no. 5, pp. 30–36, Sept. 1965. (Reprinted from *The Cornell Engineer,* Mar. 1965.) Traces the origin of the profession and of engineering education in the U.S. in very early stages, with some details of the educational programs available in electrical engineering in the nineteenth century.

[2] H. H. Skilling, ''Historical Perspective for Electrical Engineering Education,'' *Proc. IEEE,* vol. 59, no. 6, pp. 828–833, June 1971. Highlights, rather than an extensive description, of the history of engineering education, and then of electrical engineering education, with brief mentions of the major contemporary electrical discoveries along the way.

[3] R. Rosenberg, ''The Origin of EE Education: A Matter of Degree,'' *IEEE Spectrum,* vol. 21, no. 7, pp. 60–68, July 1984. Detailed examination of the status of electrical engineering education in the period 1880 to 1900, with emphasis on individuals and their roles.

[4] J. D. Ryder, ''The Way It Was,'' *IEEE Spectrum (Special Issue on Education),* vol. 21, no. 11, pp. 39–43, Nov. 1984. An informal and personal account of the development of electrical engineering education in the years between World Wars I and II.

[5] J. M. Pettit and J. M. Gere, ''Evolution of Graduate Education in Engineering,'' *Jour. Engineering Education,* vol. 54, no. 2, pp. 57–62, 1963. Historical perspective and data on graduate education in engineering in the United States. Still further data by the same authors and in the same journal is in vol. 55, no. 5, pp. 164–171, Jan. 1965.

[6] E. Weber, ''Graduate Study in Electrical Engineering,'' *Proc. IRE,* vol. 50, no. 5, pp. 960–966, May 1962. Remarks on the evolution of the objectives and patterns of graduate

education in electrical engineering, with emphasis on historical perspective.

[7] W. H. Hartwig, "An Historical Analysis of Engineering College Research and Degree Programs as Dynamic Systems," *Proc. IEEE,* vol. 66, no. 8, pp. 829–837, Aug. 1978. Quantitative analysis of several parameters (degrees granted, residence time, research budget, etc.) over two decades for the largest 20 or 25 engineering colleges in the United States.

[8] J. D. Kemper, "The Evolution of Graduate Engineering Education," *Engineering Education,* vol. 68, no. 2, pp. 176–180, Nov. 1977. Narrative and quantitative description of the five major events during the 35 year period (from 1940 to 1975) witnessing the growth of graduate level engineering education in the U.S.

[9] G. Kennedy, "Independent Consulting Engineering," *IEE Proc. (U.K.),* Part A, vol. 131, no. 6, pp. 343–354, Aug. 1984. Traces the industrial needs from the mid-1600s, through the formation of consulting engineering firms in 1800s, to the future prospects for consulting engineers.

[10] J. G. McGivern, *First Hundred Years of Engineering Education in the United States (1807–1907).* Gonzago Univ. Press, Spokane, WA, 1960. Chronological history of the 100 year period divided into six stages.

[11] G. S. Emerson, *Engineering Education: A Social History.* Crane Russak, New York, 1973. Examines the origin of technical education from the 17th century, as it occurred in various nations, influenced by their institutions and national character.

A Brief History of Engineering Education in the United States

LAWRENCE P. GRAYSON, Fellow, IEEE
National Institute of Education

Editor's Note

"Tutorial in nature" was the major recommendation that we received with the attached article. It is one of the very few that we ever have (or will) reprint from any source other than our own limited-distribution conference records. Careful thought was given to the expenditure of space; all reviewers were unanimous in their recommendation that we seek to reprint it. Our thanks to ASEE for their permission to bring it to you.

Abstract

A history of engineering education in the United States is described, beginning with General Washington's general order in 1778 calling for the establishment of a school of engineering and concluding with a discussion of the general trends in engineering education today. Emphasis is placed on the needs of a continually growing nation and on the ways in which engineering education has responded to these needs.

Manuscript received Nov. 15, 1979.

Author's address: National Institute of Education, Washington, D.C. 20208.

Engineering education in the United States began in the early 1800s for the purpose of promoting "the application of science to the common purposes of life" [1] and, indeed, its entire history parallels the changing needs of a growing nation for scientifically and technically trained manpower. It developed from a combination of borrowed and indigenous elements, which in many ways reflected the character of the country. The collegiate plan of organization and most of the traditions of the professoriate can be traced back to the older universities of England, which were reflected in the early arts colleges of the United States; engineering curricula originally were derived from French models, as was the concept of a professional school of engineering. Early approaches to teaching of shop and manual arts were based on schooling in Russia, while the emphasis and methods of research and the model for graduate education came from Germany.

Although early developments in the United States were more adaptive than creative, in at least three respects American schools have exercised conspicuous leadership. They pioneered in the introduction of laboratory instruction for the individual student as an integral part of the curriculum in the provision of distinctive training in the economic and management phases of engineering and in the development of cooperative education.

Almost from its beginnings, engineering education in the United States was in all essential aspects a form of collegiate education, instituted and directed by educators, rather than by practitioners. It was firmly established before the profession organized itself, with curricula in the various branches of engineering being developed and taught, and degrees offered before the corresponding professional societies were formed. As a result, engineering education did not evolve from apprenticeship training and only slowly replaced it, gaining the support of practitioners with considerable struggle. Even after the professional societies were formed, they initially did not assume any responsibility for creating policy or for providing guidance or control of preparatory professional education. This was directly opposite to the manner in which education for the legal, medical, and dental professions developed in the United States, as they evolved out of apprenticeship on a purely practical and technical plane, with none of the general qualities of collegiate education. These early professional schools were usually founded as institutions independent of colleges and universities, with their corresponding professional societies assuming from their beginnings a strong concern for the educational process [2]. Many of them were profit-making ventures, a form never assumed by engineering schools, and were not endowed by private funds or supported by the state. They were organized by members of their respective professions, with the curricula determined by each

Reprinted from *IEEE Trans. Aerosp. Electron. Syst.*, vol. AES-16, no. 3, pp. 373–391, May 1980.

Fig. 1. Portraits of prominent early engineering educators. (A) Jean Rodolphe Perronet (1708-1794). He established the first formal engineering school in the world in 1747, later to be named the Ecole Nationale des Ponts et Chaussées under King Louis XV of France. (B) Alden Partridge (1785-1854), first professor of engineering in the United States (appointed 1813). He founded the American Literary, Scientific and Military Academy, which later became Norwich University as the first civilian school of engineering in the United States. (C) Sylvanus Thayer (1875-1872), military engineer and educator. As superintendent of the United States Military Academy at West Point from 1817 to 1833, he reorganized the curriculum stressing honor, discipline, and achievement. He is often referred to as the "Father of Military Engineering"; also he endowed and was the guiding influence in establishing the Thayer School of Engineering at Dartmouth College. (D) Benjamin Franklin Greene (1817-1895), senior professor and first director of the Rensselaer Institute, 1847-1859. He developed the school from one of applied natural science into a polytechnic institute.

profession to meet its specific needs [3]. These differences in the ways in which education for the various professions began are still evident in the structure of higher education today.

The Beginning: 1862 and Before

The high value placed on individual achievement in the United States, as well as the British heritage of the Colonists, played a significant part in determining the role of education as preparation of the practice of engineering in early America. British, as well as American, industry in the eighteenth century and before was largely due to the inventiveness of practical and enterprising men. Education for the trades and professions in England from the time of Elizabeth to 1814 was, by law, conducted through a system of apprenticeship. This was in sharp contrast to the situation in France, in which highly developed engineering schools, with curricula based heavily on

Fig. 2. Polytechnic College of the State of Pennsylvania (1853-late 1880s), which offered the first regular curricula and granted the first degrees in mining engineering and in mechanical engineering in the United States.

mathematics, largely determined entry into engineering practice.

The Ecole Nationale de Ponts et Chaussées generally is considered to be the first formal school of engineering in the world, although a number of schools taught technical subjects prior to that time. It was founded in 1747, when King Louis XV appointed Jean Rodolphe Perronet chief engineer of bridges and highways, and gave him the authority to establish a school within the Corps de Ponts et Chaussées. Perronet, who began the school with a three-year course of study, has been referred to as "the father of engineering education"[4]. France continued its leadership in engineering education through the establishment of seveal additional schools in the late establishment of several additional schools in the late them was the Ecole des Travaux Publics, which Napoleon created in 1794 to train engineeers for public and private service in order to compensate for the poor technical training that resulted from the disorganization which followed the French Revolution [5]. This school, which later became the Ecole Polytechnique, served as the model for several of the early engineering schools in the United States.

Most colleges founded in America, from the prerevolutionary period until the mid-nineteenth century, had as their primary aim to promote some version of the Christian religion, both by providing leaders for the sponsoring religious denomination and by instructing all undergraduates in the conventional moral and religious beliefs. These colleges built their curricula around the classical subjects of Latin and Greek, history, philosphy, and religion as practiced by the founding demoninations. With strong traditions and the sense of elitism that grew around them, these institutions did not provide fertile ground for changes to occur. Yet this was the institutional climate of higher education in America when engineering education was begun. When the United States gained its independence in 1776, it had a population of about 2½ million people, who lived in a series of settlements located across 13 states. This was a pioneer age, when

Fig. 3. Sketch of West Point cadets being examined by Board of Visitors, about 1869.

Fig. 4. West Point class engaged in field exercise in military engineering, nineteenth century.

men prized versatility and practical resourcefulness above specialized training and the refinements of science. Inventions multiplied and industries developed in the hands of ingeneous individuals, while the practice of any occupation at that time, including law and medicine, was the prerogative of anyone who wanted to try. A handful of engineers trained abroad and a few self-taught Americans built the public works and satisfied the military requirements that were then needed.

As a result of the break with England and the nonimportation agreement of 1774, a number of the early colonial leaders realized the need for trained engineers to satisfy the needs of the country for the development of manufacturing, public works, and military fortifications. As early as 1778, General George Washington, while encamped at Valley Forge, issued a general order calling for the establishment of a school of engineering [6]. Washington was supported in this effort by men such as Henry Knox, Alexander Hamilton, John Adams, Thomas Jefferson, James Monroe, and Louis Duportail (who served as chief engineer in the Continental Army) [7].

The concerns of the American leaders resulted in Congress, in 1794, authorizing President Washington to raise a Corps of Artillerists and Engineers to be educated and stationed at West Point, New York. The school, whose initial curriculum included primarily subjects of a military nature, was destroyed by fire in 1796 and was not reopened until 1801. A number of the policitical leaders of the day argued that there was need for broader training, as engineers were required to serve the needs of public works, as well as of the military. James McHenry, the Secretary of War, for example, stated in 1800, "We must not conclude that service of the engineer is limited to constructing fortifications. This is but a single branch of the profession; their utility extends to almost every department of war; besides embracing whatever respects public buildings, roads, bridges, canals and all such works of a civil nature" [8].

These arguments had their effect, and on March 16, 1802, Congress established the United States Military Academy at West Point, with the provision that the engineers and cadets of the Academy were to be available for such duty and on such service as the President of the United States should direct [9]. This made the Army engineers available for performing tasks of a public as well as military nature. Although originally loosely organized and operating on meager resources, the Academy initially had no definite or consistent system of instruction or examination [10]. It was only as a result of the success of its graduates in the War of 1812 that the academy was allowed to develop a sound organizational structure, under the direction of Colonel Sylvanus Thayer, who was appointed Superintendent in 1817, following a trip to France to study its educational system. Thayer arranged the cadets into four annual classes, divided the classes into sections requiring weekly reports, developed a scale for marking, attached weights to the subjects in the curriculum necessary for graduation, instituted a system of discipline, and set a standard of high achievement. These characteristics have remained with the Academy until the present and have formed the pattern for technical education in America. The program he developed, which was largely influenced by the Ecole Polytechnique, even to adopting many of the texts used at the French institution, was based on a civil engineering curriculum with work in the design and construction of bridges, roads, canals, and railroads [11]. Although no degrees were granted until 1933 [12], West Point is considered to be the first engineering school in America. Of the civil engineering graduates engaged in public works prior to 1840, a sizable fraction were West Point graduates, with at least 30 percent of them serving as chief engineers of important projects on railways, canals, or other nonmilitary activities [13].

Alden Partridge was among the early West Point graduates, receiving a commission as first lieutenant of engineers in 1806. He was assigned to the Academy as an instructor, being promoted to captain in 1810, and then being appointed professor of engineering on September 1, 1813 [14], apparently becoming the first person to hold this title in the United States. After serving as acting superintendent of the Academy from 1815 until Thayer's appointment in 1817, he resigned from the Army and, in 1819, established the American Literary, Scientific, and Military Academy at Norwich, Vermont. The institution was patterned after West Point, with the exception that unlike the changes

Fig. 5. Electrical engineering laboratory at Stevens Institute of Technology around 1889.

instituted at the latter academy by Sylvanus Thayer, Partridge did not require a uniform course of study for all students or a uniform length of time to complete the studies [15]. The school's first catalog in 1821 states that, in addition to a large number of military subjects, instruction was offered in "Civil Engineering, including the construction of Roads, Canals, Locks, and Bridges," making this the first civilian school of engineering in the country. In 1834, the institution became Norwich University, awarding its first engineering degrees in 1837, when two students received the Master of Civil Engineering after completing a three-year course of study [16].

The pragmatism of American society was reflected in all of the early schools of technical education in this country. In 1822 Robert H. Gardiner established the Gardiner Lyceum in Maine as a school of practical science. One of its purposes was to provide a curriculum preparatory to the higher study of agriculture, mechanics, arts, and engineering for young men of the laboring class. Its two-year course of study, later expanded to three years, included, among other courses, instruction in mathematics surveying, navigation, natural philosophy, and astronomy, with civil engineering offered as an elective. In many ways, its features anticipated the important aspects of the Morrill Land Grant Act of 1862, including state financial support, short courses for those not enrolled full time, and an experimental farm operated by the students, as well as a curriculum with a uniform freshman year [17]. Due to financial difficulties, the school was closed in 1832.

Stephen Van Rensselaer, a leading public figure of his day, established the Rensselaer School in 1824 at Troy, New York, "to qualify teachers for instructing the sons and daughters of farmers and mechanics, by lectures or otherwise, in the application of experimental chemistry, philosophy, and natural history, to agriculture, domestic economy, the arts and manufacturers" [18].

Amos Eaton, a lawyer, geologist, botonist, and educational pioneer, was the school's first director and dominant influence from its founding until his death in 1842. While Partridge's academy clearly was an engineering school from its beginning, Rensselaer offered only a general technical course of one year's duration during its first decade [19]. Instruction in engineering was introduced into the curriculum in gradual stages. In 1825, courses included "land surveying, mensuration, measurements of the flow of water in rivers and aqueducts," while the term "civil engineer" first appeared in the catalog of 1828 [20]. The New York State Legislature in May 1835 authorized Rensselaer Institute, as it was then known, to establish a department of mathematical arts for instruction in "engineering and technology," which led to a newly authorized degree of Civil Engineer being granted for the first time in 1835 to a class of four. No similar degree was granted in America or Britain at this time [21]. Thus by 1835 the school had acquired a new engineering orientation which would dominate and obscure its original purpose [22].

In 1835 the territorial growth of the continental United States was completed, when James Gadsden purchased from Mexico a strip of land that occupies the present southwestern part of the country. The previous 50 years had been marked by the acquisition of a vast amount of land and expansion of the borders of the United States. Now the time was right to settle and develop that land. The movement of people and resources to the western part of the country depended upon the development of transportation and communications and required engineering skills to a far greater degree than existed before.

B. Franklin Greene succeeded Eaton as director of the Rensselear Institute in 1846 and proceeded to subject the problems of education for the technical profession to thorough investigation and analysis. He conceived of a polytechnic institute providing "the most complete realization of true educational culture" [23], an idea that received major emphasis in the United States after World War II. In 1849 the school was reorganized along the lines of the leading French technical schools, being particularly influenced by the programs at the Ecole Polytechnique and the Ecole Centrale des Arts et Manufactures. Greene began referring to the school as the Rensselear Polytechnic Institute, a name which did not become official until 1861. Its purpose became "the education of the Architect, the Civil Engineer, Constructing and Superintendent Engineers of Mechanics, Hydraulic works, Gas Works, Iron Works, etc., and Superintendents of those higher manufacturing operations, requiring for their successful prosecution strict consideration of the scientific principles involved in this respective processes." [24] A curriculum was developed to carry out the new aims, with the distinguishing feature of a parallel sequence of

Fig. 6. Lecture hall at Stevens Institute of Technology, around 1889; Dr. Henry Morton, first president of Stevens is on stage; note projector, in front of stage, and the large amount of demonstration equipment.

Fig. 7. Surveying class at the University of Illinois, 1890.

humanistic studies, mathematics, physical sciences, and technical subjects. This form of curriculum, which was new to the country at the time, but has marked American engineering curricula to this day, was intended to "contribute to the education of the man of science and the man of action, whatever be his prospective professional pursuits". The time requirement for the new program increased from one to three years. Experience, however, soon showed that the new curriculum was in advance of prevailing preparatory education. To remedy this, a "preparatory division" of one year was provided, but the distinction disappeared in 1862, leaving the integrated four-year curriculum which became the accepted norm.

In 1862 there were about a dozen engineering schools in the United States. These included the United States Military Academy; Norwich University; Rensselaer; Union College, which established a school of civil engineering in 1845; the U.S. Naval Academy, which offered courses in steam engineering from its founding in 1845 and created a separate department of engineering in 1866 when it returned to quarters at Annapolis, Maryland, at the end of the Civil War; the Chandler Scientific School at Dartmouth, founded in 1851; the Sheffield Scientific School at Yale, which was founded in 1847 but did not begin to offer engineering until 1852; the University of Michigan, founded in 1852; the Polytechnic College of the State of Pennsylvania, which was founded in 1853 and apparently offered the first curricula and granted the first degrees in the United States in mechanical engineering in 1854 and in mining engineering in 1857 before it closed in the late 1880's [25]; New York University and the Polytechnic Institute of Brooklyn, both of which began offering programs in 1854 and were to merge their engineering organizations into the Polytechnic Institute of New York in 1973; and Cooper Union, founded in 1857 as a polytechnic school "for the advancement of science and art." In addition, the Lawrence Scientific School had been founded at Harvard in 1847, but does not appear to have offered engineering in any substantial way until 1865, and the Massachusetts Institute of Technology was incorporated in 1861, but because of financial difficulties and the Civil War did not begin its first classes until 1865.

The growth of engineering education to this point had been gradual, but the lines of its development were clear. By 1866, only about 300 engineers had graduated during the past 31 years [26]; most practitioners at this time still learned engineering through on-the-job experience, but the graduates easily found employment in railroad and bridge construction. As a collegiate discipline, the poor regard for engineering and even science in the older more classical institutions can be surmised by the fact they were kept separate in scientific schools as a concession to the theological interest. Engineering, which was regarded as a technology, was not considered a respectable scholarly pursuit. In the mid-nineteenth century, students majoring in engineering or science at Harvard and Yale did not have equal status with the more elite students in the arts. Admission standards in engineering and science were lower and the curriculum was less demanding, as it only required three years to graduate in these areas but four years to obtain a bachelor of arts degree; even the dining facilities were segregated.

The Era of Growth: 1862 to 1893

The year 1862 marks a major change in the development of engineering education and of the United States as a nation, as three significant events took place. In that year the U.S. Congress passed the Homestead Act, which gave 160 acres of land free to any head of a family who worked for five years to improve the land. This prompted a large migration of people to the West of the country. Second, in 1862, the Congress granted a charter to the Union Pacific to construct a transcontinental railroad from Nebraska to California. Overcoming the difficulties inherent in spanning a region of desert wastes, wooded plateaus,

Fig. 8. Electrical laboratory at Purdue University, about 1900; note leather belts running through slots in the floor to a shaft in the basement; an Edison bipolar dc generator is in the foreground.

and precipitous mountains required an ingenuity in engineering. This opened a new chapter in engineering education. The largely self-taught engineers of earlier times, who surveyed the land, built roads, canals, and bridges, or were practical constructors of machinery were not adequate for the tasks. The railroad created a demand for engineers with a greater mastery of scientific resources, and so engineering schools arose out of simple necessity. The growth of the railroads was phenomenal. The Union Pacific was completed in 1869 and was quickly followed by the completion of the Southern Pacific and Northern Pacific railroads in 1884. Between 1866 and 1876, 40 000 miles of track were laid, bringing the total to 76 808 miles of track in operation, which rose to 167 703 miles by 1890. With the railroad came the telegraph. Supported by a Congressional appropriation of $30 000, Samuel Morse completed the first test line between Washington, D.C., and Baltimore, Maryland, in 1844 [28]. The telegraph provided a means of telecommunications that quickly spanned the country from coast to coast, with the Western Union Telegraph Company putting 108 000 miles of new wire into service between 1866 and 1876. This brought the total number of miles of telegraph wire in service in 1876 to 184 000, which increased to 679 000 miles by 1890 [29].

The third major event of 1862 was the passage of the Morrill Land Grant Act, which laid the foundations for a comprehensive system of public higher education in the United States. This Act, which was defeated in Congress in 1857, was vetoed by President Buchanan in 1859, finally passed with the absence of the Southern delegation during the Civil War [30]. It provided for the allocation of public lands by the federal goverment to each State and territory for the foundation and support of colleges, "particulary such branches of learning as are related to agriculture and the mechanic arts" for the purpose of promoting "the liberal and practical education of the industrial classes in the several pursuits and professions of life." The promotion of practical education had become extremely important for the utilization of the country's resources and the creation of the industrial activities. The effect of this single piece of legislation can be seen by the fact that in a single decade the number of engineering schools increased from about a dozen in 1862, to 21 in 1871, to 70 in 1872, a rate of expansion without equal in Americn education [31].

Notable milestones in the development of engineering education during the first decade after the passage of the Morrill Act included the establishment of the School of Mines at Columbia University in 1864, which supplied the prototype for most of the schools of mines and metallurgy that developed during the next 15 years; the founding of Worcester Polytechnic Institute in 1868, with its distinctive form of shop training in which students served as journeymen to operate a small manufacturing plant for commercial production; the establishment in 1867 of the Thayer School of Civil Engineering at Dartmouth College, which was formed as a "professional" school in which the curriculum evolved into three years of study at Dartmouth College for general subjects, followed by two years at the Thayer School for professional subjects, similar to that followed in the medical and legal professions; the creation in 1868 of Cornell University, a land grant institution which was established with the visualization that education in the mechanics arts was to be "in every way equal to the learned professions"; the opening of probably the first laboratory for engineering instruction by Robert H. Thurston at Stevens Institute of Technology in 1871; the establishment of the first summer camp for instruction in surveying by the University of Michigan in 1874; and the establishment of many of what were to become the large state universities of the Midwest—Iowa State University (chartered in 1858, classes began in 1869); University of Neberaska (chartered in 1869, classes began in 1871); Ohio State University (chartered in 1870, classes began in 1873); and Purdue University (chartered in 1874), among others.

Large-scale economic development followed the Civil War (1861-1865). In the five years from 1865 to 1870, the capital invested annually in the purchase of structures and equipment for the manufacturing industries quadrupled, rising from 0.1 to 0.4 billion (constant 1958) dollars; raw cotton used in textiles increased from 344 000 to 797 000 bales; raw sugar production went from 15 000 to 77 000 short tons and then to 1 397 000 short tons by 1880. There also was a surge in inventions. For the period from 1790, when the United States Patent Office was founded, to 1865, 60 000 patents were issued, while over 416 000 were issued from 1865 to 1890; this was almost a sevenfol increase in a third of the time [32]. Large numbers of

European immigrants, with as many as 800 000 entering the country in a single year, helped to provide the needed labor, as well as to increase the need for housing, transportation, and manufactured goods. This boom resulted in newer services, new machinery, more goods, new types of business organizations, increased markets and demand, and brought a great amount of new capital into the economy.

By 1870 the nation's rate of growth was considerable, and the stage was set for major advances in engineering education. The northern and western states were in the full swing of a vast process of industrial expansion, railroad building, city development, and land settlement. This industrialization and economic development greatly stimulated the demand for engineering services. In the older states of the East, the transition from an agricultural to an industrial society gave a strong impetus to the subdivision and specialization of the engineering profession. There was, however, a half century of delay in the economic recovery of the south, lasting until the beginning of this century, making the progress of engineering education in that region slow and difficult.

The Centennial Exposition of 1876, in Philadelphia, did much to popularize technical education, particularly in the manual arts. It was a clear display of the rate at which America's economy was expanding, of the inventiveness of its people, and of the capacity of industry to exploit ideas, inventions, and processes through technological and financial means. It was at the Exposition that Russia exhibited its methods of manual training, which influenced many of the subsequent developments in the United States in shop training. It also was here that the Corliss engine, the electric generator and motor, and the arc lamp were first displayed, and where Alexander Graham Bell introduced the telephone which within a few years would revolutionize the business routine in the large cities and became a familiar convenience in thousands of American homes. A German visitor to the Exposition remarked, "The machine is the essential element in the life of North Americans . . . . Certainly it is unlimited mastery of material which speaks for the endless machinery here; certainly it is the picture of a wild chase in which all the energies are concentrated; a chase after material gain. But who can deny there is herein greatness and power?" [33].

Engineering education was moving into a period of consistent development. The state of engineering education prior to 1870 was characterized by schools with very limited resources; engineering laboratories were unknown; textbooks were few and mainly derived from abroad; instruction was largely blackboard demonstrations prepared from texts, followed by recitation and interrogation; and the primary purpose was to train civil engineers to meet the needs of geographic expansion and urban growth [34]. After 1870 came a period of expansion in the number of schools, aided mainly by the Land Grant Act; engineers, rather than scientists, took an increasing leadership in education; an American literature in engineering was developing through the authorship by faculty members of textbooks and of articles in the journals of the newly forming technical societies; curricula began to diversify to meet the specialized needs for engineering talent; and the lecture system became widespread. In the two decades following the Civil War, the collegiate type of curriculum with its extended base of science, mathematics, languages, and social studies, though modified by the pressure of expanding engineering knowledge and technique, proved its stability and became firmly established as a basic structure for engineering education in the United States. By 1885, shop work had attained its maximum position in engineering curricula, as measured by the time allotment, and considerably overshadowed the developing emphasis on laboratory instruction. Of all the elements of progress in this period, possibly the most important was the beginning of a period of notable teaching and authorship. American engineering education turned away from European models and sources to seek its leadership and materials in its own country.

Engineering began to diversify into its major branches during this period. The first distinctive curriculum in civil, as distinguished from military, engineering was offered at Partridge's Academy in 1821, which was followed in 1828 by Rensselaer, with the latter institution granting the first degree in civil engineering in 1835. Mechanical engineering was first offered at the Polytechnic College of the State of Pennsylvania in 1854, at Yale University under the heading dynamical engineering in 1863, at the Massachusetts Institute of Technology in 1865, at Worcester Polytechnic Institute in 1868, and at Stevens Institute of Technology in 1871, a school founded specifically for the professional education of mechanical engineers. A significant assist in the development of mechanical engineering curricula resulted when the U.S. Navy went into a period of reduction in the 1870s. To utilize the educated and experienced naval engineering officers, the government legislated the assignment of these officers to teaching positions in, mainly, civilian colleges and universities to promote "knowledge of steam engineering and iron shipbuilding." The authorizing bill was passed in February 1879 and 49 appointments to 33 institutions were made in the following 17 years [35].

The diversification continued, with the introduction of mining engineering in 1857 at the Polytechnic College of Pennsylvania and then the creation of the School of Mines at Columbia University in 1864. Electrical engineering followed as an outgrowth of physics departments evolving from the scientific toward the technical rather than in the reverse direction, as did the earlier branches of engineering education. It was introduced first at Massachusetts Institute of Technology in 1882, and then at Cornell University and at several other schools in 1883, with the first department of electrical engineering perhaps being organized at the University of Missouri in 1886 [36]. The electrical industry, which had began to flourish in the preceding decade, and the popularity of the subject led Senator George Hoar of Massachusetts to remark in 1886 that "every technical school in the country has established, or is seeking to establish, a department of electrical engineering" [37]. A need for engineers knowledgeable about chemical processes was fostered by the high rate of production and consumption of oil that developed in the United States in the late nineteenth century, first for ship transport and then for the automobile, and later was given added stimulus by the enormous demand for explosive and chemicals during World War I. Chemical engineering was similar to electrical engineering in that it evolved from a scientific rather than from a technical base. It was introduced into the colleges as a series of lectures at Massachusetts Institute of Technology in 1888 and as a curriculum at the University of Illinois in 1895. It also was during this period that two distinct approaches [38] developed to the education of mechanical engineers, which resulted in the erection of contesting types of engineering colleges. One type, which included Stevens Institute of Technology, Massachusetts Institute of Technology, and Sibley College of Engineering at Cornell University (after 1885), emphasized more of the theoretical concepts in higher mathematics, research, and general science, while the second type, which included the Worcester Free Institute, Rose Polytechnic Institute, Sibley College (before 1885), and the Georgia School of Technology (now Georgia Institute of Technology) as founded in 1888, stressed practical shop work producing capable machinists and shop foreman [39]. The conflict between the two approaches reached its maximum intensity in the 1880s and persisted until after 1900 [40].

National engineering societies were established shortly after their respective curricula, indicating the growing relationship between the profession and the schools of engineering, but also showing that the profession had not yet assumed the leadership and responsibility for professional education. The founding of the American Society of Civil Engineers in 1852 was followed by the creation of the American Institute of Mining and Metallurgical Engineers (1871), the American Society of Mechanical Engineers (1880), the American Institute of Electrical Engineers (1884), and the American Institute of Chemical Engineers (1908).

During this era, America was still predominantly agricultural. The U.S. Bureau of the Census declared 1890 as the official end of the America frontier, as the census of that year showed for the first time that the increase in urban population for the preceding decade was greater than the increase in rural population. Although it was not until 1920 that urban residents exceeded rural dwellers, the trend was clearly evident. By 1890 the nation's growth was considerable. In the preceding two decades, the internal combustion engine and the spark-ignition engine were developed, as was the steam turbine, electric generator, electric motor, storage battery, voltage transformer, incandescent lamp, phonograph, and telephone. Major engineering works were proliferating. J.B. Eads completed a 500-ft center span bridge across the Mississippi at St. Louis in 1874, after 17 of the leading civil engineers of the day differed over how it should be designed [41]. John A. Roebling and his son, Washington A., constructed the Brooklyn Bridge with a record-breaking span of 1595.5 ft in 1883; while William L.B. Jenney designed and constructed the first skyscraper in 1883-1885 for the Home Insurance Company of Chicago. After Thomas A. Edison established a commercially successful electrical generating plant at Pearl Street in New York City in 1882, there was a significant development of electric power for domestic and industrial uses, for street lighting and railway transportation. Steel became available in quantity as a result of the work of Bessemer and Siemens. With the technological developments that had occurred, and with the growing emphasis on standardization and production techniques, the stage was set for the establishment of mass production.

It also was in this period that American higher education had undergone a major step in its evolution. Colleges grew in number and in the number of students who attended. The purposes of college education had begun to change. There was a definite movement to give education a greater utilitarian emphasis. Practical subjects were introduced into the curricula and many new fields of study became worthy to be taught in college. Graduation from college was connected more closely with skilled vocations and professions, which appealed to a large segment of the American population. At about the same time, graduate education had begun. Yale granted the first Ph.D. degree in 1861, and The Johns Hopkins University, which is often considered as a model for graduate education in this country, was founded in 1876. The Germanic approach in connecting graduate study and original research, with its methods of science, had taken root. By the end of this era, most elements of the structure of the American system of higher education were in place. The numbering of courses, the credit system, major fields of study and departmental organization, the lecture, recitation, and seminar mode of learning, the elective system, and the administrative hierarchy involving presidents, deans, and department chairmen had all emerged and were accepted with very little variation among institutions.

The Era of Development: 1893 to 1914

Engineering education in the United States may be said to have come into its own during this period. By its beginning, all of the major branches of engineering had developed professional societies, except chemical engineering which was yet to be founded. Engineering education, however, still did not hold a high place as preparation for the profession or as essential to the practice of the profession, and it was only after 1893 that engineering education became accepted as a distinct and worthy field of higher education.

The World's Columbian Exposition of 1893 marked a significant step in the history of engineering education in this country. In connection with the Exposition held in Chicago, a number of congresses on important branches of industry and knowledge were established. One of these was the World's Engineering Congress, which included Division E, Engineering Education, whose meetings were organized and chaired by Ira Osborn Baker, Professor of Civil Engineering at the University of Illinois. This was the first major meeting of engineering in which engineering education was recognized as an important subject and was given equal status with the branches of engineering. The success of this meeting left a general consensus that the gathering of engineering teachers should be continued in a permanent organization, resulting shortly afterwards in the creation of the Society for the Promotion of Engineering Education (SPEE). The papers presented at the Congress became volume 1 of the *Proceedings* of SPEE.

The Society held its first annual meeting at the Polytechnic Institute of Brooklyn in 1894, with an international membership of 156 people. De Volson Wood of Stevens Institute of Technology, was elected its first president. The organization quickly drew to itself the leading figures in engineering education, with its early officers and members coming from a small group of notable institutions that had sound academic standards and progressive policies. Engineering thus became the only profession with a society devoted solely to education for the profession. The membership of the society, however, which has always been significantly less than that of the major technical societies and has consisted primarily of persons in colleges and universities, was further evidence that the profession had left matters of education primarily to the engineers in academic institutions.

Sparked by activities at the World's Columbian Exposition, as well as by the first automobile exhibition held at Madison Square Garden in 1900 and the Wright Brothers first flight of an airplane in 1903, great industrial expansion took place as America

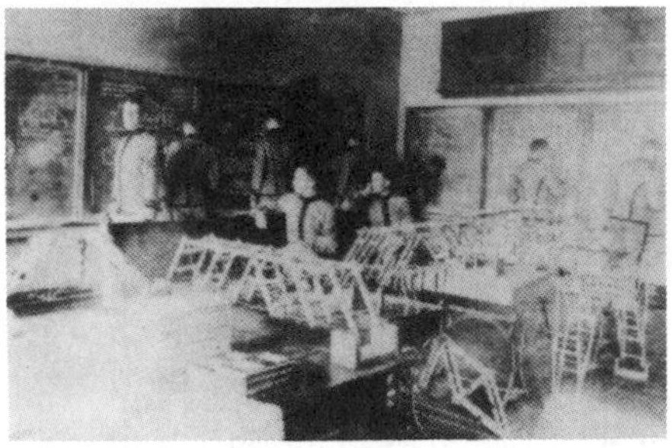

Fig. 9. Engineering class at West Point studying bridge structures, about 1914.

turned more to the exploitation of ideas and the production of goods. The country also underwent a rapid population increase. From 1900 to 1914, 13.4 million immigrants entered the country, helping to expand the population 30 percent in that time to 99.1 million people [42]. This created the necessary labor and markets for the goods produced. In 1900, for instance, there were 8 000 motor vehicles registered in the United States. This increased to 1 258 000 by 1913, the year Henry Ford set up a moving assembly line to produce the Model T. By doing so, he was able to reduce the time to assemble a chassis from 12 hours 28 minutes to 1 hour 33 minutes, and thus to greatly increase production [43], with the result that by 1918 there were 6 160 000 motor vehicles registered in the United States [44]. This enormous output of automobiles stimulated other industrial activity by increasing the demand for steel, petroleum, rubber, and plate glass.

The electrical industry was another that grew rapidly in this period. The Westinghouse Electric Company was formed in 1886, and the General Electric Company was formed in 1892 as an outgrowth of a company formed in 1882 by Thomas Edison to exploit his work in electric lighting. In 1895 a power plant was opened at Niagara Falls, with part of the electrical power transmitted to Buffalo, New York, rather than being used locally. Markets for electrical power were being extended from the source of generation. By 1902 the country produced almost 6 billion killowatt-hours of electrical energy, which increased to 43.4 billion killowatt-hours in 1917 [45]. In another area, Herman Hollerith, then working for the U.S. Patent Office, had invented in 1890 an electromechanical machine to tabulate information using cards with punched holes. These machines were first used to tabulate mortality statistics in Baltimore, Maryland, in New Jersey, and in New York City, and then were selected for compiling the census of 1890. In 1906, Hollerith formed the Tabulating Machine Company, which through mergers and growth later became the International Business Machines Corporation.

Industrial engineering was beginning to develop. H.R. Towne, in the 1870s and 1880s, had pioneered in methods to increase production through a close working of management and production after he joined Yale's company to produce pin tumbler locks. His ideas developed and spread rapidly during the two decades following 1890. Frederick W. Taylor introduced concepts of efficiency engineering, the piece rate system, and job analysis. Hugo Muensterberg coupled psychology with industrial development, and Frank and Lillian Gilbreth developed time and motion studies. Collegiate education closely followed the industrial developments. The first course in management in an engineering college appears to have been offered by Hugo Diemer at the University of Kansas in 1902, while a similar course was created at Cornell University in 1904. Diemer subsequently moved to Pennsylvania State College, where he began the first curriculum in industrial engineering in 1908 [46]. Several societies for the science of management were formed about 1912, but the trade unions resisted on the belief that efficiency meant speeding and exploitation of labor, and the movement slowed considerably.

During this period education in applied science grew in popularity, as did the number of engineering colleges. By 1896, the number of such institutions had risen to 110. Much of this expansion, however, had taken place at a fairly low level of academic quality which was viewed as a demoralizing influence and caused many complaints by the better engineering schools [47]. A classification [48] of engineering colleges at this time gave the following groupings, according to their entrance requirements:

Class A: Requiring for entrance at least algebra through quadratics, plane geometry, solid geometry, or plane trigonometry, one year of foreign language, and moderately high requirements in English — 31 colleges.

Class B: Requiring algebra through quadratics and plane geometry — 33 colleges.

Class C: Requiring less than algebra through quadratics and plane geometry — 25 colleges.

Class D: Not offering complete courses in engineering, but giving analagous work, generally under the heading of mechanic arts; many in the process of developing into true engineering schools — 18 colleges.

Class E: Having no entrance requirements as such, but doing engineering work of good grade — 3 colleges.

It was during this period more than during any other that the laboratory method of instruction in engineering grew to its dominant position. This change was due in part to the growth in importance of mechanical and electrical engineering. Early curricula had been dominated largely by the pattern of civil engineering, with its emphasis on design and field work, which in turn had been largely influenced by models of French schools. The latter were characterized by an emphasis on didactic methods of instruction and on mathematical and analytical, rather than on experimental, methods. By 1900 it was generally recognized that American laboratories and methods for the teaching of engineering were not surpassed and often not equalled in any other part of the world. This could not be claimed, however, for much of the theoretical instruction or for instruction in design.

Beginning in the latter part of the nineteenth century, the profession of engineering became deeply engaged in the industrialization of the nation, and focused its efforts on the improvement, operation, and maintenance of a growing complex of devices and processes. The growing demand for engineering was a natural consequence of the decline in the agricultural population, the increase in the urban population, and the rapid development of manufacturing. The advent of electricity, the gasoline engine and ferroalloys, and the development of the production line in the automobile industry created a demand for technical specialization. By the end of this period, engineering education was well established and concentrated upon preparing graduates who could accept jobs and be immediately useful—exactly what business and industry wanted.

The first decade of the twentieth century saw engineering education in America reach the height of its popularity relative to other divisions of higher education, as the number of engineering students increased from 10 000 to 30 300 in the period 1890 to 1910 [49]. In the second decade, agricultural education and collegiate education for business grew rapidly, a development which checked the expansion of engineering education, probably to its benefit.

Along with the growth, there was a large diversification in the types of engineering degrees offered at this time, possibly as a result of trying to prepare graduates for immediate employment. A committee of the SPEE reported that 90 different engineering degrees were offered by the colleges in 1904; 68 of the degree types were undergraduate (only 47 types were actually conferred, however) and 22 were graduate (12 were conferred). Of the 82 colleges reporting, 2393 baccalaureate degrees and 84 graduate degrees were awarded, with the major fields being civil (641 degrees), mechanical (635), electrical (414), mining (231), and general engineering (409) [50].

As a result of the Land Grant Act with its provision to promote knowledge in agriculture and the mechanic arts, agricultural experiment stations and agricultural extension grew rapidly. Similar growth in these areas did not occur in engineering, but beginnings were made. America was still an agrarian country, so that research and dissemination of information in agriculture would be viewed to benefit large numbers of people and have direct results in terms of productivity. This latter was an important objective since the country was on its way to becoming an urban, industrialized society, and higher yields from the farms with less labor would be necessary to feed the growing population of the country. Research in engineering, however, was not seen to be as necessary since great technological developments were being made by the individual entrepreneur, without the benefit of government subsidies.

Although monies were not available from the federal government, the University of Illinois established the first engineering experiment station in December 1903, and Iowa State College established the second in 1904. Eventually, all land-grant institutions having engineering programs followed this lead [51]. In the area of extension services, the growth was even slower, with the Pennsylvania State College beginning a program of industrial extension in 1907 [52].

Other prominent developments of this period included the beginnings of engineering education on the west coast, the introduction of part-time education and of cooperative education, and the passage of the first law to license engineers for practice. On the west coast, the University of California at Berkeley was founded in 1868 and Stanford University in 1891 with instruction in engineering begun from the start in each case. Throop Polytechnic Institute began offering degrees in electrical, mechanical, and civil engineering in 1908, before becoming the California Institute of Technology in 1921. The foundations for major developments in engineering education on the west coast, to be stimulated particularly by World War II, were now in place.

Part-time education took several forms, with evening undergraduate courses in engineering introducted first at the Polytechnic Institute of Brooklyn in 1904. This approach gained favor with employed people, so that by 1975 about 17 000 or 7 percent of all engineering undergraduate students were enrolled in part-time degree programs. A second approach was introduced by Herman Schneider in 1906 at the University of Cincinnati. His plan for cooperative education divided each class of students into two groups that alternated periods of full-time instruction with appropriate employment in industry. This six-year program utilized the same entrance requirements and same instructional material as the full-time, four-year curriculum. The cooperative plan spread steadily, until by 1970 there were 104 engineering schools that had adopted it [53].

States began to license engineers for practice, a development that would affect curricula, particularly in those branches of engineering that had a direct implication for the health and safety of the public. As a result of chaotic conditions that developed when homesteaders "surveyed" their own water rights and signed their names as "engineers," Wyoming passed the first state law in 1907 to register engineers.

Fig. 10. Drafting class at the University of Illinois, 1925.

Louisiana followed in 1908 with a law regulating the practice of civil engineers to protect the public from unqualified and unscrupulous persons who practiced engineering. Registration was based on professional competence and required a certain amount of experience in responsible engineering positions as well as the passing of an examination in the theory and practice of engineering before a license would be granted [49].

The Period of Evaluation: 1914 to 1940

The development of and demand for the automobile and airplane, the growing need for petroleum and electric power, and the techniques of mass production, amplified by the requirements generated by World War I, brought engineering and engineering education into a new phase. In 1914, for instance, there were 1.7 million motor vehicles registered in the United States. The healthy state of the economy and the production capability of the industry caused the number of vehicles to rise dramatically to 26.7 million by 1929, when the Great Depression began. This level remained nearly constant until the number of vehicles registered began to rise again to a total of 32.5 million in 1940. In this period from 1914 to 1940, Americans bought about three-quarters of all automobiles produced in the world [50].

A similar situation existed in the number of airplanes produced. The early adoption of the airplane by the U.S. Post Office, its use in World War I, the establishment of transcontinental service in 1921, and Lindbergh's historic flight in 1927 stimulated the growth of a new industry and aided the acceptance of the airplane by the American public. It was used for crop spraying, irrigation work, planning for towns and railways, as well as for military application. Between 1914 and 1940, the United States produced a total of 38 000 aircraft for the military and almost 40 000 additional aircraft for civilian use [56].

In line with these developments, departments of aeronautical engineering were created in the colleges. The airplane also had significant effects on other branches of engineering and technology. It called for

Fig. 11. Students studying railroad engineering at Purdue University, 1935.

precision engineering, high tensile steel, the machining of light alloys, the creation of fiber glass and plastic adhesives, new lubricants, high quality fuels, and powder metallurgy.

The demand for oil and petroleum also rose significantly during this period. America was the major producer and consumer of oil in the world, consistently producing about two-thirds of the world's oil, while its rate of consumption per capita was ten times that of any other country [57]. While the automobile created a demand for large quantities of fuel, the airplane increased the requirements on the quality of the fuel. This led to improvements in the distillation process, thermal cracking, the production of petroleum (which up to the war was mainly a waste product), and the creation of tetraethyl lead. The need for chemical engineers rose, and many departments of chemical engineering came into being.

The extreme diversification of engineering curricula that had taken place in the preceding decades was placing severe demands on engineering colleges for a variety of courses dealing with specific technical applications. There were programs in aeronautical, agricultural, architectural, automobile, bridge, cement, ceramic, chemical, civil, construction, electrical, heating, highway, hydraulic, industrial, lighting, marine, mechanical, metallurgical, mill, mining, railway, sanitary, steam, textile, telephone, topographical engineering, and engineering administration. The pressure to produce graduates who had immediate industrial usefulness, which had existed since the passage of the Morrill Act, left little time for engineering education to examine its structure or to relate the objectives of the profession to society as a whole. Fortunately, however, a number of leaders in engineering practice and education were instrumental in initiating a major study of engineering education. This move ushered in a series of periodic evaluations of engineering education, which has continued to the present time.

The Society for the Promotion of Engineering Education, at its Annual Meeting in 1907, invited the American Society of Civil Engineers, the American Society of Mechanical Engineers, the American Institute of Electrical Engineers, and the American Chemical Society to join with it in appointing a "Joint Committee on Engineering Education." This committee was charged to examine all branches of engineering education, including engineering research, graduate professional courses, undergraduate engineering instruction, and the proper relations of engineering schools to secondary industrial schools, and to recommend the degree of cooperation and unity that should exist between engineering schools. A year later, the Carnegie Foundation for the Advancement of Teaching provided funds for the effort, and Professor Charles R. Mann of the University of Chicago was appointed to conduct the study. His report issued in 1918 was the first comprehensive study of engineering education [58]. The report suggested a return to fundamentals and a movement toward the unification of engineering curricula, the need for the development of the student's intellectual capacities and discipline in his habits of work and study. Due to the conditions following World War I, the report received less attention from engineering educators than it deserved.

America's involvement in the war, however, served to direct a different attention to engineering education. The prominent place gained by engineers in the management and direction of industry, the notable contributions of engineering to the techniques of production, and the conspicuous place taken by engineers in war activities led to a marked increase in emphasis on the administrative and economic sides of engineering.

Distinct curricula to emphasize the administrative rather than the technical aspects of engineering were introduced widely. The place given to economics in all curricula was augmented, and business electives in engineering education were more generally provided. After the war, there was a noticeable trend away from specialization for undergraduates. The effort to stretch a four-year program into an all-around and a specialized training was abandoned, with a movement toward simplification of programs and a greater emphasis on general training. In order to prepare engineers to serve in the full scope of technical, administrative, and executive responsibilities, engineering schools developed general types of curricula that would be useful in a wide range of occupations. They were built on a foundation of science, of humanities, and of social relationships, rather than on the practical techniques required for specific industries or occupations. The curricula were functional in nature, similar to these in agriculture, commerce, journalism, or education, but they emphasized the professional responsibilities of the graduate much more than did these other curricula. However, few, if any, engineering schools made preparation for professional registration and practice their predominant objective, as is the norm in law and medicine.

The second major evaluation of engineering education, and the first carried out by members of the profession, was conducted in 1923 to 1929 [59]. Under the sponsorship of the Society for the Promotion of Engineering Education, a committee headed by W.E. Wickenden performed the most comprehensive evaluation of engineering education conducted to date. The committee's report examined virtually every aspect of engineering education in the United States, including its historical development, how it compared with engineering education in Europe, curricula, faculty preparation, relationships with industry, opinions of past graduates, and other aspects. The second part of the report, published in two large volumes, is devoted to the less than baccalaureate programs to train engineering technicians as supporting personnel. As might be expected, the report had a profound effect on engineering education and set the tone for future evaluations.

The principal recommendation of the report was carried out when the Engineers' Joint Council for Professional Development was established in 1932 and given authority to accredit engineering curricula. The number of schools with accredited curricula grew rapidly, increasing to 125 in 1940, to 159 in 1960, and to 234 in 1976. By October, 1976, there were over 1161 basic level and 78 advance level curricula accredited by ECPD [60]. Accreditation is voluntary and much sought.

It was in this period that graduate education in engineering made rapid progress. Although Rensselaer encouraged persons to enroll who had completed a baccalaureate degree program from another college, and Norwich granted master's degrees as early as 1837, the first graduate programs which followed a four-year undergraduate education can be traced to the 1890s. The Lawrence Scientific School at Harvard offered a graduate program in electrical engineering in 1893, while Massachusetts Institute of Technology offered a master's degree program in civil engineering in 1894 and graduate programs in chemical, mechanical, and sanitary engineering within the next ten years.

In this early period, graduate work was limited to comparatively few students, and often was no more than a fifth year of undergraduate instruction in subject matter, method, and status. In 1904, for instance, 114 engineering colleges reported a combined enrollment of 15 004 undergraduate and 249 graduate students [61].

Fig. 12. Engineering student at lathe, Purdue University, 1937.

Engineering generally accepted the M.S.-Ph.D. pattern of the arts and sciences, and the administration of engineering programs came within the scope of established graduate schools instead of developing as separate professional schools. From 1921 to 1930 graduate enrollments in electrical engineering increased ninefold, in mechanical engineering sevenfold, and in chemical engineering sixfold. Much of this growth, however, was in part-time and evening study, particularly in the metropolitan areas, and most often terminated with the master's degree.

The Scientific Era: 1941 to 1968

The United States' involvement in World War II, from 1941 to 1945, brought out weaknesses in engineering education, particularly in the electrical and electronics areas, which stimulated a change from a strong emphasis on practical subjects having an immediate utility to industry to a stress on the scientific principles underlying the technology. This movement was further advanced as a result of the Soviet-American tensions of the 1950s and 1960s which resulted first in a Cold War and then in a race to dominate outer space and to land a man on the moon.

Fig. 13. Students in forge and welding shops at Purdue University, about 1940.

It became evident as early as the spring of 1940 that the number of engineering and science graduates from the country's colleges would not be sufficient to meet the critical needs of national defense. In cooperation with engineering colleges the federal government therefore initiated a specialized training program "to supplement the regular curricula of engineering colleges with short, intensive, college-level courses having as their objectives the preparation of trainees to perform specific industrial jobs, the retraining of graduate engineers to perform new or more specialized tasks needed in the defense effort, and the building up of a large supply of technicians, draftsmen, inspectors, testers, and engineering assistants" [62]. The program included to a lesser extent the disciplines of physics, chemistry, and production supervision. From its inception in October 1940, to its conclusion in June 1945, 227 colleges participated in the Engineering, Science, and Management War Training program, offering over 31 000 engineering courses to more than 1.3 million people [63]. The program was highly successful in meeting the war time needs for technically trained manpower and was so conducted that upon its conclusion it did not disrupt the regular engineering college programs. Further, it provided a large group of people who after the war had a desire to obtain a more complete and systematic education in engineering and who were able to receive that schooling under the G.I. Bill of Rights. The veterans saw things with a much more mature view than the more usual and younger college student and proved to be very capable students, although their presence on campus was transitory. An awareness was developing at this time among the middle class and skilled workers that some form of college was necessary for economic betterment, so that the number of college students, which increased significantly after the war, never decreased again to the prewar levels.

The war lead to numerous technological developments that had implications beyond this immediate military application. Developments in the electronics field occurred in radar, microwaves, sophisticated control and navigation systems, and electronic instrumentation; in the aeronautical and mechanical fields in research on high speed aerodynamics, structures, and fatigue of airframes, the development of gas turbine engines, new fuels, and lubricants, as well as new techniques in forging, pressing, milling, and testing high tensile light alloys; in metallurgy in the development and use of duralumin, beryllium, titanium, iridium, and high temperature steel; and in high speed production. The Allied forces in World War II used as much petroleum in a single day as the entire Allied armies used in the entire period of World War I [64], causing major research and development in catalytic cracking and petroleum production. The loss of Malaya stimulated research in synthetic rubber, while efforts were expanded to produce asphalt for runways and roads, synthetic fibers such as nylon, plastics and textiles, petroleum products, and a wide variety of medicines. These military-directed advances caused corresponding developments in the theory and technology of electronics, metallury, applied mechanics, aerodynamics, and chemical operations and processes.

The war time demand for new and advanced knowledge in almost every branch of engineering and science showed the shortcomings of engineering education and brought engineering colleges into graduate education and research in a serious way. While engineers made many contributions to the development of ships, tanks, planes, and armament, it often was the physicist with advanced fundamental training who took the initiative in creating new devices and systems. The war produced and increased awareness of the importance of academic research, which, coupled with increased amounts of money from the Army, Navy, and Air Corps, lead to the establishment of research programs conducted by faculty members and graduate students.

Before the war, a major fraction of all graduate enrollments in engineering was in part-time courses conducted in a few of the larger centers of population. However, this changed. While many part-time opportunities still existed, and in fact expanded, full-time enrollments grew. This was aided by the establishment, with federal government funds, of graduate scholarships, assistantships, and fellowships, and by expanded research activities. The undergraduate curriculum was redesigned to serve a twofold purpose of preparing some graduates for immediate employment and others for graduate study.

These movements received another important thrust in 1958, when the Russians launched Sputnik. The resulting Soviet-American race for technological leadership, and the national goal to land a man on the moon, gave rise to a scientific and technological boom. Funds for science and engineering education suddenly became available from private foundations and from federal and state government appropriations. Faculty salaries rose, facilities and equipment improved, and research grants became plentiful. Academic standards were raised, and course requirements became more stringent. Scientists and engineers assumed new roles in government leadership, particularly during the Kennedy administration. This lasted almost until the end of this period, assuring that research was fully integrated with the purposes of engineering colleges. The funds available to academic institutions increased significantly in the 1950s and 1960s, with the result that funds expended at universities for research and development increased from $334 million in 1953 to $2.6 billion in 1968 [65]. It was only after 1966, when America had landed a man on the moon, and the nation's priorities began to change, that the amount of support for research declined. The number of graduate degrees awarded annually in engineering had a parallel rise, with the number of master's degrees increasing from about 1300 in 1940 to 4800 in 1950 to over 15 000 in 1968, while the number of doctors' degrees increased from about 100 to 500 to 3000 over the same period [66].

The west coast schools and industries, particularly in the electronics field, grew dramatically as a result of the war. A small group of entrepreneurs, many of whom were graduate students of faculty members at Stanford University, created a number of companies to develop and manufacture electronic equipment in the two decades preceeding World War II. Litton industries, Hewlett-Packard, and, after the war, Varian Associates were begun in this way. These companies became important contributors to the defense effort and, in the process, grew significantly in assets and sales. A major university-industrial relationship developed on the west coast, primarily centered around Stanford. This was similar to the relationships that had developed a few decades earlier on the east coast, with Massachusetts Institute of Technology and Harvard providing a stimulus and source of ideas for a large number of electronics firms that developed in the Boston area. An indication of the benefits and effect that this industrial relationship has had on Stanford University is that its government sponsored research program in electrical engineering increased from about $20 000 per year prior to World War II to almost $6 million in 1974-1975 [67]. In addition, the California Institute of Technology and the University of California at Berkeley, which had become centers of activity in chemistry in the 1920s and 1930s and in nuclear physics in the 1930s, and had developed relationships with research laboratories established in California by several oil companies in the 1920s [68], expanded their engineering departments after the war.

The evaluations of engineering education that had begun with the Mann report continued. In 1940, an SPEE committee, under the chairmanship of H.P. Hammond, issued the report, "Aims and Scope of Engineering Curricula" [69]. The committee found that although engineering was taught in a wide variety of types of institutions, engineering curricula were very similar throughout the nation. They recommended that there should be much more diversification and variation among curricula in order to prepare engineers for a wide range of technical, administrative, and executive responsibilities. They also made a recommendation that engineering curricula be developed along a parallel sequence of scientific-technological and humanistic-social. In 1944, an SPEE Committee on Engineering Education after the War, with a largely similar menbership, and also chaired by Hammond, restated these same objectives and elaborated upon methods for their achievement [70].

Fig. 14. Students taking measurements at steam turbine, Purdue University, about 1940.

In 1946 the Society for the Promotion of engineering Education changed its name to the American Society for Engineering Education (ASEE), with the continuation of its primary purpose to improve engineering education in all of its facets. In 1955 an ASEE committee, under the chairmanship of L.E. Grinter, issued the "Report on Evaluation of Engineering Education" [71]. As with the similar groups that preceded it, the committee was charged with recommending ways in which engineering education could keep pace with the rapid developments in science and technology, as well as how future engineers should be educated to provide the professional leadership needed over the next 25 years. The report reemphasized the conclusions of the Hammond

Fig. 15. Dean A.A. Potter observing Navy students studying electrical engineering in V-12 program at Purdue University, 1943.

reports and set specific objectives for the profession in both the technical and social areas. It also discussed in detail how the objectives could be implemented and suggested broad guidelines for dividing engineering curricula into stems that included humanities and social sciences, mathematics and basic sciences, engineering sciences, engineering specialty subjects, and electives. In addition, the report recognized that an undergraduate curriculum must serve a twofold purpose of preparing some students for graduate study and others for immediate employment.

These reports, coupled with the rapid accumulation of new knowledge of all kinds, the accelerated pace of technological advances, and the growing complexities of social, technical, and economic relationships in modern society, had major effects on the development of engineering curricula during this period. The general result is unique in American higher education. Engineering curricula attempt to provide within the confines of a four-year program both a broad general education and a specialized technical education of great and growing complexity.

The engineering programs of this period incorporated the natural sciences, social sciences, humanities, and communication arts into a strong core of mathematics, engineering science, and analysis, and tried to bring these intellectual disciplines and fields of knowledge to bear on real and contemporary problems of society. On the one hand, there was an emphasis on fundamentals and an acceptance of a uniform first year to provide the engineer with a basic technical knowledge that would allow him to practice in a variety of occupations. On the other hand, there was a tendency to broaden the content of engineering programs in all branches, including some training in economics, management, social and humanistic studies, statistics, and computer programming.

In order to achieve both breadth and specialization, the developing general trend was to develop flexible undergraduate programs of four-years duration that could be followed by a year of more of graduate study. Indeed, the growth of graduate study was the most significant trend of the period. In the latter part of the period, graduate enrollment grew faster than undergraduate enrollment.

In addition to the education of engineers, opportunities were enlarged, particularly at the end of this period, for the education of engineering technologists and technicians, and for providing continuing education. Programs for engineering technologists developed along both two-year and four-year lines. The first engineering technology baccalaureate program was accredited by the Engineers' Council for Professional Development in 1967, and by 1976, there were 57 institutions that had a total of 150 curricula accredited. Over the period 1967 to 1975, the full-time enrollment in ECPD-accredited engineering technology programs, both two-year and four-year, in-creased from 23 669 students at 55 schools to 58 002 students at 176 schools [72].

Many engineering colleges expanded the established pattern of high-quality, part-time, advanced degree programs for on-campus study by employees of nearby industry and government and began to utilize new techniques and arrangements for extending advanced engineering education, usually through the use of telecommunications, to persons employed so far from the campus as to make commuting difficult or impossible. Larger companies developed their own training programs, sometimes having them videotaped for offering at widely scattered plants, while many engineering schools began to cooperate more closely with industry, government, and the professional engineering societies to provide a variety of opportunities for continuing studies on the part of engineers.

Present State: The Era of Social Involvement

Engineering education is moving into a new era of development. The United States has begun to shift its priorities away from defense toward the human and social problems of the nation. The country has landed men on the moon and is, at this time, no longer in a race with Russia for the conquest of space. There has been a decrease in the federal funds available for defense and space activities and an increase in funds for applying science and technology to domestic problems, such as housing, transportation, health care, education, pollution control, and energy. It is clear that new engineering graduates will have to be more concerned not only with technical developments, but also with the impact of those developments on society.

It has been usual for the engineer to concern himself primarily with the technical aspects of his job, leaving its moral and sociological implications to politicians, social scientists, and others. This, however, is changing. While the design of a system to meet a set of requirements and specifications may be fairly straightforward technically, the human setting into which the system will be placed can have profound implications for the design of the system. Engineers in the future will have to deal with important nontechnical constraints that arise out of legal, social, economic, aesthetic, and human considerations and will have to take into account the interaction of technology with the social and physical environment.

Engineering has always been a form of public service, aimed at meeting the needs of society by conceiving, developing, and implementing solutions to technical problems of concern to society. While the technical role of the engineer in American life has been largely unchanged, the social role received added prominence when Herbert Hoover, an engineer, became President of the United States in 1929. Hoover called attention to the unplanned effects that technology was having on society, to the waste and inefficiencies in industry, and to the plundering of the natural resources that were occurring as a result of industrial development. Under his successor, Franklin D. Roosevelt, the Tennessee Valley Authority was started in 1933, as the largest effort in social engineering ever attempted in America. The Authority, which has been a major contributor to the development and economy of its region, is engaged in flood control, soil conservation, fertilizer development, reforestation, public health, construction of an inland waterway, and generation of hydroelectric power. Today, it is thc largest single consumer of coal and the largest producer of electricity in the United States, with an installed generating capacity of 26.7 million kilowatts and an additional 21.2 million kilowatts of capacity under construction or planned [73].

While the work of engineers has traditionally been concerned with human problems, it is the degree of interaction and the greater perceived impact that engineering solutions are having on people that have increased dramatically in the past few years. Technology is affecting virtually every aspect of life in the United States, and indeed in the world, from the gross national product, foreign trade, and balance of payments, to the growth of cities, production of goods, and national defense, to the tallying of individual credit balances, and tax returns. The general public is being directly affected by the technologies involved in pollution and environmental control, housing, health care, the development of major transportation systems including the supersonic transport, the desirability of new missile defense systems, the development of data banks, and the production of nuclear and other forms of energy. More than ever before engineering must be viewed as a service to society, taking into account the interaction of the technical with the social, physical, cultural, and political environments.

"Goals of Engineering Education," the most recent of the periodic assessments of the profession, completed in 1968 under the auspices of ASEE, recognized this trend [74]. Among its numerous observations and recommendations on virtually all aspects of engineering education, the report made the following points: society's needs in the decades ahead will call for engineering talent on a scale never before seen in the United States or elsewhere; the engineer of the future will be called upon to play an increasing role in the solution of complex social problems; and the future engineer will need greater technical competence to cope with the complexities of technological endeavors.

Engineering schools are becoming more aware that the engineer of tomorrow must become more conscious of the social consequences of his work, and that he will have to work in much closer concert with sociologists, economists, industrialists, psychologists, physicians, politicians, and even theologians. With financial help from the federal government and from private foundations, programs are being developed at numerous universities, some at the senior year of the undergraduate level, others at the master's degree level, to acquaint the student with political and social processes and values, the structure of public systems, economics, social sciences, law, public health, rural, urban and international development, civil welfare, business administration, environmental design, and public policy. These programs currently involve faculty from many parts of the university and often incorporate problem-oriented, interdisciplinary courses. They all, however, are attempting to make the student more conscious of the obligation the engineer has to society and to enlarge his outlook to include the human aspect of his future work.

With all of the change that has taken place in the last century and with the changes that are being called for, engineering curricula, in one sense, have maintained a remarkable stability. In their more essential qualities, engineering curricula are today what they were 100 years ago—a distinctive type of college program based primarily on the principles and applications of physical sciences and mathematics, with associated studies in humanities and social sciences, intended to precede and supplement, but not supplant, an extended professional apprenticeship. College training is assumed to provide the more general, the apprenticeship the more specific, preparation for an engineering career. The numerous changes in details that have occurred in the course of 100 years do not invalidate these characterizations.

The stability of the broader features of engineering curricula through a century of revolutionary changes in other realms of higher education may be ascribed to the fundamental soundness of engineering programs. There are few, if any, branches of higher education that are more severely tested by objective realities, nor are there any in which the content of curricula is more constantly and critically scrutinized for its relevancy, in consequence of the rapid advance of knowledge and technical practice and the enforced limitaions of time.

Engineering curricula aim to provide a thorough grounding in the principles of science and the methods of engineering, togther with elements of liberal culture intended to enrich the personal life of the student and fit him for a worthy place in human society. They do not provide a complete professional training, if they ever did. Whatever their original purposes, the majority of engineering graduates actually put their training to the uses of a general education. These curricula therefore should be appraised as a type of college training, aiming at larger general values and intended to lay only the broader rather than the specialized foundations for a professional career. The present trend in engineering education is toward an emphasis on general, rather than technical, educational values in undergraduate college work, followed by one or more years of graduate work for specialized training, and then periodically supplemented by continuing education throughout one's professional carrer.

References and Bibliographic Notes

[1] This phrase originally was used by Count Rumford in 1799 in a prospectus for The Royal Institution of Great Britain; see P.C. Ricketts, *History of Rensselaer Polytechnic Institute.* Wiley, 1934, p. 4. Steven Van Rensselaer used it in a letter of November 5 1824, to Samuel Blatchford defining the purpose of the Rensselaer School; see S. Rezneck, *Education for a Technological Society.* Troy, N.Y.: Rensselaer Polytechnic inst., 1968, p. 3.

[2] This is brought out clearly in articles dealing with the founding of the American Medical Association, the American Society of Dental Surgeons, and the American Bar Associaton; see J.G. McGivern, *First Hundred Years of Engineering Education in the United States (1807-1907).* Gonzaga Univ. Press, 1960, pp. 107-108.

[3] Ibid., p. 57.

[4] W.K. LeBold, "Engineering education," in R.L. Ebel, Ed., *Encyclopedia of Educational Research,* 4th ed. New York: Macmillan,1969, pp. 435-443.

[5] McGivern, *First Hundred Years,* p. 10.

[6] General Washington's order, issued on June 9, 1778, did not result in the establishment of an engineering college, but is considered to be the genesis of the United States Army Engineer School, currently located at Fort Belvoir, Va.

[7] E.L. Armstrong, ed., *History of Public Works in the United States 1776-1976.* American Public Works Assoc., Chicago, Ill.1976, p. 666.

[8] Committee on History and Heritage of American Civil Engineering, *The Civil Engineer — His Origins.* New York: American Society of Civil Engineers, 1970, p. 37.

[9] Ibid.

[10] *The Centennial of the United States Military Academy at West Point,* Washington, D.C.: U.S. Government Printing Office, 1904, p. 263.

[11] McGivern, *First Hundred Years,* pp. 37-38.

[12] The degrees were made retroactive to 1802 for all graduates of the Military Academy under a law drafted by General Douglas MacArthur.

[13] C.W. Cullum, *Biographical Register of Officers and Graduates of the United States Military Academy,* vol. 3. Boston, Mass.: Houghton Mifflin, 1891, p. 2300; see also A.M. Wellington, "The Engineering Schools of the United States III," *Engineering News,* vol. 27, p. 318, Apr. 2, 1892.

[14] L.A. Webb, *Captain Alden Partridge and the United States Military Academy 1806-1833,* Northport, Ala.: American Southern, 1965, p. 23; see also A. Johnson, and D. Malone, Eds. *Dictionary of American Biography,* vol. 7, pt. 2. New York: Scribner's 1959, p. 281.

[15] W.A. Ellis, Ed., *Norwich University, 1819-1911, Her History, Her Graduates, Her Roll of Honor,* vol. 1. Montpelier, Vt.: Capital City Press, 1911, p. 2.

[16] G.S. Emerson, *Engineering Education; A Social History,* New York: Russak, 1973., p. 142.

[17] McGivern, *First Hundred Years,* pp. 46-48.

[18] From letter of Stephen Van Rensselaer of November 5, 1824, to Samuel Blatchford (see [1]) included in the act incorporating the Rensselaer School, passed by New York State Legislature, March 21, 1826.

[19] D.H. Calhoun, *The American Civil Engineer, Origins and Conflict.* Cambridge Mass: The Technology Press, distributed by Harvard Univ. Press, 1960, p. 45.

[20] W.E. Wickenden, "A comparative study of engineering education in the United States and in Europe," in *Report of the Investigation of Engineering Education 1923-1929,* vol. 1. Lancaster, Pa.: Society for the Promotion of Engineering Education, Lancaster Press, 1930, pp. 811-812.

[21] Emerson, *Engineering Education,* p. 146.

[22] Rezneck, *Education for a Technological Society,* p. 44.

[23] B.F. Greene, "The true idea of a polytechnic institute," originally published in 1849, reprinted by Rensselaer Polytechnic Institute, 1949, p. 56.

[24] Rezneck, *Education of a Technological Society,* pp. 53-84.

[25] J.P. Wickersham, *A History of Education in Pennsylvania.* Originall published in 1886, reprinted by Arno Press, New York, 1969, pp. 431-432; see also McGivern, *First Hundred Years,* p. 87.

[26] Wickenden, *Report of the Investigation of Engineering Education,* p. 542; also known by the name of the study director as the Wickenden report.

[27] P. Doty, and D. Zinberg, "Science and the undergraduate," in C. Kaysen, Ed. *Content and Context.* New York: McGraw-Hill, 1973, p. 155.

[28] *World Book Encyclopedia,* vol. 19, 1975, p. 75.

[29] *Historical Statistics of the United states, Colonial Times to 1970,* pt. 2. Washington, D.C.: Government Printing Office, 1975, ser. Q. 321, p. 731, and ser. Q 47, p. 788.

[30] W.H.G., Armytage, *A Social History of Engineerng.* New York: Pitman, New York, 1961, p. 178; see also, McGivern, *First Hundred Years,* p. 92.

[31] R. Fletcher, "A quarter century of progress in engineering education," *Proc. Society for the Promotion of Engineering Education,"* vol. 4, 1896, pp. 31-50 at 37; see also Wickenden, *Report of the Investigation of Engineering Education,* pp. 750-1015 at 816.

[32] *Historical Statistics of the United States,* pt. 2, ser. W 99, p. 959; ser. P 110, p. 683; ser. P 228; p. 689; ser. P 232; p. 691; ser. P 265; p. 694.

[33] Emerson, *Engineering Education,* p. 276.

[34] R. Fletcher, "A quarter century of progress in engineering education," pp. 31-50 at 37.

[35] F.M. Bennett *The Steam Navy of the United States.* Pittsburg, Pa.: 1896, pp. 733-736.

[36] F.E. Terman, "A brief history of engineering education," *Proc. IEEE,* pp. 1399-1407, Sept. 1976.

[37] J.W. Oliver, *History of American Technology.* New York: Ronald Press, 1956, p. 361.

[38] Calvert viewed this as "a deeper struggle between two cultures—school and shop—for control of the whole process of socialization, education, and professionalization of mechanical engineers in America." See M.A. Calvert, *The Mechanical Engineer in America 1830-1910*. Baltimore, Md.: The Johns Hopkins Press, 1967, p. 62.

[39] Ibid., pp. 56-57; see also J.E. Brittain, and R.C. McMath, Jr., "Engineers and the new south creed: The formation and early development of Georgia Tech," *Technology and Culture,* vol. 1, no. 1, 1977, pp. 175-201 at 176 and 189.

[40] Calvert, *The Mechanical Engineer,* p. 281.

[41] Armytage, *Social History of Engineering,* p. 175.

[42] *Historical Statistics of the United States,* pt. 1, ser. A 29, p. 10; sec. C. 89, p. 105.

[43] Armytage, *Social History of Engineering,* p. 175.

[44] *Historical Statistics of the United States,* pt. 2, ser. Q 152, p. 716.

[45] Ibid., pt. 2, ser. S 32, p. 820.

[46] H.P. Hammond, "Promotion of engineering education in the past forty years", *Proc. Society for the Promotion of Engineering Education,* vol. 41, 1933, pp. 44-66 at 57.

[47] Wickenden, *Report of the Investigation of Engineering Education,* p. 548.

[48] "Entrance requirements for engineering colleges," Report of the Special Committee, *Proc. Society for the Promotion of Engineering Education,* vol. 4, 1896, pp. 101-173 at 103-104.

[49] McGivern, *First Hundred Years,* p. 147.

[50] W.T. Magruder, "Report of the Committee on Statistics of Engineering Education," *Proc. Society for the Promotion of Engineering Education.* vol. 14, 1906, pp. 94-96.

[51] H.L. Plants, and C.A. Arents, "History of Engineering Education in the Land-Grant Movement," *Proc. American Association of Land-Grant Colleges and State Universities,* vol. 2, Centennial Convocation, 1961, p. 93.

[52] Ibid., p. 93.

[53] H. Schneider, "Coopertive course in engineering at the University of Cincinnati," *Proc. Society for the Promotion of Engineering Education,* vol. K, 1907 p. 391-398; see also "Two years of the cooperative engineering course at the University of Cincinnati," vol. 16, 1908, pp. 279-294.

[54] McGivern, *First Hundred Years,* pp. 159-160.

[55] Armytage, *Social History of Engineering,* p. 175.

[56] *Historical Statistics of the United States,* pt. 2, ser. Q 153, p. 716; ser. 566 and 567, p. 768.

[57] Armytage, *Social History of Engineering,* p. 206.

[58] C.R. Mann, *A Study of Engineering Education,* Bull. 11, Carnegie Foundation for the Advancement of Teaching, 1918.

[59] Wickenden, *Report of the Investigation of Engineering Education,* 1923-1929.

[60] *44th Anual Report,* vol. 1 Engineers Council for Professional Development, 1976, Table 1, p. 10.

[61] "Report of the Committee on Statistics of Engineering Education," *Proc. Society for the Promotion of Engineering Education,* vol. 10,, 1902, pp. 231-257 et 239.

[62] H.H. Armsby, *Engineering Science, and Management War Training, Final Report,* Washington, D.C.: Bull. U.S. Office of Education, U.S. Government Printing Office, 1946, p. 45. This program was initially called Engineering Defense Training, then Engineering, Science, and Management Defense Training, before assuming its final title of Engineering, Science, and Management War Training.

[63] Ibid., p. VIII.

[64] Armytage, *Social History of Engineering,* p. 270.

[65] *Historical Statistics of the United States,* pt. 2, ser. W, pp. 114-117.

[66] A.B. Bronwell, "Enrollment in engineering colleges," *J. Engineering Education,* vol. 39, pp. 1-19 at 7, Feb. 1949; see also S.J. Armore, and H.H. Armsby, "Engineering enrollments and degrees in ECPD-accredited institutions: 1957," *J Engineering Education,* vol. 48, pp. 415-432 at 420, Feb. 1958; see also *Projections of Education Statistics to 1983-84, 1974 Edition.* Washington, D.C.: U.S. Government Printing Office, 1975, Table 24, p. 52, and Table 25, p. 55.

[67] Terman, "Brief history of engineering education," p. 1406.

[68] A.L. Norberg, "The origins of the electronics industry on the Pacific coast," *Proc. IEEE,* pp. 1314-1322, Sept. 1976.

[69] "Aims and scope of engineering curricula," also known by the name of the study director as the (H.P.) Hammond Report, *J. Engineering Education,* vol. 30, Mar. 1940.

[70] "Report of Committee on Engineering Education after the War," written by a committee of the SPEE chaired by H.P. Hammond, *J. Engineering Education,* vol. 34, May 1944.

[71] *Report on Evaluation of Engineering Education (1952-1955),* also known by the name of the study director as the L.E. Grinter Report, American Society for Engineering Education, June 1955.

[72] *Engineering and Technology Enrollments,* fall 1967 ed. and fall 1976 ed., Engineering Manpower Commission.

[73] T.K. McCraw, "Triumph and irony—The TVA," *Proc. IEEE,* pp. 1372-1374, Sept. 1976.

[74] *Final Report: Goals of Engineering Education,* committee chaired by E.A. Walker, American Society for Engineering Education, Jan. 1968.

A Brief History of Electrical Engineering Education

FREDERICK E. TERMAN, FELLOW, IEEE

Abstract—Electrical engineering curricula made their first appearance in the U.S. in the early 1880's as options in physics that aimed to prepare students to enter the new and rapidly growing electrical manufacturing industry. As this industry developed, so did electrical engineering education, and within a decade made a place for itself as an equal among the older engineering departments. The curricula that evolved followed the needs of the industry, and before World War I were concentrated largely on the properties of dc and ac circuits and equipment and associated systems of power distribution.

Before World War I, little graduate work was carried on, and what passed in academic institutions for "research" was typically advanced testing. The standard career pattern was to receive a B.S. degree and then obtain a job where one could learn how practical electrical work was done. After World War I, developments in broadcasting and communication led to the appearance of communication options within electrical engineering departments. Concurrently, students having a special interest in teaching or in research were increasingly encouraged to obtain the master's degree. However, the numbers who did so were small, and practically no electrical engineers sought a doctor's degree. For example, at the Massachussetts Institute of Technology in 1925 there was only one member of that large faculty who held an earned doctorate, while the background of about half of the faculty consisted of a bachelor's degree plus practical experience. Under these circumstances research performed in academic institutions was in most cases superficial, although here and there some significant work was carried on by an unusual professor.

When World War II came along and brought into being such new electrical and electronic techniques such as radar, microwaves, control systems, guided missiles, proximity fuses, etc., the electrical engineers were caught unprepared. As a group they had neither the fundamental knowledge required to think creatively about these new concepts, nor the research experience to carry through. Thus most of the great electrical developments of the war were produced not by engineers, but rather by scientists, particularly physicists who had turned engineers for the duration.

In the decade after the war, electrical engineering education went through a complete transformation. Prewar courses were drastically revised. Increased emphasis was placed on fundamentals, including particularly emphasis on physical and mathematical principles underlying electrical engineering. These results were achieved by reducing the time devoted to teaching engineering practice, by eliminating subjects such as surveying that were of little concern to electrical engineers, and by reducing the concentration on 60-cycle power. In addition, master's programs were developed that were direct extensions of the revised bachelor's program, and in time the master's degree became the recommended degree goal of the student who desired to follow a career in technical engineering.

Concurrently, the doctor's degree became the objective of those who planned a career in academia or of research in industry, or who wanted training superior to that of their many classmates working for the master's degree. With government funds available, programs of student-faculty research developed on many campuses that were the equal of the research being carried on in the best industrial laboratories.

The combined effect of curriculum changes, more students carrying on graduate work, the existence of university research laboratories of the highest caliber with this research led by well-trained faculty aided by doctoral and master's candidates, has completely changed both the character and intellectual level of electrical engineering on the campus. This is illustrated by the fact that in a 1969 survey of a representative group of major high technology firms, 82 percent agreed with the statement that "Engineers now learn enough science and mathematics so that they can adequately fill positions once occupied only by physicists." If another world emergency should arise, the electrical engineers will this time be ready to carry their share of the leadership.

Manuscript received January 1, 1976.
The author is with Stanford University, Stanford, CA 94305.

Reprinted from *Proc. IEEE,* vol. 64, no. 9, pp. 1399–1407, September 1976.

Introduction

THE HISTORY of electrical engineering education parallels the development of the electrical industry, particularly of the electrical manufacturing industry. The electrical experimenters, inventors, and innovative entrepreneurs such as Edison, Morse, Weston, Brush, Bell, Sprague, Westinghouse, Thomson, etc., who developed the early practical applications of electrical phenomenon, were either trained in related disciplines such as physics, chemistry, mechanics, etc., or were self-trained resourceful tinkerers possessing elements of genius. However, once industrial applications had been developed to the point where there were electrical installations to be designed and electrical equipment to be manufactured and sold in substantial volume, a need existed for trained electrical engineers to design, test, and improve this equipment as well as to supervise production, installation, and maintenance. Thus the history of electrical engineering education over the years has paralleled the developments taking place in electrical manufacturing.

The Beginnings of the Age of Electricity

The first important practical application of electricity was the telegraph, invented by Samuel F. B. Morse, who was an artist by profession. The key date is 1844 when telegraph service between Baltimore and Washington was inaugurated. Also, more or less with the introduction of the telegraph, electrical systems came into use for such applications as fire and burglar alarms and for railway signaling. Important as these events were, they did not create much of a demand for electrical engineers since the instruments to be manufactured were simple, and were inexpensive as compared with the value of the outside plant.

In the mid-1870's, Alexander Graham Bell, a speech teacher, began to experiment with the electrical reproduction of sound, and in March 1876 he was issued the basic patent on the telephone. An early model of his telephone system was exhibited in Philadelphia at the Centennial Exhibition in June 1876.[1] In spite of patent litigation and the need for making further technical improvements, some 778 telephones were in service by August 1877, and another application of electricity had been found.

In 1884 a group of 71 individuals who were active in the application of electricity to useful ends gathered together in New York City and formed the American Institute of Electrical Engineers. These charter members included Weston, Brush, Sprague, Edison, Thomson, Bell, Sperry, and Professor Cross.

The first industrial application of electrical *power* was the illumination of streets, auditoriums, and other large spaces by the electric arc. A commercially successful arc light system was developed in the period 1875–1879 by Charles Brush of Cleveland, culminating in the installation in 1879 by the California Electric Company of San Francisco of two dynamos supplying a total of 22 arc lights. This was the first electric central station in the world, and it was an immediate commercial success. Within six months additional equipment had been installed that supplied over 50 arc lights. During the next two years Brush central stations were established in a number of American cities including New York, Boston, and Philadelphia. Additional companies quickly entered the field; of these the most successful was the Thomson–Houston Electric Co. of Lynn, MA.[2]

During the period 1877–1880, Edison developed an incandescent lamp system[3] as an alternative to the previously used gas lights and kerosene lamps. In 1882 a central station embodying the Edison principles was completed on Pearl Street in New York City, and sold electric lighting on a commercial basis that was competitive with gas lighting. This installation was a financial success, and incandescent lighting was soon being employed in an increasing number of cities. The firm created to exploit the Edison system was the Edison General Electric Co.

The resulting widespread availability of electrical power provided by the Edison lighting system created a market for electric motors, a market which Frank Sprague was the first to exploit in a major way beginning in 1884. The electric motor soon made it unnecessary for industrial plants to use a steam engine, or to be located near water power.

The development of satisfactory electric motors opened up the possibility of electric traction. The first entirely satisfactory electric railway system was that built by Sprague in Richmond, VA.[4] It became fully operational in 1888 and made horse cars forever obsolete.

Around 1885 attention began to be given to the possibility of using alternating current (ac) instead of direct current (dc) in systems of electrical power. The transformer had been invented a few years earlier and although there was some initial confusion as to the best way to use it, there was an appreciation of the fact that the ability to transform voltage held promise of economies. The ac motors were being developed, including commutator motors, and also Tesla's induction motor for which Westinghouse had purchased the American patent rights. A meter for measuring ac power was also invented around this time by Shallenberger. The essentials for an ac system of electrical lighting and power accordingly became available.

In 1886, George Westinghouse formed the Westinghouse Company as a spinoff from his Union Switch and Signal Company for the special purpose of concentrating on ac possibilities. This was in spite of the fact that Westinghouse's advisors almost to a man felt that ac had little future in competition with dc.

Under the encouragement of Westinghouse, William Stanley installed an ac incandescent lighting system in Great Barrington, MA, in 1886, which involved transformers that were considered the heart of ac long-distance transmission. In the following years additional ac central lighting systems were installed by Westinghouse Electric Company in various parts of the United States, and also by the Thomson–Houston Company. The Edison interests reacted vigorously against the use of ac but were unsuccessful in their efforts. The controversy between ac and dc systems subsided in 1892 when the Edison General Electric Company united with the Thomson–Houston Company, already in the ac business, to form the General Electric Company. Thomas Edison felt so strongly on the subject of ac that after the merger he resigned as director of the General Electric Company and for the rest of his life had nothing to do with it.

[1] See Hounshel paper, this issue.

[2] Other pioneer developers of arc lighting systems included Edward Weston, founder of the instrument company bearing his name, and Elmer Sperry of gyroscope fame.

[3] This employed the Edison 3-wire direct-current system.

[4] See Condit paper, this issue.

Around 1890 the subject of how to utilize the power that was potentially available from Niagara Falls became a lively topic. After a number of studies the decision was made in 1893 to use ac for transmitting power to Buffalo, NY, 20-mi distant, and bids were put out in October 1893 for the first three generators. Westinghouse insisted on the use of 25 Hz rather than the originally specified $16\frac{2}{3}$ Hz and won the initial contract. Later orders were divided between Westinghouse and General Electric. The first generators went into service in 1895. As things developed, much of the power was used locally for electrochemical industries that sprang up adjacent to the power houses, but 5000 kW of the initial power was transmitted to Buffalo in a two-phase 2200-V system.[5]

Thus by 1900 an electrical industry had come into being and was a part of life in the United States. There was the telegraph and the telephone. The country's streets, stores, homes, and buildings were being illuminated by electric lighting, using either arc or incandescent lights as the occasion required. Power distribution systems had been developed that made possible the economic transmission of electrical energy over substantial distances, and electric motors were coming into use in large numbers. The electric street car was in common use. Some of the enthusiasts of the time exclaimed that the age of steam was over, and the age of electricity had arrived!

Early Electrical Engineering Curricula

The flowering of the electrical industry in the decade 1875–1885 not only established electrical engineering as a challenging profession, but also created the need for educational programs that would prepare young men for careers in this new and exciting field of activity.

The first educational program in the U.S. designed to train young men for a career in the new electrical industry was established at the Massachusetts Institute of Technology (M.I.T.) in 1882. It was under the friendly sponsorship of Physics Professor Charles Cross, head of the Physics Department, who had become interested in the applications of electricity. The 1882–1883 M.I.T. catalog describes it as "an alternative course in physics · · · for the benefit of students wishing to enter upon any of the branches of electrical engineering." In 1884 this course of study was renamed electrical engineering, although still under the sponsorship of the Physics Department where it remained until 1902, when a separate Department of Electrical Engineering was established at M.I.T.

Similar programs quickly followed at other institutions. In 1883, Cornell University announced a program in electrical engineering sponsored by Physics Professor William Anthony. Subsequently, in 1885 when Thurston became head of engineering at Cornell, he took an interest in electrical engineering and worked cooperatively with the Physics Department. In time a separate department of electrical engineering came into existence.

In 1886 an electrical engineering department was organized at the University of Missouri. The University of Wisconsin organized such a department in 1891. When Stanford University enrolled its first freshman class in 1891 the catalog stated that students interested in electrical engineering should enroll in mechanical engineering, but the 1892–1893 catalog shows a functioning but very small (one man) separate electrical engineering department.

In 1881, the year before the M.I.T. course in electrical engineering was first announced, only four of all the graduates that M.I.T. had produced since its first commencement in 1868 were working in the field of electrical engineering. Thus the first electrical engineering programs were created more in anticipation of what was expected to develop than to meet an already existing need. However, events quickly justified the supporters of these programs, and by the 1890's enrollment in them was as great if not greater than in the older fields of civil and mechanical engineering. Thus at M.I.T. 27 percent of all the institute graduates in 1892 were electrical engineers. Again, at Stanford the "pioneer" class of 1895 included more electrical engineers than either mechanical engineers or civil engineers.

The electrical content of the early electrical engineering curricula was minimal. Engineering knowledge about electrical phenomena was limited, there were few if any textbooks, and laboratory facilities were meager. For example, Harris J. Ryan, long-time head of electrical engineering at Stanford, once stated that when he entered Cornell as a freshman in 1883 the electrical engineering laboratory of the university was "little more than the electrical section of the physics laboratory of that day." The "little more" was one direct current generator built by Professor William Anthony in 1874 and exhibited at the Centennial in Philadelphia. At M.I.T. the laboratory situation was only slightly better until the completion of the 40 000 square-foot Augustus Lowell Laboratory of Electrical Engineering in 1902 financed by a memorial gift of \$50 000 made by the sons and daughters of Augustus Lowell.

M.I.T.'s 1882 curriculum for electrical engineers is given in Table I and clearly shows its close relationship to physics. It is interesting to note the absence of electives and the considerable number of required courses in the humanities and social sciences. During the following years, until well after World War I, the general pattern of electrical engineering curricula gradually changed with emphasis on dc and ac circuits, on the characteristics of motors, generators, transformers, distribution systems, etc., and on the measurement of electrical quantities. A few courses were commonly available as professional electives dealing with such subjects as communication systems, batteries, electrical railways, illumination, etc. Some schools offered a course in "wireless" telegraphy, but this was the exception rather than the rule.

Undergraduate Curricula Between the Wars

After World War I, new factors began to influence electrical engineering. The vacuum tube had become a device that could not be ignored. The broadcasting industry came into being and grew rapidly. Radio communication expanded as the possibilities of the higher frequencies became understood, and water cooled tubes were developed that could produce substantial power, including power at these "short-wave" frequencies. Furthermore, the telephone industry exploited new possibilities created by the vacuum tube, and not only steadily increased the technological level of its activities, but became of growing importance as an employer of electrical engineers.

As a consequence of these new factors, communication options (sometimes formal, often informal) began to appear in electrical engineering curricula in the 1920's and were selected by an increasing number of students. These communication programs were typically built around the interests of one or two younger faculty members, many of whom had been radio

[5] See Belfield paper, this issue.

TABLE I
ELECTRICAL ENGINEERING CURRICULUM M.I.T. 1882

FIRST YEAR

First Term	Second Term
Algebra continued.	Plane and Spherical Trigonometry.
Solid Geometry.	General Chemistry.
General Chemistry.	Qualitative Analysis.
Chemical Laboratory.	Chemical Laboratory.
Rhetoric.	English History.
English Composition.	English Literature.
French.	French.
Mechanical Drawing.	Mechanical Drawing.
Free Hand Drawing.	Free Hand Drawing.
Military Drill.	Military Drill.

SECOND YEAR

First Term	Second Term
Physics, Lectures.	Physics, Lectures.
Physical Laboratory, General Laboratory Work and Experimental Acoustics.	Physical Laboratory, General Laboratory Work, Acoustics, Simple Applications of Electricity.
Analytic Geometry.	Differential Calculus.
Shopwork, Carpentry; Wood and Metal Turning.	Shopwork; Wood and Metal Turning.
Descriptive Astronomy.	Physical Geography.
English History and Literature.	English History and Literature.
German.	German.
	General Physics, Theoretical Acoustics.

THIRD YEAR

First Term	Second Term
Physical Laboratory, Special Methods in Photometry.	Physical Laboratory, Electrical Measurements and Testing.
General Physics, Electricity, Photometry.	General Physics, Electricity.
Integral Calculus.	Advanced Physics, Memoirs, etc.
Applied Mechanics.	History of Physical Sciences.
Mechanical Engineering, Theory and Practice of Steam and other Engines.	Applied Mechanics.
Mechanical Laboratory, Use of Dynamometers, Indicators, etc.	Mechanical Engineering.
Constitutional History.	Mechanical Laboratory.
	Political Economy.
	German.

FOURTH YEAR

First Term	Second Term
Physical Laboratory, Electrical Testing and Construction of Instruments.	Physical Research.
General Physics, Applications to Telegraph, Telephone, Electric Lighting, etc.	General Physics, Applications of Electricity.
Photography.	Advanced Physics, Memoirs, etc.
History of Physical Science.	Principles of Scientific Investigation.
Mechanical Engineering Laboratory.	Advanced Mathematics.
Applied Mechanics, Thermodynamics, Hydraulics, etc.	Note.--The student is advised to take Advanced German.

hams in their earlier days, and who entered into the rapidly expanding field of what we now call electronics with an enthusiasm which they transmitted to their students. The result was that the communication options in electrical engineering grew steadily in popularity through the 1920's and 1930's.

GRADUATE STUDY IN ELECTRICAL ENGINEERING 1882-1945

Graduate study in electrical engineering beyond the bachelor's degree developed only very slowly in the period before World War I. This is illustrated by the data in Table II. The general attitude during this period was that upon obtaining a bachelor's degree the electrical engineering student should find a job and get practical experience. In fact, until well into the 1920's there was little in the way of organized instruction in electrical engineering beyond the bachelor's degree available on the typical campus, and the question could legitimately be raised as to whether many of the professors of that era were really qualified to offer *bona fide* graduate work.

Before World War I the large manufacturing concerns, notably General Electric and Westinghouse, had developed special programs for the initiation of college graduates into the world of electrical engineering. These company-sponsored activities were regarded by students as highly desirable stepping stones in the development of careers in electrical engineering. In public utilities, fresh college graduates were commonly assigned to the drafting board or to construction projects and thereby gained practical experience in a different manner. During this period many of the better organized employers felt that a college man with a master's degree was less useful to them than a man with a bachelor's degree, since in their opinion the former had wasted a year by hanging around college and thereby avoiding facing up to the real world.

TABLE II
MASTER'S AND DOCTOR'S DEGREES AWARDED IN SOME REPRESENTATIVE INSTITUTIONS

	Master's Degrees[1] Total in 5-year periods					Doctor's Degrees Total in 5-year periods				
Period	MIT	Stanford	U Cal (B)	Cal Tech	Cornell	MIT	Stanford	U Cal (B)	Cal Tech	Cornell
1900-04	2	1	0	+	6	0	0	0	+	1
1905-09	4	3	0	+	8	0	0	0	+	1
1910-14	10	4	0	+	7	2	0	0	+	1
1915-19	27	6	5	+	9	4	1	0	+	1
1920-24	127*	34	8	0	15	2	0	0	0	0
1925-29	291	66	8	15	13	4	3	1	5	0
1930-34	256	50	33	62	29	16	8	0	12	6
1935-39	215	41	31	51	5	19	5	5	15	3
1940-44	156	52	9	34	2	6	11	1	6	5
1945-49	337	200	46	99	25	16	24	1	12	17
1950-54	546	329	133	93	57	65	67	19	26	13
1955-59	665	418	133	172	46	69	94	26	17	11
1960-64	820	670	328	197	123	137	185	72	33	31
1965-69	1109	873	584	141	364	204	252	158	46	68
1970-74	602	827	630	88	300	231	242	202	43	72

1 Some of numbers contain a small proportion of pre-doctoral, post-master degrees.

* In 1920, 1921, 1922, 1923, and 1924, degrees were 7, 4, 37, 45, 34, respectively.

\+ CIT did not function as a collegiate institution until 1921.

In the pre-World War I period almost no doctor's degrees were awarded by engineering schools, as is apparent from Table II. In this period few industrial employers would have known what to do with a man who had a doctor's degree in electrical engineering, beyond ignoring the fact that he was "overeducated."

After the end of World War I, the situation began to change. By this time two new factors had entered the picture. First and most important was the growing importance of the communications field, particularly of the vacuum tube. This technology was sufficiently complex that a year of graduate work added very substantially to the competence of a young man in the communication field. Furthermore, the teachers in communication were typically young and vigorous faculty members, who were themselves exploring and developing the field of communication, and so had interesting projects for bright students who stayed in school beyond the bachelor's degree. A second factor was that by the end of World War I even the older fields of electrical engineering had matured sufficiently to provide subject matter of solid worth—material that was important in the real world but which could not be added to an already crowded four-year undergraduate curriculum. The combined result of these factors was that the period 1920–1942 saw a gradual expansion of enrollment of graduate work in electrical engineering (see Table II).

A careful examination of Table II shows that something obviously happened to M.I.T.'s master's degree program in 1922. The "event" was the graduation of the first class completing M.I.T.'s cooperative program. This was an imaginative arrangement devised through extensive discussions between M.I.T. and several thoughtful leaders at the Lynn works of the General Electric Company, including Elihu Thomson. The cooperative students were a selected group who at the end of their sophomore year were enrolled in a three-year program (including summers) that involved alternating periods of study at M.I.T. and work assignments at the Lynn plant. At the end of five years (including the freshman and sophomore years on campus) these "co-op" students received the bachelor's and master's degrees simultaneously. A unique feature of this program was that during their work periods the co-op students carried at least one regular M.I.T. course taught in the evenings by an M.I.T. faculty member or a General Electric engineer. The reasoning was that in the real world engineers would need to continue to study and they had better start developing the habit of doing so as soon as possible.

This program was extremely successful. In five years of elapsed time it gave the student a far better training than he could get in a four-year bachelor's degree program plus a year and a half of real world experience. Further, it generated enough income to more than finance the additional year involved in the program. In time M.I.T. developed cooperative arrangements with companies in addition to General Electric, and the program continues down into the present.

Although successful from every point of view, the M.I.T. type of cooperative course terminating with the master's degree was not copied by other institutions. The reasons are not clear, but perhaps lie in the fact that General Electric under Elihu Thomson's influence took a special interest in this particular program. Cooperative programs elsewhere developed in the pattern originated by Herman Schneider at the University of Cincinnati in 1906, which provided a five-year undergraduate program leading to a terminal bachelor's degree.

Although graduate study before 1942 normally meant study for the M.S. degree, an interest in doctoral study began to develop during the 1920's and 1930's. The doctoral students were few in number, but those institutions that had successful master's programs began to accommodate the exceptionally bright and ambitious student who wanted a better foundation in mathematics and fundamental sciences than was provided for in the bachelor's and master's programs, and who wished to obtain research experience. The number of such individuals was not large, as illustrated by Table II, but it nevertheless was a fairly steady trickle that slowly expanded with the years. It will be noted that California Institute of Technology was the first institution in Table II to place emphasis on doctoral studies in electrical engineering.

The typical electrical engineering teacher of the early post-World War I period combined a bachelor's degree in electrical engineering with some practical experience. Very few held a master's degree, and almost none had a doctorate in engineering earned at a U.S. institution. As an illustration, when the author was a graduate student at M.I.T. in 1922–1924, only one member of the electrical engineering faculty of that institution possessed an earned doctor's degree, and only several of those with the rank of Assistant Professor or higher held a master's degree. Of those who held the rank of Instructor (equivalent to today's Assistant Professor), over half held no degree beyond the bachelor's. With the passage of time an increasing number of young Instructors began to work toward the doctorate. Nevertheless, the number of electrical engineering teachers with doctor's degrees was very limited until after World War II.

Academic Research Before World War II

Very little research in electrical engineering was performed on campuses during the earlier days of electrical engineering education. The reasons for this are several: 1) this was a practical age in which acquisition of practical experience was regarded as more important than seeking new knowledge, 2) professors were in general not trained in the basic science of electricity, but rather in the applications of electricity, 3) the "publish or perish" syndrome had not yet been invented, 4) universities had a minimum of laboratory facilities and money for the support of research, 5) interest was heavily on electrical power, and this involved availability of machines of substantial size and cost, 6) there was a lack of graduate students to collaborate and help with the research.[6]

With the passage of time the research situation slowly improved. Islands of real research developed here and there around individual professors. Specialized laboratories dealing with high voltage were established at several institutions and collaborated with power companies on electrical power transmission problems. The developments in communication around the time of World War I and later, including particularly the growing importance of the vacuum tube, gave opportunities for research that were particularly suitable for academic work. Broadcasting was coming up over the horizon, long-distance telephony was increasingly important in communication, public address systems were in common use, talking pictures had arrived, etc. This led to opportunities for new kinds of research, and in addition made knowledge of fundamentals as important as practical experience. Even then it took time to change academia; it was not until after World War II that a substantial fraction of the electrical engineering teachers were regularly engaged in research.

The situation with respect to electrical engineering research on the campus in the mid-1920's is indicated by a survey that the author carried out in 1927.[7] This study showed that in the six-year period 1920–1925 inclusive, there was an average of nine technical papers of college origin per year appearing in the *AIEE Transactions*. This represented virtually the total research output of the nation's teachers of electrical engineering, and of their students, that was deemed of more than temporary value.[8] Approximately seven of these nine papers originated in five institutions; the remaining 100 or more departments of electrical engineering together produced a total of less than two publications per year during this period.

A similar analysis of the *Proceedings of the IRE* for the same period showed less than five publications per year of college origin, and of these over half were credited to physics departments rather than electrical engineering departments. Only one electrical engineering department in the country had more than two papers published in the *Proceedings of the IRE* during the six-year period.

If one considered that a college professor was a productive research worker if he and his students together turned out one technical paper of professional quality every two years, it is found that in the six-year study period there were a total of eight productive research workers on the faculties of the electrical engineering departments in the entire U.S., and that these eight men and their students produced over half of the university research in electrical engineering! Three of these were on the faculty of M.I.T., while the other five were distributed one to a school. Only one of the eight published anything in the *Proceedings of the IRE*.

This may sound almost incredible by present standards, but it highlights the fact that as of the early 1920's teachers taught the existing art of electrical engineering, but did very little to extend that art. It was a situation that offered a marvelous opportunity for an ambitious young faculty member with good training to make a showing. All he needed to do was to write a couple of papers that got published, and he became an important man in his EE department.

Post-World War II

World War II made profound changes in the education of electrical engineers. The war developments such as radar, microwaves, pulse technology, sophisticated control systems, electronic navigation systems, new types of electronic instrumentation, etc., added dimensions to the electrical (electronics) industry that did not die out at the end of the war, but rather continued as permanent additions to the field of

[6] Students receiving the bachelor's degree were commonly required to carry through a project and write a report, but while this gave students valuable "hands-on" experience, it was seldom true research that represented an addition to knowledge.

[7] F. E. Terman, "The electrical engineering research situation in American universities," *Science*, vol. LXV, pp. 385–388, Apr. 22, 1927.

[8] At this time in the history of electrical engineering, the *AIEE Transactions* and the *Proceedings of the IRE* were the only technical journals having substantial professional standing.

electricity. Furthermore, the technological impetus generated by the war continued into the post-war period, and led to such post-war developments as the transistor, integrated circuits, magnetic recording, computers and calculators, guided missiles, communication satellites, the laser, etc. Television displaced radio as the most popular medium of mass entertainment, to be followed by color television.

The result was a virtual explosion of the electrical (electronic) industry. Innumerable new products and devices found a ready reception in the marketplace, and new companies sprang up, first by the hundreds and then by the thousands. Moreover, the tight patent monopoly that had been maintained in the electronics industry through the 1920's and 1930's by RCA, General Electric, Westinghouse, etc., was loosened by the war developments, and the field became essentially open to all comers on reasonable terms.

The exciting developments of the war that triggered off the new electronics were largely the work of physicists temporarily turned engineers. The typical electrical engineer trained in the pre-World War II pattern did not know sufficient fundamental science and mathematics and did not possess the research seasoning to contribute in the creative electrical (electronic) developments of World War II. Engineers were relegated to working out design details, and to following the new equipment through production, test, and installation, but as a group played only a secondary role in the process of generating new ideas.

Those electrical engineering educators who participated in the war developments recognized this situation, and upon returning to their institutions at the end of the war, were forces for upgrading the education of electrical engineers. The times were favorable for doing this. War veterans were anxious to obtain systematic training in the war-time developments. Young men who had worked on war projects were available as teachers, and were not only qualified to teach the new subject matter, but were eager to do so. A scattering of middle-aged faculty members who had participated in the war activities were available to provide leadership for change. Finally, but by no means last, very shortly after the end of the war the government began to support basic research at universities in these new areas of electronics.

As a result of these influences the undergraduate electrical engineering curriculum began gradually but steadily to increase the emphasis on the fundamental science aspects of electrical engineering, particularly physics and mathematics. This was achieved by reducing the time devoted to teaching engineering practice, cutting out subjects that were of little concern to electrical engineers such as surveying, by reducing the intensity of the concentration on 60-Hz power, and by revising the content of many courses.

Master's programs in electrical engineering were developed as direct extensions of the revised bachelor's program, thus making five years of coordinated training available to turn out a well-rounded engineer. The circumstances caused the master's degree gradually to become the degree goal of the student who desired to follow a career in technical electrical engineering, and who sought training that would enable him to work with new ideas that kept coming into electrical engineering.

This new role of the master's degree was accompanied by a change in the character of the associated curriculum. Once the master's degree almost invariably required a thesis project that typically occupied one-third or more of the student's time. However, when the objective became to provide a strong general technical background, most institutions either made the thesis optional or did away with it entirely. The time thereby released was filled with additional course work at graduate or advanced undergraduate level, selected to broaden and strengthen the student's technical and scientific background.

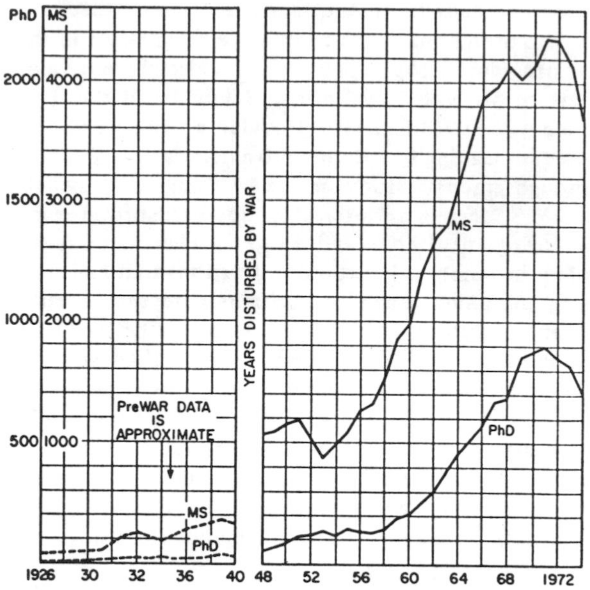

Fig. 1. Graduate degrees awarded in electrical engineering 1926-1974. Data for the pre-World War II years are fragmentary and in some cases estimated, but of the correct order of magnitude.

The doctor's degree then became the degree sought by those who wanted training superior to that of their many classmates working for the master's degree, or who planned a career of research in industry, or whose goal was to be a faculty member at an educational institution that had a graduate program.

Under these new conditions, faculty–student research became concentrated at the doctoral level, where the student had the time to perform a really important piece of research, particularly in view of the fact that before embarking on his dissertation the doctoral student would already possess the background provided by the MS course work. This greatly raised the level of faculty–student research being performed on campus.

As shown in Fig. 1, the number of master's and doctor's degrees grew steadily after the war until by the early 1970's approximately a third of the bachelor's degree students went on to a master's degree and approximately 8 percent followed their studies through to the doctor's degree. In contrast, in the early post-war period approximately 10 percent of the bachelor's degree recipients carried their studies to a master's degree, and only a few percent followed through to the doctorate. At the same time, the academic curricula in electrical engineering, including the bachelor's degree program, became steadily both stronger and broader, and by pre-war standards more difficult.

The availability of government research grants and contracts beginning immediately after the end of World War II has had a major impact on electrical engineering education. Such funds have provided the creative faculty man with the resources he needs to work effectively on sophisticated problems of contemporary importance in the real world. They also have enabled him to build up research teams of bright and eager graduate students, who are supported through employment as part-time research assistants but who simultaneously are also trainees. Research funds likewise have made it possible for the faculty man to work on his research full time during the summer, while receiving summer pay from his research project; this increases both his scholarly productivity and his income.

Thus government sponsored research makes it possible for both faculty and students to perform at a higher level than would otherwise be possible, with corresponding effect on the intellectual tone of the electrical engineering department. In addition, sponsored research supports graduate students while simultaneously giving them a unique and valuable educational experience.

The combined effect of curriculum change, more students carrying on graduate work, and the existence in the university laboratories of electrical engineering research of the highest caliber with participation in this research by doctoral, master's, and sometimes even bachelor's degree candidates, has completely changed both the character and intellectual level of electrical engineering on the campus. This is illustrated by a 1969 meeting of the corporate associates of the American Institute of Physics at which it was postulated: "Engineers now learn enough basic science and mathematics so that they can adequately fill positions once occupied only by physicists." Of the corporate associates present, 82 percent agreed with this statement.[9] It is clear that should another national emergency such as Pearl Harbor occur, electrical engineers will not be found unequal to the challenge as was the case in 1941.

The last twenty-five years have seen increasing interactions develop between universities with strong engineering and industry in geographical proximity to such universities, particularly in electronics. This is not an entirely new phenomenon, as opportunities have traditionally existed for young electrical engineers to improve their competence through enrolling in night courses at nearby institutions and for professors to gain supplementary income through consulting. The difference today is the magnitude of the activity, and the levels at which the interactions take place.

The growing importance of the master's degree has resulted in the widespread development of part-time degree programs structured to suit engineers who have full-time jobs. Various arrangements are used, including courses in the evening, courses offered in the early morning, and courses made available during working hours through live television or videotape. Today practically every center of industrial activity has available some arrangement by which a master's degree in various fields of engineering can be earned by employed engineers.

Faculty interaction with industry has also progressed beyond individual consulting. There are today a number of electrical engineering faculty members around the country who have helped found successful firms, and in some cases have given up teaching to become industrial executives. Furthermore, it is commonplace for high technology companies to have one or more faculty members on their Boards of Directors. In some cases, faculty members have even helped start companies in the role of advisors and/or investors; today there are a number of campuses on which there are one or more unobtrusive electrical engineering professors and ex-professors who are millionaires as a result of knowing which of their graduate students to back with a few thousand dollars worth of consulting assistance, often taken out partially in stock.

The importance of today's electrical engineering departments to high technology industry results largely from the high level of training of electrical engineering faculty members, combined with the opportunities faculty have to sharpen their expertise through well-financed research projects which often generate useful ideas. For example, at Stanford the government sponsored research program in electrical engineering in 1974–1975 accounted for an expenditure of approximately $6 million. This is to be compared with an expenditure on electrical engineering research of less than $20 000 per year in the era before World War II, about one-fourth of which was in electronics. No wonder today's electrical engineering faculty operate at a high level and turn out doctorate students who are impressively competent.

The oldest university-industrial complex of importance exists in New England, and is built around M.I.T. and Harvard. Many of the old-line manufacturing companies in the vicinity of Boston have a history that clearly shows important inputs of both personnel and ideas from these institutions. However, the most spectacular university–industrial development is probably that which has grown up on the San Francisco Peninsula around Stanford since the end of World War II. In this case the interactions are quite clear, since they have come into being in a relatively short time and most of the original actors are still alive. Here the contributions that electrical engineering at Stanford has made to the industrial community are quite evident, as are the contributions both intellectual and financial that the contiguous industry has made to the development of a strong electrical engineering department at Stanford.[10]

The best electrical engineering departments of today possess a sophistication and a diversity that gives them much in common with a high technology industrial complex. Each partner in this arrangement benefits from the presence of the other. As a result, one finds that high technology electronic industry tends to be increasingly associated geographically with educational activities.

If one looks at the present world broadly, it is apparent that electricity in its various manifestations is becoming increasingly involved in almost every aspect of our technological civilization and of our daily lives. Today's electrical engineer is being trained in ways that enable him to capitalize on this situation, with the result he can choose between many interesting alternative ways of spending his life. Furthermore, these alternatives characteristically interact with other disciplines. Electrical engineers today are involved not only with such traditional electrical activities as telephony, telegraphy, "wireless," and the generation and distribution of electrical power, but such matters as new sources of energy, optics as represented by the opportunities made possible by fiber optics and by the laser, the mysterious properties of semiconducting materials, medical electronics, computers (from giant computers to pocket calculators), pulse and digital techniques, instruments of almost unbelievable complexity that not only calculate the answers, but even plot the results, etc., etc., seemingly without limit.

Where this will lead is difficult to see, but one thing is certain. This is that electrical engineering is not going to stop advancing. Educators will have to continue to run fast in order

[9] See A. A. Strassenburg, "Supply and demand for physicists," *Physics Today*, vol. 23, pp. 23–28, Apr. 1970.

[10] See Norberg paper, this issue.

to keep even with their field of specialization, and practicing engineers are going to have to spend a certain fraction of their time studying the new knowledge that is being generated by our research oriented academic and industrial laboratories if they are not to become technologically obsolete as they grow older. And never again will electrical engineering have to turn to men trained in other scientific and technical disciplines when there is important work to be done in electrical engineering. Finally, the electrical industry and electrical engineering education is no longer focussed primarily on electrical energy of sinusoidal wave form at 60 ± 0.0000 Hz.

Acknowledgment

Acknowledgment is made to the following persons for helpful discussions and/or information: Arthur Norberg, Coordinator of Science and Technology Project, University of California (Berkeley), Karl Wildes, Massachusetts Institute of Technology, John R. Pierce, California Institute of Technology, and Donald F. Berth, Cornell University.

References

[1] K. L. Wildes, "Electrical engineering at the Massachusetts Institute of Technology," unpublished manuscript, 1971.
[2] H. C. Passer, *The Electrical Manufacturers: 1875-1900*. Cambridge, MA: Harvard Univ. Press, 1953.
[3] F. E. Terman, "Electrical engineers are going back to science!" *Proc. IRE*, vol. 50, pp. 955-956, May 1962.
[4] —, "Electrical engineering education—1912 vs. 1962," *IRE Student Quart.*, May 1962.
[5] —, "The development of an engineering program," *Eng. Education*, vol. 59, pp. 1053-1055, May 1968.
[6] B. Berelson, *Graduate Education in the United States*. New York: McGraw-Hill, 1960.

Part II
Overall Goals of Engineering Education

Scope and Purpose

ESTABLISHING the goals of engineering education is important because (i) the goals provide a purpose and direction to the instructional activity, (ii) the goals can be used to identify and validate the instructional activities that are in support of the goals, and (iii) the success of an enterprise can be measured only by reference to its desired goals. Although the goals can be stated at many different levels of detail, there are two major varieties. The instructional objectives for a single classroom session, or for a single course in a particular subject, are typically prepared by the instructor; the skills for their preparation are the subject of Part III of this volume. In contrast, the educational goals for an entire engineering curriculum, or for all of engineering education, are usually established through the collective wisdom of many educators and engineers, and require an extensive background, a breadth of understanding, and a long time for reflection. Naturally, the wider the scope of applicability of goals, the lesser the specificity in the goal statement, and global goals applicable to entire engineering education can be expected to have the flavor of broad philosophical statements. These are the subjects of this part.

A topic which is closely related to the educational goals is the accreditation of engineering curricula. The broad goals determine the curriculum, defined as a systematic group or sequence of courses or educational experiences that is either offered or prescribed by a school or under a program, or that is required for graduation, certification, or as a preparation in a field, trade, or profession. Ideally, the curriculum would be designed to meet the goals in the most efficient and effective manner possible. In practice, the translation of the broad goal statements into an actual curriculum is a non-trivial process, involving many constraints, uncertainties, and decisions. As a result there is no consensus among educators as to what constitutes a good curriculum, nor a universally accepted curriculum, but only broad curricular guidelines, enforced through the means of accreditation.

The accreditation of an educational program is a voluntary procedure whereby the involved individuals (practitioners, educators, scholars) and institutions (professional societies, educational institutions) review, ensure, and certify that the program meets some minimum criteria of quality and effectiveness, and this certification serves the community by encouraging continual improvement, updating and renewal in existing programs, and by providing students and employers with a means of distinguishing acceptable programs from unacceptable ones. Accreditation criteria and procedures in professional fields like engineering are much more developed than in higher education in general, and the accreditation of engineering programs in the U.S. is carried out by the Accreditation Board for Engineering and Technology (ABET). Part II includes a very brief introduction to this accrediting body and to the accreditation procedure.

The Issues

Some of the issues of interest to those involved in the development of the goals of engineering education, and accreditable engineering curricula, are as follows.

- What roles do engineers play in society, and what educational preparation is needed for each of those roles?
- What characteristics, attitudes, and training distinguish engineers from technicians and craftsmen on the one hand, and from applied scientists and researchers on the other hand?
- To what extent should the training of an engineer for technical competence be supplemented by training for other skills, such as business management and communication abilities?
- How can the inevitable changes in technology over time be accounted for in designing training programs for engineers?
- In what way should engineering education prepare students to combat technical obsolescence in their professional life?
- What is a suitable length of the educational program needed to prepare competent engineers?
- What level of specialization in individual technologies must be attained in a basic, first degree program?
- What should be the qualifications and experiences of the faculty who teach engineering?
- What should be the balance between the study of the details of specific engineering devices, processes, and techniques on the one hand, and the study of the global picture and organizing principles on the other?
- Where should engineering education stop and on-the-job training take over?

The Reprinted Papers

The following two articles (and a newspaper column) are reprinted in this part:

1. "IEEE Position Statement on the Goals of Engineering Education Report," E. E. David, Chairman, *IEEE Spectrum,* vol. 5, no. 12, pp. 81–83, Dec. 1968.

2. "What is ABET?" *IEEE The Institute,* vol. 6, no. 7, p. 10, July 1982.

3. "Accreditation: Perspectives and Procedures," I. C.

Peden and M. E. Van Valkenburg, *Proc. IEEE,* vol. 66, no. 8, pp. 849–854, Aug. 1978.

The most ambitious attempt to produce a single document stating the broad goals of engineering education to date has been that by the American Society for Engineering Education (ASEE). The ASEE has a long history of involvement in the preparation of comprehensive written goals of engineering education. The latest of these attempts was in the mid-1960s, which led to the so-called "Goals Report" in 1968. The final goals report is about 68 pages long, and therefore not reprinted here in its entirety. Instead, a highly condensed summary of the recommendations of the report, along with the position statement of one engineering society (IEEE), is reprinted here to give the reader a flavor of the original [1]. Lest one think that it is possible to have a consensus among the various constituencies and individuals interested in this task, the preliminary version of the ASEE goals statement generated several dozen printed comments, criticisms, and reactions; an incomplete bibliography of these comments is in [4]. Numerous individual opinions on the desirable goals of engineering education can also be found in the literature, containing forceful arguments, examples from personal experience, and interesting points of view, and usually emphasis on a single aspect or goal; see for example [5].

The next reprinted paper, by Peden and Van Valkenburg, is devoted to the subject of accreditation in engineering. The process, and the body responsible for it, are described in this paper. In addition, the paper briefly states the rationale behind accreditation, and the limitations of using accreditation as a means of ranking universities and controlling enrollments. Although some changes of detail have taken place in accreditation since this paper was written, including the replacement of ECPD (Engineers' Council for Professional Development) by ABET, the paper still accurately reflects the perspectives and spirit of the process. The details of accreditation, such as the accreditation criteria and a list of accredited engineering programs in the U.S., is available from the ABET headquarters (345 East, 47th Street, New York, N.Y., 10017). In addition, specific guidelines and recommendations for some of the engineering disciplines are available from the corresponding engineering societies (e.g., electrical engineering guidelines from IEEE).

The Reading Resources

[1] ASEE Goals Committee, *Goals of Engineering Education.* American Society for Engineering Education, Washington, D.C., 1986, 68 pp. This is the final report of the goals committee, as adopted by ASEE.

[2] L. E. Grinter, "Report on Evaluation of Engineering Education (1952–55)," *Jour. Engineering Education,* vol. 46, no. 1, pp. 25–63, Sept. 1955. An earlier ASEE goals report, widely quoted until the 1968 report.

[3] E. A. Walker and B. Nead, "An Interpretation by the Chairman, ASEE Goals Committee. The Goals Study," *Jour. Engineering Education,* vol. 57, no. 1, pp. 13–19, Sept. 1966. Comments by the chief author of the ASEE goals report on the preliminary version of the report.

[4] W. K. LeBold, W. E. Howland, and G. A. Hawkins, "Reactions to the Preliminary Report of the ASEE Goals Study," *Jour. Engineering Education,* vol. 57, no. 6, pp. 437–444, Feb. 1967. Contains a bibliography of the many published comments and reactions to the preliminary version of the ASEE goals report.

[5] N. H. Crowhurst, "The Objective of Engineering Education," *Proc. IEEE,* vol. 52, no. 2, pp. 202–203, Feb. 1964. An illustrative example of how the objectives of engineering education stated by individuals tend to emphasize single issues; this one emphasizes the need to teach adaptability for new, future situations.

[6] G. Beuret and A. Webb, "Goals of Engineering Education. Engineers—Servants or Saviours," *Matrix Tensor Quarterly,* vol. 33, no. 4, pp. 69–87, June 1983. Formulates the goals of engineering education based on a survey of several hundred employed British engineers and their employers; goal statements are contained in an appendix.

[7] G. Burnet, "Recent Developments in Accreditation," *Engineering Education,* vol. 66, no. 2, pp. 175–178, Nov. 1975. Describes changes in accrediting bodies, the historical background of ECPD, and some procedural matters followed in 1975.

[8] E. W. Ernst, "IEEE and Accreditation of Engineering Programs," *IEEE Trans. Education,* vol. E-22, no. 1, pp. 3–6, Feb. 1979. A description of the current accreditation criteria and procedures, and of the role of a professional society (IEEE) in this process.

IEEE position statement on the ASEE Goals of Engineering Education Report

Foreword

During the last 50 years engineering education has been the subject of several comprehensive self - investigations undertaken with the view of defining objectives, identifying methods, clarifying common problems and needs, and focusing the attention of the engineering community upon the processes by which people become engineers. Perhaps the best remembered of these among engineering teachers of today's generation is the "Grinter Report" of 1952–1955.[1]

In 1961 the Engineers' Council for Professional Development, believing the time was appropriate for another look, suggested that the American Society for Engineering Education undertake another study in this sequence of self-appraisals. This suggestion was accepted by the ASEE Executive Board and General Council in November 1961. Funding for the project in the amount of $307 000 was granted by the National Science Foundation in 1963. After 2½ years spent on an outstanding job of information gathering, analysis, and interpretation, the Goals Committee issued a preliminary report of its findings and recommendations.[2]

The Preliminary Goals Report was intended to sound out the reactions of engineers, educators, and practitioners alike, and to generate comment that might be fed into further refinement of the report itself. Members of the Goals Committee held several open discussions with interested persons at engineering society meetings in various parts of the United States during 1965 and 1966. Hundreds of unsolicited letters were received from individual engineers who wished to express their own thoughts. Formal position statements were generated by some groups and engineering organizations, including IEEE (these have not been published). An interim report, considerably modified as a result of these interactions, was issued in 1967[2]; it excited little additional comment. In January 1968 the Final Goals Report was published, the culmination of what certainly must have been the most comprehensive effort of its kind in the history of engineering education.[2]

The IEEE statement on the Final Goals Report, presented in its entirety in the following, was authorized by the Educational Activities Board. The statement was prepared by E. E. David, Jr., with consultation and assistance. It has been endorsed by the Board of Directors and the Executive Committee as representing the official position of the IEEE on the recommendations contained in the Final Report.

John N. Shive, Chairman
Educational Activities Board

REFERENCES

1. Grinter, L. E., "Report on evaluation of engineering education 1952–1955," *J. Eng. Educ.*, vol. 46, p. 25, 1955.
2. Available from American Society for Engineering Education, 2100 Pennsylvania Avenue, N.W., Washington, D.C. 20037.

As the Goals Report itself states, it is "neither a detailed evaluation, nor a consensus of current suggestions for improvements, but an effort to delineate significant trends. . ." in relation to the "directions which engineering education must take if it is to meet the demands of the future." Thus, in taking action on the Report and its recommendations, educators must interpret its generalities in terms of their local situations. As far as the IEEE is concerned, the influence of the Goals Report on the profession as a whole is at issue. Again, determining this influence requires interpretation of the recommendations. The scope and generality of the Report make this a considerable task at best. The purpose of this position statement is to bring the Report one step closer to its ultimate local applications and to delineate a position concerning its influence on the profession as represented by IEEE. The Goals Report and this position statement relate to engineering education in the United States, though some of the matters addressed may be more broadly relevant.

This statement condenses the Goals Report to eight points, each of which subsumes several recommendations of the Report. This condensation is believed to be representative of the Report as a whole, but of course all of the Report's nuances and subtleties are not represented explicitly. Actually, the condensation includes a paraphrase of all Goals Report recommendations. These paraphrased recommendations are intended merely for ease of reference to the complete versions in the Report itself, and the position stated here is based on the complete versions. To aid the reader, the recommendations contributing to each of the eight points in the condensation are identified by the number of the page on which they are found in the Goals Report. The IEEE position on the eight points is stated separately for each.

Goals recommendations and IEEE comments

1. *Broaden the scope of engineering education*
   (a) by including high-quality social science and humanities courses (p. 11)
   (b) by studying further the role of communications, humanities, and social sciences in the education of engineers (p. 11)
   (c) by increasing the flexibility of the curriculum to

Reprinted from *IEEE Spectrum,* vol. 5, no. 12, pp. 81–83, December 1968.

encourage a wider range of subject matter at both undergraduate and graduate levels (pp. 19, 41)

(d) to recognize engineering careers in design, development, management, and all engineering functions as desirable educational objectives (p. 36)

(e) so that master's degree programs with design and new pedagogical techniques are broadly available (p. 38)

(f) to use degrees intermediate between the master's degree and the doctorate more extensively (p. 42)

(g) to provide part-time advanced-degree programs (p. 53)

(h) to recognize continuing education as a distinct function (p. 59)

IEEE comment. Many electrical and electronic engineering educators have been adventurous in introducing new subjects and curricula, including interdisciplinary programs and broader elective choices for students. This flexibility is highly appropriate since the relative importance of various specialties continues to change rapidly, and new specialties are to be expected as technology and science advance and as social problems begin to influence technical components of curricula. Furthermore, students are individuals. Engineering education should capitalize on this individuality by permitting each to tailor a program suited to his own interests and talents.

Also, society is demanding an increased level of social consciousness by engineers in setting goals that are socially and technically acceptable. Emphasis on high-quality social science and humanities courses, aimed at developing a student's sense of social and esthetic values, would be welcome. A study supported by the Carnegie Foundation entitled "A Study of the Role of the Humanities and Social Sciences in Engineering Education" is being carried out. The report on this project may be a valuable guide to action in curriculum development.

The engineer's education is not complete when his formal course work is ended. It is essential to the individual, to his client or employer, and to society that his education be a continuing, lifelong process. Employers of engineers should stimulate this process by providing opportunities for continuing study. The broad responsibilities and more flexible perspectives indicated by these recommendations are to be endorsed and encouraged.

A key to attaining these goals is faculty competence and involvement, which, in turn, hinge upon the criteria for faculty promotion. Today these criteria are slanted toward research output and publication. A better balance is needed between these and accomplishment in pedagogy, as well as social involvement, public service, and engineering practice. Broader faculty horizons are a requisite for successful broader educational programs. Also, if broader faculty competence is to be more effective in the education of students, a close student–faculty relationship is essential.

2. *Lengthen the formal phase of engineering education*

(a) so that most engineering graduates have the opportunity for one year of graduate study (p. 13)

(b) so that basic engineering education (the education ideally expected for entry into the profession) includes one year of graduate study (p. 14)

(c) to satisfy the desires of individuals for higher education (p. 25)

(d) to meet the challenge of the future needs of society (p. 18)

(e) to develop fully the national resource of able students (p. 29)

(f) to live down the concept that engineering is an undergraduate discipline only (p. 35)

(g) and publicize that graduate study and advanced degrees are an integral part of engineering education (p. 35)

IEEE comment. The scope of activities encompassed by IEEE interests implies that no single educational level will be uniformly appropriate for professional standing. The level will depend upon the engineer's function—for example, research, design, development, teaching, or operating. Clearly there are many engineering tasks that require only a four-year degree, supplemented by on-the-job education and training. In fact, in many practice-oriented instances or instances in which highly specialized knowledge is required, the on-the-job aspect is unique to the engineering environment and at best can be imperfectly imitated in academic environments. These situations will not disappear in the foreseeable future. Whether these tasks are of professional stature or not at any given time depends upon the standards of the profession. The IEEE believes the situation today and for the foreseeable future justifies bestowing professional standing on the graduate from a four-year accredited engineering curriculum. In the many cases where further qualifications are desirable, selective hiring and directed educational programs should be the mechanisms rather than a blanket increase in educational requirements for professional standing.

The IEEE recognizes that increasing the educational level of the engineers is a vital goal for the future. However, lengthening the usual curriculum to five years does not assure the result. What is required is a critical examination of the educational goals to be achieved and repackaging of the relevant subject matter to attain them; in other words, modernization of the curriculum. Overall, although length and content do have a positive correlation, there is not a one-to-one correspondence. A five-year curriculum is justified only when the combination of educational goals and curriculum values so dictates.

3. *Expand the number of engineering graduates*

(a) by attracting larger number of students and providing adequate facilities, and by retaining a larger fraction of students who matriculate in engineering (p. 9)

(b) by establishing new graduate schools of high quality (p. 30)

(c) by developing new or enlarged graduate-degree programs in a deliberate time scale to maintain quality (p. 56)

IEEE comment. All signs point to the need for more, better-educated engineers, although better management and clearer goals for engineering effort would partly satisfy this need. The public conception of the nature of engineering may be a major impediment to increasing the numbers of high-quality engineering graduates. Merely expanding facilities will not necessarily result in more graduates. Steps to improve understanding about engineering careers among precollege students, faculties, and guidance counselors are vital to maintaining an adequate flow of excellent graduates to fill the broad range of engineering opportunities.

There is evidence from some parts of the United States that today there are not adequate numbers of students to fill existing and planned schools. This situation may become more general because of the new student draft rulings. Educators should proceed slowly in creating new graduate schools until existing ones have enough students and the flow of additional qualified students becomes assured. However, there may be a need for new programs and schools to provide opportunities in new or expanding disciplines not now adequately covered.

4. *Increase the flexibility of accreditation*
   (a) and accredit either the bachelor's or master's degree, or both, and accredit either individual curricula or an overall college, school, or department (p. 20)
   (b) but only the first professional degree (p. 57)

IEEE comment. Accreditation is intended as an assurance of quality education for both students and employers. Flexibility in accreditation is desirable, and we endorse the proposal to accredit *both* the master's and the bachelor's degree. Accreditation of one should not preclude the other, however.

The present practice of accrediting individual curricula permits accreditation to be based upon the degree to which the specific goals of a curriculum are achieved. Larger academic units necessarily have broader, more diffuse goals, and the accreditation process for these becomes correspondingly less specific and more difficult. It is vital that curricula be examined critically and individually before they are accredited as acceptable.

5. *Maintain and augment faculty competence*
   (a) by favoring doctorate-level degrees for new faculty (p. 44)
   (b) by providing opportunities for self-improvement through diverse opportunities beyond the home institution (p. 44)

IEEE comment. Important aspects of engineering instruction can best be imparted by practicing engineers or engineers whose background is practice. In acquiring new faculty, competence in the subject should be the criterion. Since degree level and competence are not synonymous, the former is not a guarantee of the latter. The tendency to consider the doctorate exclusively as a qualification for faculty membership is unfortunate.

Opportunities for nonacademic engineering experience are to be encouraged and expanded in scope wherever possible. A more effective symbiosis between engineering education and engineering practice is clearly desirable to achieve "more learning in practice" and "more practice in learning."

6. *Increase and broaden the base of research support*
   (a) by encouraging increased government support (p. 48)
   (b) by encouraging industry to finance a larger fraction of research support (p. 49)

IEEE comment. Achieving these goals may not be either necessary or realistic considering the budget situation in the U.S. and the increasing skepticism on the part of government, industry, and the public toward research as a panacea. Meanwhile, there is need for increased public understanding of what research can reasonably be expected to do. In order to develop this understanding it is necessary for people and organizations to present the case for research in a responsible way, particularly where the risks are high and the subject matter new or unconventional. *Promises* of results and impacts within an unrealistic time frame, and consequently not likely to be fulfilled, should not be made. It is important, too, that engineering researchers identify with the problems of society, and communicate their concern with real rather than pseudoproblems. Finally, academicians should be certain their research involvement does not detract from, but augments, their role as educators.

For the long-term future, we need to know how much research per faculty member is required to maintain a dynamic and exciting educational program. The amount will certainly be influenced by the faculty's consulting activity and the extent to which they are involved in important social problems. As faculty and students become more concerned with problems of modern significance in engineering, increased dollar support of university research (apart from inflation) may not be needed.

7. *Study technician education (p. 10)*

IEEE comment. The problems of continuously upgrading technician education to keep step with engineering education are of great moment, and must receive continuous attention by engineering educators if the supply of technical assistants is to be adequate in quality and quantity.

8. *Develop new aids to pedagogy*
   (a) particularly using electrical communication and new information displays to reach points remote from campuses (p. 54)

IEEE comment. Communication technology and computer-based systems offer many possibilities for education independent of location. However, techniques for assessing the educational values achieved are required before rational choices can be made between the many techniques available. A necessary goal should be to achieve a high educational level within realistic economic constraints.

Additional IEEE comment

The foregoing commentaries concern specific recommendations of the Goals Report. Some other matters vital to engineering education that need emphasis are

1. The necessity for continuing intellectual growth and broadening of interests of individual engineering faculty members so that new and promising subjects will be introduced within the academic framework. For example, need exists for treating broad interdisciplinary subjects, such as systems engineering.

2. The necessity for evaluating the long-term significance of newly appearing engineering specialties and providing appropriate changes in educational programs. One example is recognition of digital computation and software not only as aids to problem solving in engineering and technician education but also as engineering disciplines in themselves.

3. The responsibility of engineering education to contribute to general education so that the engineering culture and mission are understood by the nonengineering student and by other professions, the government, and the general public.

These matters, too, must be of concern to engineering educators.

The Goals Report expresses high aspirations for engineering education as a dynamic and evolving enterprise. In these aspirations the IEEE enthusiastically concurs.

E. E. David, Jr., Chairman
Ad Hoc Committee to Review
Goals of Engineering Education Report

What is ABET?

In 1932, seven engineering societies joined in establishing a new organization, the Engineers' Council for Professional Development (ECPD), formed for the "enhancement of the professional status of the engineer." How best to achieve such a sweeping goal was not immediately clear, and in its early days, ECPD invested its efforts in many different activities, from drafting an ethics code to compiling a reading list of classics for engineers. ECPD decided at its inception that the main problem it had to tackle was the quality of engineering education. In 1935, the Council began accrediting engineering programs; programs at The Massachusetts Institute of Technology, Princeton University, and Yale University were among the first to be accredited. By 1938, accreditation became nationwide. It was placed on the back burner during the war years, becoming fully operational again in 1948.

The Council altered its programs over the years by expanding its accreditation activities to cover technology programs; by refining criteria for evaluating educational programs; and changing its name to the Accreditation Board for Engineering and Technology (ABET). However, the philosophy behind accreditation has not changed in half a century: accreditation, according to the ABET charter, must be a peer review process. Programs are evaluated by engineers on the basis of standards set by the profession through the engineering societies.

How does accreditation work? ABET sends a self-evaluation form to the dean of engineering asking about a program's goals and curriculum, faculty, facilities, and finances. Once the form is returned to ABET, a team of engineers from industry and academia (suggested by the IEEE for electrical or computer engineering programs) visits the school to get a first-hand look at the program. The visiting team reports back to ABET. ABET then gives the school an opportunity to respond to the facts at issue. ("What do you mean we have no faculty in thermodynamics? We have an excellent person who happened to be in the hospital the day you visited.")

The next step is a meeting of the ABET Accreditation Commission, comprising members from all the societies, where the decision on whether or not to accredit is made. When the Commission reaches a consensus—often a long and heated process—the school is notified of the decision. The school can appeal a decision not to accredit at a special hearing attended by engineers and administrators. No lawyers are allowed because accreditation is a peer-review process.

The role of professional societies in the process is great. Though ABET sets the general criteria for accreditation (stating, for example, that students should have at least half a year of basic sciences), it is the societies that give guidelines as to what these general rules mean to a given field (for instance, civil engineering students should take courses in inorganic, organic, and physical chemistry, and so forth).

Societies are constantly modifying their guidelines to keep up with the state of the art, and the IEEE has often been a leader in keeping guidelines current. The IEEE was, for instance, the first society to introduce guidelines requiring students to use computers in advanced course work. Other societies have since followed suit.

IEEE Vice President for Educational Activities Edward Ernst believes additional changes in ABET's criteria are required. "We need more specific language in the criteria," he says. For example, they now state that 'Appropriate laboratory experience *should* be included in the program of each student.' Why don't we say '*must*'?

"The criteria have their origins in the days when we were a less litigious society. In the early days, if I believe the stories my predecessors tell me, ABET would simply suggest to a school that it make a change, and the school would try. But today schools are less compliant because they are up against hard times. So the criteria must be stated more firmly."

Changing criteria is more difficult than changing guidelines, since it requires consensus among all the professional societies. Says Dr. Ernst: "I have the feeling that we're going to see more changes in the next few years than we've seen in awhile. Still, it is an evolutionary, not revolutionary, process."

Reprinted from *IEEE The Institute,* vol. 6, no. 7, p. 10, July 1982.

Accreditation: Perspectives and Procedures

IRENE C. PEDEN, FELLOW, IEEE, AND M. E. VAN VALKENBURG, FELLOW, IEEE

Invited Paper

Abstract—The historical operation of the engineering accreditation process in which IEEE participates is described, stressing recent changes in procedure. Some key steps in the accreditation process are described, as well as some changes that might come in the future.

Introduction

ACCREDITATION is a highly charged word in the eyes of many IEEE members at this time. That fact is interesting, in itself, to those who have long been associated with accreditation of engineering programs. Just a few years ago minimal interest was aroused in the topic, and that was confined essentially to people directly affected by its actions inside academic institutions. A small band of dedicated volunteers carried out the work for IEEE in that era. Their numbers were replenished from time to time with newcomers who had demonstrated interest, qualifications, the capabilities, and suitable time schedules to discharge the responsibilities without undue hardship to other key volunteers involved in the complex chain of accreditation events that includes visitation, evaluation and report writing—all against rigid deadlines. Selections were made by a small group of IEEE volunteers experienced in accreditation affairs. They were largely educators themselves, due to the fact that little interest or committment was expressed by the remainder of the electrical engineering community. This long period of relative disinterest on the part of the general membership coincided with growth of the economy, availability of a wealth of engineering jobs, and with exciting technical advances in the field of electrical engineering on a number of fronts.

Manuscript received April 1, 1978.

I. C. Peden is with the Department of Electrical Engineering, University of Washington, Seattle, WA 98195.

M. E. Van Valkenburg is with the Department of Electrical Engineering, University of Illinois, Urbana-Champaign, IL 61801.

Reprinted from *Proc. IEEE,* vol. 66, no. 8, pp. 849–854, August 1978.

The last several years have seen a downturn in the economy, together with related upheavals that included threats to individual job security. These years have been associated with the emergence of considerable pressure for change. One target has been the accreditation programs. The process and the largely academic group of IEEE volunteers who once carried the responsibilities are objects of contention and complaint. Such activity appears motivated by the fact that accreditation is seen by some as a mechanism for potential control of the entry of new electrical engineers into the job market. In addition appropriate concern for professional quality motivates a new interest in the accreditation of electrical engineering programs. This concern is well received by the academic community, which endorses quality in engineering education as strongly as any other interest group.

Present Procedures

Who Has the Accrediting Authority?

At present, the central accrediting body in the United States is the Council on Post Secondary Education (COPA). It is concerned for uniformity and quality of the accreditation process, for seeing that the number of accrediting agencies does not proliferate, and for determining that no conflict of interests exists among them. The Engineer's Council for Professional Development (ECPD)—an umbrella organization in which numerous engineering societies hold membership—belongs to COPA. IEEE is part of ECPD, as are other participating bodies such as the American Society of Civil Engineers, the American Institute of Chemical Engineers, the National Society of Professional Engineers, the American Society for Engineering Education, The American Nuclear Society, and more. There are also regional, interdisciplinary accreditation bodies with COPA membership (i.e., North Central, Southern, etc.) which accredit entire program offerings of institutions. In this relationship, an entity such as ECPD is satisfied that programs external to engineering on which it depends—such as mathematics, physics, chemistry, etc.—are meeting general academic standards.

ECPD, through its membership in COPA, is the government-recognized body for the accreditation of all engineering programs. Because of its membership in ECPD, IEEE presently has the authority for electrical engineering and electrical technology. Should IEEE withdraw from ECPD for any reason, the government-recognized authority to accredit electrical/electronics programs would remain with ECPD.

What Does Undergraduate Accreditation Accomplish?

It is central to the concept of accreditation, as presently and historically practiced, that a determination is made as to whether or not an educational program meets certain *minimum criteria.* Briefly, they include the content of the engineering curricula being measured—as shown in Table I—and the adequacy of other key academic areas—mathematics, pure sciences, humanities, and social sciences—to support engineering; the nature and extent of the laboratory equipment, instructional space, support services, computational and library facilities; and the quality of the student body and of the faculty, and also the adequacy of their numbers to cover the curriculum and to meet the educational goals of the program. No grades are given to the individual institutions; rather, an accreditation decision relates only to compliance with *minimum* standards. Potential trouble spots are assessed, together with the time and institutional resources which might be required in order to bring them to adequate levels of functioning, before decisions and final recommendations are made.

There may be readers who believe that all electrical engineering programs in this country can be rank-ordered, the best being first and the rest following in a ladder formation, with a horizontal line drawn to delineate those programs which meet minimum ECPD standards from those that do not. According to this model, raising minimum standards would move up the dividing line and considerably reduce the number of electrical engineering programs to receive accreditation. This potentially would discredit the future graduates of the remainder, in the eyes of potential employers. The appeal of this model to those individuals seeking to limit entry into the profession is apparent. The total number of ECPD-accredited electrical engineering programs currently stands at somewhat more than two hundred; so we have a large number of schools, and numerous different educational missions being carried out. It is difficult to imagine how criteria might be established to permit a unique rank-ordering of these educational programs. Even if they were more similar than they actually are, it would be necessary to recognize further that there are no known measurement methods that could cause a unique rank-ordered list to be generated. Furthermore, it is often not understood nor acknowledged that reducing the number of newly trained engineers, below the needs of prospective employers, could cause new blood—desired by industry and government—to be trained and hired through other educational programs, i.e., physics, technology, computer science, etc. Furthermore, the use of a "ladder and magic line" model is only one possible choice. Using a different model which might be equally valid, we could hypothesize that people will do whatever is required in order to reach their goals. Raising standards would then cause no change in numbers of electrical engineers available to the job market; however they would all be better trained!

Who Benefits from the Traditional Minimum-Standards Process?

Prospective employers of the students of a particular institution are reassured that graduates will meet or exceed a certain level of performance. Students and their families are assured that the chosen institution offers a program that is validated, in this sense, by a nationally recognized body representing the engineering profession. Those students whose goal is professional registration after graduation are assisted by the fact that their educational program was ECPD-accredited.

Universities certainly benefit from accreditation; otherwise, they would not continue to seek the voluntary periodic reviews associated with it nor pay the fees required to implement it. Benefits are derived from the recognition, and from nationally available listing of all ECPD-accredited programs. A higher quality of student body can normally be attracted, thereby enhancing the reputation of the institution. Further, federal grants of some kinds are not available to the unaccredited program units within a university. Prime values are derived from periodic peer review and the self-examination preparatory to ECPD evaluation. The potential exposure of related weaknesses may become a powerful tool in obtaining needed resources in a timely manner. Engineering education is costly. Central administrators must cut back wherever they believe possible in these times of inflation and shrinking budgets. Even if tangible resources were not an issue, a well-

TABLE I
IEEE GUIDELINES FOR ANALYZING DISTRIBUTION OF COURSES
(BASIC-LEVEL PROGRAMS)

Curricular Area	IEEE Minimum
Mathematics (beyond Trigonometry)[a]	12.5%
Basic Sciences	12.5%
Engineering Sciences	25.0%
Engineering Design	12.5%
Humanities / Social Sciences	12.5%
Other Required Technical Courses	
Other Required Courses (non-technical)	
Other Technical Electives	
Other Free Electives	
Total of "Other" (maximum)	25.0%
Engineering Science Outside EE	one course
Meaningful Laboratory Experience[b]	yes
One Computer Language	yes
Computer Usage Integrated in EE Courses	yes

[a]IEEE minimum requirements in mathematics are differential and integral calculus with some analytic geometry and differential equations. It is highly desirable that one additional advanced mathematics course be included.

[b]In addition to this minimum requirement laboratories should emphasize good experimental procedures.

qualified and perceptive reviewer can provide valuable insights and good professional advice. The values of accreditation are highly regarded by educators.

What Is Advanced Level Accreditation and Where Does It Fit In?

A mechanism for accrediting master's level program was made available by ECPD for the first time in 1972. As we near the end of the first six years of trial, this mechanism—called advanced level accreditation (ALA)—has proven controversial; actually it has been adopted by so few academic institutions that its future is now in doubt.

Officially endorsed by IEEE through its Board of Directors, ALA criteria address themselves exclusively to the master's curriculum, specifying a minimum design content of one-third of the total number of credit hours devoted to pursuit of this degree.[1] Intended to provide guidelines for a terminal degree that would lead to industrial employment, the strong design focus has caused the accredited program to come into conflict with the traditional master's level program geared for later pursuit of the research-oriented Ph.D. degree. Electrical engineering in particular draws a significant number of graduate students from other engineering disciplines and from mathematics, physics, and computer science. All are excluded from entry to the accredited master's program unless they are willing to make up those deficient undergraduate hours needed to fulfill basic electrical engineering accreditation requirements. For the same reasons, foreign students are usually excluded. The potential elimination of a sizeable group of graduate degree candidates may be viewed with jubilation by those who are unilaterally dedicated to limiting the number of available electrical engineers; but it must not be forgotten that the field evolved from other disciplines and has been nourished—to become the challenging, exciting technical area it is today—due to the ongoing contributions of creative individuals trained in ancillary fields. Virtually each school having accredited master's level programs at this time also offers side-by-side unaccredited programs to accommodate the broad spectrum of student interests traditionally in pursuit of graduate engineering. Although this parallel format increases the record-keeping of the institutions, it is viewed as necessary by most planners.

Many of the arguments favoring ALA are extensions of those favoring basic undergraduate level accreditation; they will not be repeated here. Additionally, advocates argue for the value of providing a minimum level of quality in advanced design, and incentives for strengthening design-oriented programs. They correctly point out that ALA offers the only current mechanism for accrediting institutions with advanced level programs, and tend to believe that it discourages proliferation of master's level programs. Antagonists can be heard to argue essentially the opposite of some of these points. They believe that ALA encourages proliferation, that schools are forced to have at least two master's degree level programs in each field in order to meet the needs of students and the profession, and that loss of the vitalizing effect of graduate students with baccalaureate degrees in mathematics, physical sciences, etc., is counterproductive, as is the exclusion of many research-oriented programs. They believe that current curricular guidelines are too constraining for long-range viability and that they will not effectively implement quality control.

ALA's acceptance level stands at approximately 15 percent at this time. It is geographically confined essentially, although not exclusively, to the southeastern region of the country.

HOW IS ACCREDITATION ACTION TAKEN?

Procedures for accreditation actions were codified into their present form by a Task Force on Accreditation, appointed by the IEEE Educational Activities Board in 1977. This group produced a detailed documentation of IEEE accreditation policies and procedures now in use. This Task Force was chaired by E. W. Ernst with members Edwin C. Jones, Jr., Harry Venema, William T. Sackett, Earl L. Steele, and Carl R. Wischmeyer; many subcommittees assisted this group in providing policies that were seen as being responsive to concerns on the part of IEEE membership regarding accreditation. Some features of the new policies will next be outlined.

Who May Be on an Accreditation Team?

Each year, EAB selects members to be on the official IEEE *Ad Hoc* Visitors List, giving appropriate consideration to balance of industry and academic affiliation, geographical location, area of specialization, etc. Only those on this list may serve on an accreditation team, and then only after suitable instruction in procedures has taken place. An EAB committee suggests visitors for each program to be visited, taking the many factors of balance into account.

[1] For a complete statement of advanced level criteria, see "Criteria for accrediting programs in engineering in the United States including objectives and procedures," available from ECPD Headquarters, 345 East 47th Street, New York, NY 10017.

TABLE II

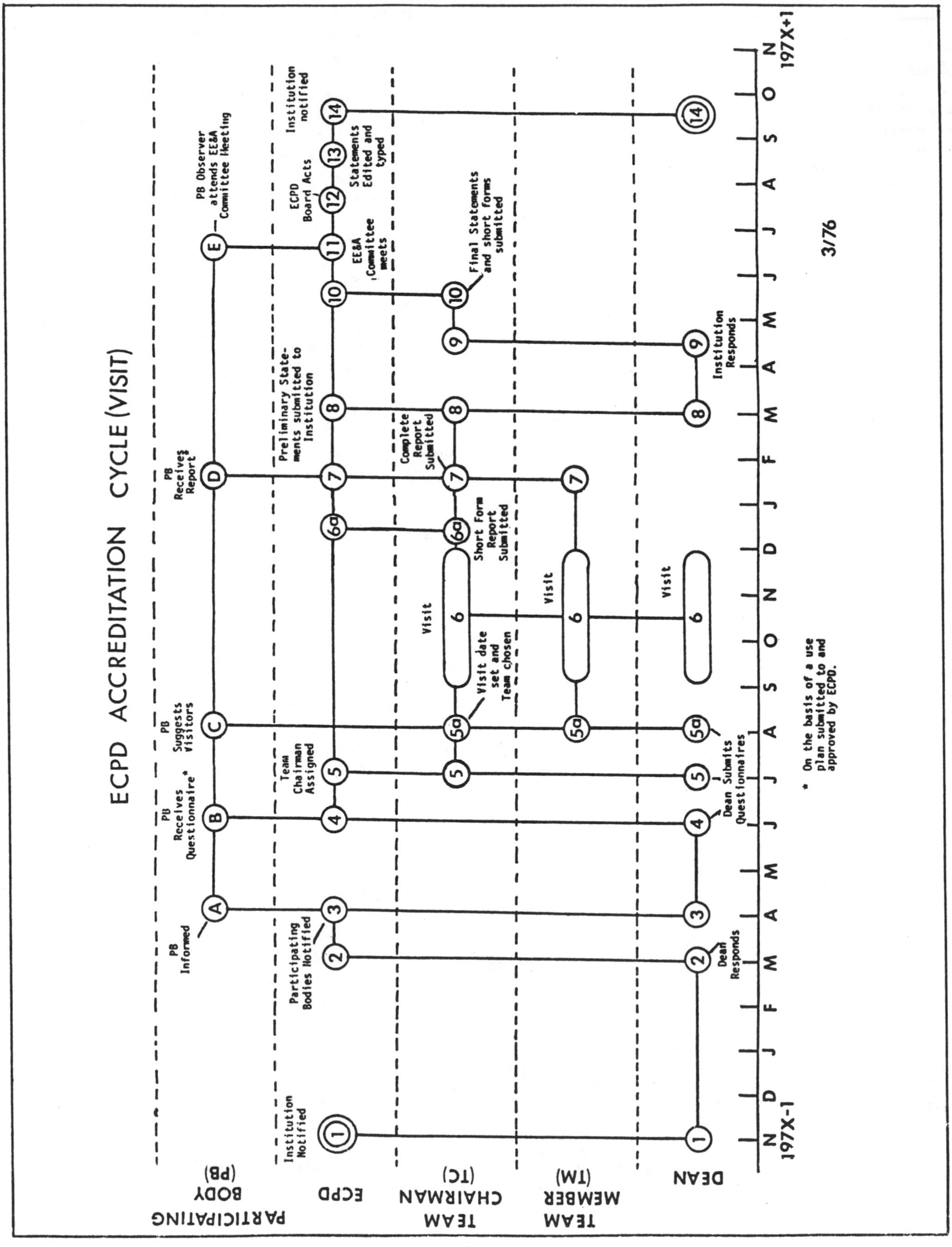
ECPD ACCREDITATION CYCLE (VISIT)
PARTICIPATING BODY (PB)
ECPD
TEAM CHAIRMAN (TC)
TEAM MEMBER (TM)
DEAN
Institution Notified
Participating Bodies Notified
PB Informed
Dean Responds
PB Receives Questionnaire *
Dean Submits Questionnaires
Team Chairman Assigned
PB Suggests Visitors
Visit date set and Team chosen
Visit
Short Form Report Submitted
PB Receives Report
Complete Report Submitted
Preliminary Statements submitted to Institution
Institution Responds
Final Statements and short forms submitted
EE&A Committee meets
PB Observer attends EE&A Committee Meeting
ECPD Board Acts
Statements Edited and typed
Institution notified
N D J F M A M J J A S O N D J F M A M J J A S O N
197X-1
197X+1
* On the basis of a use plan submitted to and approved by ECPD.
3/76

How Does the Team Function?

The team to visit a particular university is composed of a chairperson who has ECPD responsibility for a particular geographical section of the United States, and an *ad hoc* visitor drawn from the professional society responsible for each curriculum. The visit takes most of two days, and involves a searching evaluation of conditions as observed. Prior to the actual trip to the campus, each member of the team will have received extensive documentation prepared by faculty from the program to be evaluated, as well as criteria and guidelines from ECPD and the appropriate professional society. During the visit, transcripts are examined, faculty interviewed, laboratories inspected, students interviewed, etc. The observations of the visitors are reported in detail and form the basis for accreditation action. The complex schedule by which this is accomplished is shown in Table II.

Post Visit Activities

Within a month, each visitor sends a detailed written report to the team chairperson who synthesizes from them a complete proposed statement to the institution, adding the chairperson's report on administrative policies and procedures, and other general matters related to the quality of the engineering and support programs. The complete report goes to ECPD, where it is edited before its release to the institution. The Dean of Engineering receives the preliminary statement about mid-March. He has approximately one month to review it with program heads and faculty before responding. This so-called "due process response" is intended to provide the institutions with the opportunity to correct errors of fact, or misunderstandings, and in some cases to provide documentation to ECPD that supports program changes responsive to the team's suggestions and which could be implemented in the time elapsed since the visit. Responses are reviewed by the team chairman, appropriate team members, and officers of the ECPD EE&A Committee; some rewordings and/or changes in recommendations may result, before final actions are taken by the ECPD Board of Directors in the summer. A final statement goes to the institution by early fall.

Accreditation Decision-Making

ECPD presently has two time intervals over which accreditation may be extended, namely three years and six. Follow-up information is obtained in three ways, when the term is restricted to three years. Specifically, these are: a report submitted by the institution to ECPD; another visit by a different committee; or a combination of the two. A representative, though incomplete, set of examples follows.

1) Six years and a visit (6V): The program meets or exceeds minimum standards on all counts.

2) Three years and a report which may become the basis for an additional three-year extension (3R3): This action is normally recommended when the restriction was based on factors that can be unambiguously documented, i.e., acquisition of new faculty in a program whose staffing was borderline at the time of the visit; revision of catalog materials when these were previously at variance with actual program details; reallocation of required hours in the curriculum to reflect strengthening in one or more areas viewed as borderline by ECPD or IEEE.

3) Three years and a visit, which may become the basis for another three year extension (3V3): A likely recommendation in borderline cases involving factors based on subjective judgements, i.e., faculty morale and/or stability. Another visit by a different team would be preferred to a report written by administrators, for reasons of credibility.

4) Three years and show cause: The program fails to meet minimum standards on such counts as deficient curriculum; number of faculty too small to cover an adequate academic program; instability in faculty group; number or quality of students too low for continued existence of the program, etc. It then becomes the responsibility of the institution to show ECPD why accreditation should not be removed from one or more of the programs. The institution has the option to decide whether or not it can and will comply; if not, its accreditation lapses by default in three years, whereas a positive decision yields a report documenting compliance with ECPD minimum standards. If the report is deemed responsive, a new team will be sent to review the program (s) after the indicated three year "shape up" interval has passed. Virtually never is an institution so unrealistic as to believe it can implement a decision to comply when it, in fact, cannot do so. For this reason, ECPD rarely if ever needs to take the initiative to remove accreditation—the institutions take this themselves. ECPD would in no case deem a report unresponsive and remove accreditation without the courtesy of a final visit.

What Changes Are Ahead

Those who work in engineering accreditation perceive the ECPD operation as being finely tuned and polished—thanks to dedicated volunteers. The changes ahead will have only minor effects on day-by-day operation, but there will be changes. One of these relates to the way in which ECPD interfaces with the particpating bodies such as IEEE. As expressed by elected officers, IEEE is anxious to have a greater role in the various decision processes. Since other societies such as IEEE express similar objectives, it seems clear that changes will be made, perhaps as early as 1980. There is also the basic decision to be made as to the point at which a graduating engineer enters the profession. Traditionally, this has been after a four-year program in engineering. Engineering alone now has dual levels at which graduates are considered to enter the profession. Can we continue to have dual levels of accreditation, implying that some enter after four years and others after five?

IEEE guidelines for accreditation require that the student have a "meaningful laboratory experience" and that "laboratories should emphasize good experimental procedures." Whether such guidelines are satisfied is now judged by the individual *ad hoc* visitor, and most will agree that amplification of the guidelines will be useful. Can we formulate guidelines that will insure that engineering graduates are experienced in the use of modern equipment, and that the laboratories in which he or she works have no obsolete equipment. Can we replace cook-book kinds of laboratories which often require the student to verify well established laws with more open-ended experiences, called project laboratories by some? Finally, what if anything can we do to encourage a system in which faculty members are given as much credit for designing creative and meaningful laboratory experiences for their students as they now receive for publications?

Many urge that the engineering curriculum contain courses in such subjects as ethics, professionalism, writing, proposal preparation, legal aspects of engineering, etc. While few disagree that such information will be useful to the engineer, there remains the problem of designing meaningful courses. One solution to such a problem is that we accredit by the satisfaction of "end objective." At present, accreditation is accomplished, in part, by counting courses as in Table I (referred to as "bean counting"). An "end objective" approach would measure whether the student at graduation had adequate information of both technical and "professional" subjects. The manner in which the information was acquired would not be open to question.

Accreditation continues to play an important role in upgrading the quality of engineering education. Although important guidelines are provided by both IEEE and ECPD to the *ad hoc* visitor, these individuals themselves comprise the key component of the process. To be effective, each must be selected for current technical expertise, mature judgment, and the ability to make fair decisions under considerable pressure. Ideally, each functions in both evaluative and consultative modes, providing helpful advice as well as objective judgments regarding program quality. Clarity of purpose is required, to avoid emphasizing either mode in favor of the other. In taking on the primary role in both selection and assignment of its own visitors, IEEE has assumed a significant responsibility in engineering education—one that is appropriately assumed by the largest and most resourceful engineering organization of its kind in the country.

Part III
Instructional Objectives and Syllabus

Scope and Purpose

THE analysis of the broad goals of engineering, and the deduction from them of the content of an entire set of courses and educational experiences that is prescribed and required for graduation in an engineering degree program (''the curriculum'') is typically carried out by the faculty acting as a group, often requires some formal approval within the university, and is intimately related to the process of accreditation of the degree program. Since this is a collaborative process, and one that is crucial to the institution, a beginning teacher of engineering, if involved in this process, is likely to get a great deal of assistance from senior colleagues and the institution. For this reason, the subject of curriculum development is not taken up in this volume at all.

By contrast, the content of a course (''the syllabus'') is mostly the responsibility of the course instructor, and although it may have been broadly defined by an official course description, there is usually much latitude in the selection of course content, particularly in the upper level courses. It is not unusual for a beginning teacher to be assigned to teach a course, with no guidance as to the content, other than a description of the course printed in the university bulletin. How does the teacher then decide what to teach? In practice, this decision gets made in one of the following three ways:

(1) The Least Effort Method, where the course contents are governed by the time constraints, either because the instructor devotes no time to the decision making, or because the decisions are made at the last minute, leaving little time to prepare for implementation. As a result, the instructor relies on, and selects from, content that can be taught with minimal effort, that he happens to be familiar with, or that he already has instructional materials and notes for.

(2) The Borrowed Organization Method, where the instructor follows a syllabus based on a textbook, or the past history of the course, or a colleague's advice, or his own recollection of what his teachers taught in a similar course, or some other such basis. In effect, he borrows the framework and content from someone else who has had to worry about the adequacy, validity, consistency, completeness, organization, sequencing, depth and breadth of the course material. This is not necessarily a poor *modus operandi*, because often the borrower may lack the time, expertise, and experience to plan a significantly better course.

(3) Systematic Planning and Design, which requires a sizable initial investment of time, but pays significant dividends in the long run, both in the students' learning, and the instructor's convenience. This design process is the main topic of Part III of this volume.

The systematic analysis and design of instruction begins with the preparation of a detailed statement of the purpose of each sub-unit of instruction (''the instructional objectives''), and the selection of instructional framework (''the syllabus''). The statement of objectives in advance of instruction is important because it ensures that each instructional activity serves a definite purpose and is not just a time-filler, guards against inadvertent omissions, and greatly influences the outcome of instruction. These objectives can be arranged in a hierarchy in which skill at lower level objectives is necessary in order to attain the complex higher level objective. Experience shows that when the different kinds of objectives are not explicitly recognized, the instruction usually tends to be heavily weighted in favor of one or two of these objectives, and often at the lower levels of the hierarchy.

The syllabus is important because—unlike general education, where the goal is the development of some general skills such as the ability to think logically, and the actual content of the course that serves as the vehicle for learning those skills is not necessarily important in itself—engineering is a professional field, where education is expected to impart a body of knowledge in addition to the general skills. Operationally, the syllabus of a course is determined primarily by what the student does (and is therefore greatly influenced by the textbook, homework problems, and examinations), rather than by a list of topics or objectives prepared by the instructor.

The Issues

The preparation of instructional objectives, and the definition of course content, raises numerous issues such as the following:

- What are the essential constituent parts of a discipline?
- What kinds of skills must the student develop in the subject matter?
- Which skills are prerequisites to, and therefore of lower level than, other skills?
- What makes an objective teachable and testable?
- What kinds of evidence can be used to ascertain the accomplishment of a given objective?
- What factors, other than technical content, determine the suitability of a textbook for a course?
- What are the appropriate purposes of homework—drill, or extension of classwork, or individual exploration, etc.?
- How does one efficiently search for suitable homework problems that are new, challenging, or introduce variety?

- Is the teacher's time spent in grading homework problems well spent, compared to other instructional improvement activity in which the teacher could otherwise engage?
- What alternative types or mechanics of homework might be used and what purposes might be served by this?
- What is an appropriate amount of homework?
- What type of homework problems are most suitable in a given course?
- What motivation may be provided to the students to encourage them to do the homework?

The Reprinted Papers

The following three papers are reprinted in this part:

1. "A First Step Toward Improved Teaching," J. E. Stice, *Engineering Education,* vol. 66, no. 5, pp. 394–398, Feb. 1976.
2. *Instructional Objectives,* R. J. Leuba, American Society of Engineering Education, Washington, D.C., 1980, 15 pp.
3. "What to Teach: Understanding, Designing, and Revising the Curriculum," M. S. Gupta, *IEEE Trans. Education,* vol. E-24, no. 4, pp. 262–266, Nov. 1981.

Two kinds of consideration are needed to define the contents of a course, and are the themes of the papers reprinted here. First, one must determine the kinds of behaviors that the learner is expected to demonstrate (e.g., the student should be able to define, or compare, or recognize), and state them as instructional objectives. Instructional objectives are of many different types. In particular, an objective is called a behavioral objective if it satisfies three conditions: (a) the learning outcome is measurable, (b) the conditions under which the outcome is to be measured or demonstrated is specified, and (c) the minimum performance level which defines the achievement of the objective is specified. The movement toward preparation of behavioral objectives in education is a relatively recent one, having gathered support in the 1960s, when a large number of books and articles on this subject appeared in print. The first two reprinted papers are devoted to the skill of developing the behavioral objectives of instruction by an instructor. The paper by Stice is a practical one, particularly valuable to beginning teachers. The article by Leuba is a more detailed guide to the preparation of instructional objectives with engineering educators in mind.

Next, one must consider the subject matter to determine what is significant, teachable, and essential in that discipline. This consideration is discipline-dependent, and requires subject matter expertise. The third reprinted paper by Gupta suggests a method of analyzing the content of a discipline into three essential elements: domain specification, principles, and problems. This explicit recognition of the three elements serves several purposes. First, it ensures that each element receives attention in instruction, and none is inadvertently omitted. Second, it helps understand the evolution of curricula with time, as shown in the paper. Finally, when detailed behavioral objectives have not been prepared, attention to the three elements will ensure that the behavioral objectives at several different levels of the hierarchy of objectives are being reached.

The Reading Resources

A. Instructional Objectives

[1] R. F. Mager, *Preparing Instructional Objectives.* Fearon Publishers, Belmont, CA, 1962. A widely read and cited small book on the subject, written from a practical, classroom teacher's point of view. Later followed by two other books, *Goal Analysis* (1972), and *Measuring Instructional Intent* (2nd ed., 1973) by the same author and publisher.

[2] B. S. Bloom, Ed., *Taxonomy of Educational Objectives. Handbook I: Cognitive Domain.* David McKay Co., 1956. This short book is the original source of the well-known Bloom's taxonomy of objectives; widely quoted in the literature; has numerous illustrative examples from different disciplines. A subsequent volume on affective domain is D. R. Krathwohl, B. S. Bloom, and B. B. Masia, *Taxonomy of Educational Objectives. Handbook II: Affective Domain.* David McKay Co., 1964.

B. Instructional Design

[3] *IEEE Transactions on Education,* vol. E-22, no. 2, May 1979. Special Issue on Curriculum Development in an Era of Rapid Change. Many issues and opinions on curriculum design; proposed model curricula; curricula in various subjects and at different institutions; goals and procedures of curriculum design.

[4] R. B. Waina, "System Design of Curriculum," *Engineering Education,* vol. 60, no. 2, pt. 1, pp. 97–100, Oct. 1969. Expresses the objectives of a course in terms of definitive problems that the learner should be able to solve; proposes the use of these objectives for the guidance of the educational process, and for evaluation purposes.

[5] L. P. Grayson, "On a Methodology of Curriculum Design," *Engineering Education,* vol. 69, no. 3, pp. 285–295, Dec. 1978. A thorough, step-by-step discussion of the procedure of curriculum design, starting from the statement of goals.

[6] L. D. Feisel and R. J. Schmitz, "Systematic Curriculum Design," *Engineering Education,* vol. 69, no. 5, pp. 409–413, Feb. 1979. Proposes five performance objectives (define, compute, explain, solve, and judge) for engineering courses; describes an iterative procedure for analyzing the syllabus to arrive at a hierarchical chart of prerequisites.

C. Homework

[7] J. D. Horgan, "Engineering Reality in Single Answer Problems," *IEEE Trans. Education,* vol. E-21, no. 2, pp. 65–68, May 1978. Proposes the use of current professional and technical literature for designing new classroom engineering problems, and gives examples.

[8] M. W. Milligan and R. L. Reid, "Homework: Its Relationship to Learning," *Engineering Education,* vol. 64, no. 1, pp. 32–33, 65, Oct. 1973. Reports a single-trial experiment which showed that the collection and grading of homework made no significant difference in learning, as measured by tests, and that this conclusion is not affected by whether the students are non-majors or majors in the field of the course.

James E. Stice
University of Texas at Austin

A First Step Toward Improved Teaching

Ten characteristics of good teachers have come out of studies of effective teaching.[1] The good teacher knows the subject matter, is competent; presents well-prepared lectures; relates the subject to life; encourages students' questions and opinions; is enthusiastic about the subject; is approachable, friendly and available; is concerned for students' progress; has a sense of humor; is warm, kind and sympathetic; and uses teaching aids effectively.

Some of these characteristics are personal qualities and are difficult to change. One's mastery of his or her subject is assumed in this article. Attitudes toward students are the result of a complicated interplay of emotions, personal values, and prejudice; if a teacher does not like his students or himself, it may take intelligent introspection and even therapy over long periods of time in order to improve his attitudes. But those characteristics that deal with content, organization and presentation of material are subject to improvement and the time required is not prohibitive.

When a teacher asks me for help in improving his teaching, I always begin to work with his methods and procedures, because it is in this area that I believe the quickest results can be obtained in the shortest period of time. I am not a psychological counselor, and lack the expertise to embark on a lengthy program aimed at changing someone's personal qualities or attitudes. My approach nonetheless has worked well. To describe it, I will start with the question of instructional objectives.

What Is An Instructional Objective?

A lot has been written about instructional objectives. Probably the best known sourcebook is Robert Mager's *Preparing Instructional Objectives*,[2] a paperback which may well have sold millions of copies. Although the book is somewhat simplistic, it is well-written and easy to read, and the author certainly makes his point. According to Mager, "an objective is an *intent* communicated by a statement describing a proposed change in the learner—a statement of what the learner is to be like when he has successfully completed a learning experience. . . . When clearly defined goals are lacking, it is impossible to evaluate a course or program efficiently, and there is no sound basis for selecting appropriate materials, content, or instructional methods." Accordingly, a Magerian instructional objective must:

1) Describe what the learner will be doing when demonstrating that he has reached the objective.

2) Describe the important conditions under which the learner will demonstrate his competence.

3) Indicate how the learner will be evaluated, or what constitutes acceptable performance.

Put another way, a well-written instructional objective causes the teacher to ask, "Where am I going? How shall I get there? How will I know I've arrived?"[3]

Let's use an example of an instructional objective as an illustration. A reasonable objective for a class in thermodynamics might be, "Given the absolute pressure in a steam main, and the temperature indicated by a throttling calorimeter which is operating in steady-state, the student will be able to calculate the quality of the steam in the main, using data from the steam tables. He will also be able to estimate the quality from the Mollier chart." When the teacher writes an objective like this, he is stating that he wants students to be able to use the steam tables to calculate or estimate the quality of steam; he feels this is sufficiently important to be included in his course.

What does a student have to know to achieve this objective? Well, he has to know that in a steady-state operation the change in internal energy of the system is zero; that a throttling calorimeter is adiabatic; that changes in potential and kinetic energy in a throttling calorimeter are negligible; that there is no work done on or by the system; and applying this information to the general energy equation, he concludes that the expansion that takes place in a throttling calorimeter occurs at essentially constant enthalpy. He must also know that the small cylinder downstream of the throttling valve which contains the thermometer is

Reprinted with permission from *Engineering Education,* vol. 66, no. 5, pp. 394–398, February 1976.

open to the atmosphere. Using the temperature obtained from the calorimeter and the atmospheric pressure he can determine the enthalpy of the steam leaving the calorimeter. He can then enter the saturated steam tables, and using the value of enthalpy he has obtained and the specific enthalpies of liquid and vapor at the pressure in the main, he can write the algebraic equation which allows him to solve explicitly for the quality of the steam in the main.

This objective, in fact, is rather global. It not only applies to straightforward cases, it also covers unusual or "tricky" situations, such as the steam leaving the calorimeter with a quality less than 100 percent, or the case where the steam in the main is saturated vapor or even superheated vapor.

How will the learner demonstrate that he has achieved the objective? At the least he should produce two answers and perhaps a sketch. Most teachers really would like to see more than just the answers; we would like to know whether the student understood the concepts involved, and whether he correctly applied them to the particular problem at hand. So, if the student is test-wise, he will show that he knows the expansion is isenthalpic, he will put down the enthalpy of the steam leaving the calorimeter, he will show the algebraic equation used to solve for the quality, using the enthalpies of liquid and vapor at the pressure in the main, and will then indicate his answer. His sketch will be a reasonable representation of the actual process on an H-S diagram, and the quality he reads from the Mollier diagram will be a satisfactory approximation. If his method of solution is correct and he gets the right answer, the teacher can infer that the student understands how to determine the quality of steam using a throttling calorimeter, and could correctly solve a variety of such problems.

Next, under what conditions is the student expected to demonstrate his competence? The conditions implied in the statement of the objectives were that the student would have available to him a set of steam tables and a Mollier chart. The conditions can be more elaborate. For instance, if the teacher feels that students should be able to solve this kind of problem in five minutes or less, then that restriction can be added to the statement of the objective. In other situations one can specify the kind of equipment which will be available to students or the sorts of reference materials which they can use.

Finally, how is the student's performance to be evaluated, that is, what is acceptable performance? The above objective does not say anything about that, so the implication is that acceptable performance is to get the correct answer for the steam in the main. One can be more precise by saying, "Give three reasons why . . ." or "Work any six of the following eight problems."

I feel a number of authors are unnecessarily rigid in their prescription of good instructional objectives. For instance, Walbesser *et al*[4] suggest that one use only nine action verbs to indicate what the student is supposed to do to demonstrate he has achieved the objective. These nine verbs are: to name, identify, order, describe, construct, distinguish, demonstrate, state a rule, and apply a rule. The aim is to adopt a set of words which describe behaviors everyone can agree on, so that objectives will not be ambiguous and everyone can agree on what the objective will measure. But, they leave out many good words which I use all the time, such as "derive," "explain," "calculate," and "estimate." I certainly know what I mean by these words, and my students evidently do —at least, no student has complained that he did not understand what was expected of him when I used these words. But if limited to nine verbs, I can see myself getting bound up in all kinds of twisted and confusing verbiage, and coming out with some dandy grammatical horrors. I think we will achieve better communication if we use words that are common in a discipline, the good, old, comfortable words that everybody in that discipline accepts.

> "A sizeable mythology surrounds teaching and learning, and although some of the myths contain elements of truth, many reflect a kind of 'gut reaction' from a number of otherwise intelligent people."

Reasons for Using Instructional Objectives

There are at least two reasons why one would go to the trouble to prepare instructional objectives for a course. The first, and most important, is that the teacher will subject the course to a thorough analysis, and select *on purpose* what he expects students to learn in that course in the time they will spend with him. The second, and only slightly less important, is that the teacher can let students in on the secret, letting them know what is expected of them. What we are after is clear communication of intent; the teacher is "getting his head together" about what he is trying to teach, and writing down on paper what he expects his students to be able to do when the course is over, how he is going to measure whether they have done it, and what he will accept as evidence that they have achieved what he expected.

Three Types of Objectives

Most of the literature I have read on the subject of instructional objectives recognizes three different types of objectives: cognitive, psychomotor, and affective. Cognitive objectives emphasize intellectual outcomes, such as knowledge, understanding, and thinking skills. Psychomotor objectives emphasize motor skills, such as drafting, operating a lathe, and manipulation of laboratory equipment. Affective objectives emphasize feeling and emotion, such as interests, attitudes, appreciation, and methods of adjustment. (Descriptions adapted from Gronlund, ref. 5.) Engineering education is heavily oriented toward cognitive learning, although psychomotor skills also are acquired by our students along the way. Affective learning is *consciously* stressed very rarely in engineering classes. Whether this is a desirable state of affairs is beyond the scope of this paper.

Cognitive objectives, the kind nearly all engineering teachers use, are the easiest to write. Next in order of difficulty are objectives dealing with psychomotor skills. Affective objectives are difficult to write, and

Table 1. Categories in Bloom's Taxonomy of Educational Objectives, Cognitive Domain.

1.00	*Knowledge.*	The remembering of previously learned material. Represents lowest level of learning outcome.
		1.10 Knowledge of specifics
		1.11 Knowledge of terminology
		1.12 Knowledge of specific facts
1.20	Knowledge of ways and means of dealing with specifics	
		1.21 Knowledge of conventions
		1.22 Knowledge of trends and sequences
		1.23 Knowledge of classifications and categories
		1.24 Knowledge of criteria
		1.25 Knowledge of methodology
1.30	Knowledge of the universals and abstractions in a field	
		1.31 Knowledge of principles and generalizations
		1.32 Knowledge of theories and structure
2.00	*Comprehension.*	The ability to grasp the meaning of material. Represents the lowest level of understanding.
3.00	*Application.*	The ability to use learned material in new and concrete situations. Represents a higher level of understanding than comprehension.
4.00	*Analysis.*	The ability to break down material into its component parts so that its structure may be understood. Represents a higher level than application, because understanding of both content and structural form are required.
5.00	*Synthesis.*	The ability to put parts together to form a new whole. Creative behaviors are stressed, with major emphasis on the formulation of new patterns or structures.
6.00	*Evaluation.*	The ability to judge the value of material for a given purpose. The highest level of intellectual activity, because elements of all other categories are contained, plus conscious value judgements based on clearly defined criteria.

sometimes very nearly impossible to prepare. It is not so difficult to write down *what* you want your students to learn ("I want my students to learn to think"), but it is often extremely difficult to measure their degree of attainment of the objective. It is hard to tell from the way a student behaves whether he or she has learned how to think, or to appreciate, or to value. But affect often is the most valuable thing students learn in a course, and just because we cannot measure it very accurately does not mean it is not important. If we do our best to write down what we want our students to accomplish in our course, then we are more likely to design some instruction to help them achieve these goals, and more likely to design test items that effectively measure their attainment.

The Taxonomy of Educational Objectives

In the early 1950s a committee of educators began to try to identify and define instructional objectives, their aim being to agree on a classification scheme which would permit educators, psychologists, and measurement and evaluation personnel to discuss educational objectives intelligently, using the agreed set of definitions and categories. The result of their efforts was Volume I of the *Taxonomy of Educational Objectives.*[6] Table 1 above gives the categories they developed for the cognitive domain. There is not sufficient space to reproduce the entire cognitive taxonomy, but the subheadings in the Knowledge category are given to reveal the extensive nature of the taxonomy which Bloom and his colleagues developed.

Bloom's *Taxonomy* is not an easy book to read, but it is a very useful one. It discusses and describes in considerable detail the progressively higher levels of cognition and, more important, it gives a large number of examples chosen from a wide variety of disciplines. After studying Bloom, and with a little practice, you can become pretty adept at classifying the cognitive skills you wish to develop in your students. I highly recommend that you become acquainted with Bloom's *Taxonomy.* If you do not have a copy, borrow one from an older colleague. If he hasn't heard of it, I suggest you give him a short session with the knout and go directly to the dean. If he hasn't heard of it get a lock of his hair and give it to your friendly neighborhood juju man, together with whatever instructions you feel are appropriate, and head for the library. (Actually I don't mean to be smug about this; it was only eight years ago that I discovered instructional objectives, and at that time I had already been a college teacher for 13 years. But converts tend to get carried away.)

The value of the *Taxonomy* to us, as classroom teachers, is that it allows us to determine the level at which we are teaching our courses. I am sure most of us feel we are teaching thorough, challenging, and fairly difficult courses. That may or may not be the case, but you can check up on it if you are interested. Dwight Scott, from the University of New Brunswick, showed me how to do this. He took one of his final examinations in his thermodynamics course and used the *Taxonomy* to determine the cognitive level of the various questions on the exam. He was dismayed, because he did not have a single question higher than the third level (application)—only halfway up the cognitive ladder. After that he orchestrated his examinations to cover a wider range of cognitive categories, and he is much more satisfied with those examinations now. Subjecting one of my examinations to the same kind of analysis, I too was dismayed. Not a single question was higher than the third level, either. I had deceived myself into thinking that I was teaching a challenging course, because I assigned a lot of homework and almost no one finished my exams early. But time-consuming and demanding do not necessarily mean challenging, and I was not even scratching the surface of the analysis, synthesis, and evaluation categories—the higher levels of intellectual activity. You cannot deal with a problem until you know it exists, and becoming familiar with the *Taxonomy* told me something about myself. Now I do things a little differently.

Volume II of the *Taxonomy* has

also appeared; edited by Krathwohl,[7] it deals with the affective domain. A third volume discussing the psychomotor domain is apparently in preparation, but I know of nothing published on psychomotor objectives at this time. There is, however, an unpublished project report on this topic.[8]

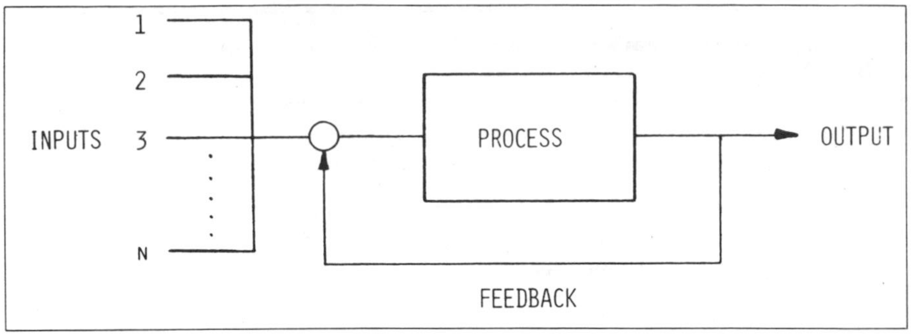

Figure 1. A simple system.

System Design

The subject of instructional objectives is only one facet of a much broader topic, that of educational system design. Figure 1 is a simple sketch of an elementary system, generalizable to almost any system—an educational system, a control system, an information system, or an economic system come immediately to mind.

If the system portrayed in the figure is thought of as an educational system, then the inputs become the attitudes, skills and knowledge the students possess when they come into a course; the attitudes, skills, and knowledge the teacher has; the materials available to the students and the teacher, such as the textbook, library reference materials, and laboratory facilities; the classroom facilities supplied, and so on. In general there is not much one can do with these inputs; they are what they are. The output is the spectrum of skills, attitudes, and knowledge the students and teacher possess when they leave the course. The educational process is whatever teacher and students do together to change the input into the output. Feedback is the information the teacher obtains through personal observation, examination results, student course evaluations, or any other means which, hopefully, he will use to make some changes in the system so that it will more effectively and efficiently produce the desired learning.

Instructional objectives are the means by which we specify the output, that is, the specific learning outcome we seek to achieve. They do not define the process, but the product. In any real design problem the quality and the amount of the product are specified, and the designer's job is to devise a process that will achieve the desired results. Further, the designer generally has various options by which he can achieve those results, and he chooses the method which is most appropriate from the standpoint of economy, ease of maintenance, ease of operation, or whatever criterion is most important. For instance, a distillation column may not be the most desirable equipment to separate products in a chemical plant. Depending on the situation, it may be better to use liquid-liquid extraction, crystallization, absorption, or evaporation.

Theoretically, the same holds true for instructional design. One chooses teaching methods and materials only after instructional objectives are known; the teacher employs those methods and materials best suited to help the learners achieve those objectives. As a practical matter, there are some constraints in applying this model to educational design. Educational resources and funds are limited, and there are many demands on the time and energies of engineering teachers. Many educators are committed to the lecture system; they are good at it; they feel comfortable with it, and they are not going to change. That really doesn't matter. Instructional objectives are still the most important tool in the teacher's kit, because they specify the outcome of a course.

When I wrote my first set of instructional objectives, it was for a course I had taught eight or ten times by the lecture method. It took considerably longer than I had expected, and I spent about two days of concentrated effort going through the textbook to decide what topics were of paramount importance and what topics were "nice to know" but not essential. When finished, I was not very well satisfied with the results and laid them aside. About a week later I hauled them out again and worked on them some more. Finally obtaining a list of objectives I thought I could live with, I was a little surprised at the results. Several topics were omitted that I had always spent time on before, but which were not prerequisite information for the following course in the sequence. I ruthlessly excised some of my favorite "war stories" (I have been known to tell them) which were fun to relate, but which really did not have anything to do with the course.

The omission of the "nice to know" material and the irrelevant yarns yielded about three weeks of extra time, which I was able to use in covering material that was important and was prerequisite for the following course, but which I had never had time to cover in the past. Using these objectives in teaching the course the following fall, I knew where I was going and why I was going there, and most of us got there. The students reported that the course was coherent and logically developed, and they appreciated knowing what was expected of them. They were better oriented and were able to use their study time more efficiently. They knew what I wanted them to learn, and therefore they knew what they did not need to spend time on. They learned a lot of the "nice to know" information anyway, but they didn't worry about it—and they did a better job of learning the things I felt they had to know. Incidentally, I gave the students the objectives, but not all at once. I figured that if I gave them the objectives for the whole semester all together, it would very likely scare the hell out of them. So they got lists of objectives for each chapter, with a new list every three or four weeks; this seemed to work very well.

Engineers are used to the idea of design, and therefore engineering educators are probably better equipped for educational design than almost any other group. The philosophy and process of design are pretty well developed; we teach it in many of our courses. We know it works and we believe in it. It works equally well in education, and if we do not apply it there, we are not practicing what we are preaching.

Pro and Con

Having spent some time extolling the virtues of instructional objectives, it seems appropriate to list some things college teachers have had to say about them, both pro and con. I am indebted to Susan Hereford, associate director of the University of Texas Measurement and Evaluation Center, for the following list of statements she has gleaned from consultation with faculty members and from workshops.

The Cons

Specifying instructional objectives seems like a good idea, except it—

- Discourages creativity on the part of the teacher and the learner.
- Takes the "challenge" out of studying.
- Is not worth the amount of time and effort required.
- Leads to concentration on the specific details of a subject, while the "big picture" is missed by students.
- Makes test scores meaningless, since students know their learning goals in advance.
- Insults the students' intelligence.
- Seems mechanistic and dehumanizing.

The Pros

On the other hand, specifying objectives—

- Forces an instructor to critically evaluate the relative importance of topics and the allocation of instructional time.
- Can contribute to a more open and candid classroom atmosphere, and more positive and honest teacher and student relationships.
- Focuses the student's attention on learning tasks rather than on "psyching out" the instructor.
- May promote rather than discourage creativity through the reduction of anxiety about tests and grades.
- Causes the teacher to appreciate and make good use of individual differences in teaching and learning styles. It specifies the *product* and allows intelligent choice of the *process* by which an individual teacher or learner progresses toward the goal.

A sizable mythology surrounds teaching and learning, and although some of the myths contain elements of truth, many reflect a kind of "gut reaction" from a number of otherwise intelligent people who are not familiar with the facts. Similarly, most people used to believe we live in an earth-centered universe because of the evidence available, and they had no way of knowing the true state of affairs. The first seven statements above (the cons) are incorrect for most students and most teachers; given the variability of human beings, some of the statements are true for a small minority of students. The five statements listed under the pros represent the current conclusions of a large body of educational research.

Summary

There are a number of characteristics which effective university teachers possess. Many of these depend on personal qualities and attitudes, and changing them requires considerable time and often outside assistance. Course content and organization are areas in which a teacher can improve, with little or no outside assistance, at little or no monetary expense, and with reasonable expenditures of time. Preparing instructional objectives gives the teacher a clearer idea of the purposes and goals of his course, and introduces him to the concept of instructional design and the attendant benefits which can accrue to his teaching activities. Well-written objectives also aid students by providing a more logical and coherent framework for a course, by letting them know what is expected of them, and by allowing them to use their time more efficiently. Instructional objectives are particularly useful in the highly structured kinds of knowledge which make up most engineering courses, although they also have been proved very helpful in such fields as music, drama, history, Elizabethan literature, and philosophy.

If you want to begin to improve your teaching today, preparing and using good instructional objectives is the place to start.

References

1. Sheffield, Edward F., *Teaching In The Universities—No One Way,* McGill-Queen's University Press, 1974, pp. 206-218.

2. Mager, Robert F., *Preparing Instructional Objectives,* Fearon Publishers, 1962, p. 52.

3. Mager, Robert F., *Developing Attitude Toward Learning,* Fearon Publishers, 1968, p. vii.

4. Walbesser, Henry H., and Edwin B. Kurtz, Larry D. Goss, and Richard M. Robl, *Constructing Instruction Based on Behavioral Objectives,* Engineering Publications, Oklahoma State University, 1971, p. 41.

5. Gronlund, Norman E., *Stating Behavioral Objectives for Classroom Instruction,* Macmillan Co., 1970, pp. 18-25.

6. Bloom, Benjamin S. (ed.), *Taxonomy of Educational Objectives. Handbook I: Cognitive Domain,* David McKay Co., 1956.

7. Krathwohl, David R., Benjamin S. Bloom, and B. B. Masia, *Taxonomy of Educational Objectives. Handbook II: Affective Domain,* David McKay Co., 1964.

8. Simpson, Elizabeth Jane, *The Classification of Educational Objectives, Psychomotor Domain,* unpublished progress report, University of Illinois, Urbana, 1966.

Instructional Objectives

RICHARD J. LEUBA

Introduction

SINCE Ralph W. Tyler, at the University of Chicago, enunciated his concept of instructional objectives in 1949, literature on the subject has mushroomed. There are literally hundreds of published books and articles on instructional objectives. Some of the effort is research, relating use of objectives to teaching effectiveness. Much of the writing is polemical. An increasing body of literature is elucidative, intended to help the teacher employ objectives for improved instruction.

A milestone is Robert Mager's now quite famous little book, *Preparing Instructional Objectives,* 1962. A major undertaking is the cooperatively edited, three-part handbook, *Taxonomy of Educational Objectives,* begun in the 1950s. These earlier works have been followed with books by Cohen, Kibler, McAshan, and others. Chapters on instructional objectives now appear in educational psychology texts—for example, Anderson and Faust.

In engineering, frequent references to objectives may be found in the writings of such perennial contributors to *Engineering Education* as Gordon Flammer of Utah State University, Lee Harrisberger of the University of Alabama, and Charles E. Wales of West Virginia University.

This guide is for the educator who may have heard of objectives and who would like to know more. It should also serve the instructor who is using, or has used, objectives and is troubled by some aspects. (Throughout the paper, "he" and "his" are used to represent the correct, but cumbersome, "he or she" and "his or her.") So, let us get on. When you have completed your study of this paper you should be able to

1) Spot the following weaknesses in an illustrative set of defective objectives: a) failure to express in action, or behavioral, terms, b) failure to specify conditions, and c) failure to define the required level of performance.
2) Correct a defective instructional objective, using your own words.
3) List, and give examples of, the three requirements of an instructional objective as given in (1) above.
4) Recall and discuss at least six commentaries on practicalities and philosophy of instructional objectives.
5) Give three reasons for instructional objectives, with specific illustrations of each.

Instructional Objectives are Specific

Instructional objectives have to do with the goals of education, and as we shall quickly see, to be most useful, they must be quite *specific*. But before turning to specifics, let's take a look at rather broad goals and see their weaknesses. Some broad goals of education are to teach students to think, increase students' knowledge, develop students' potentialities, teach salable skills, teach professional attitudes and develop students' self-confidence. Laudable and important as these goals are—and you could list more—they are too vague. There is vast latitude in the way an instructor might plan and execute a particular course using such goals. Widely different, conscientious interpretations of these goals by different instructors would have different outcomes in terms of what students learned. This disparity raises doubts about what such goals say. Criticism is directed not at the apparent spirit of such goals, but at their lack of specificity.

In engineering education, generally speaking, our goals are not so diffuse: We want students to learn computer programming, introductory electrical engineering, statics, and so forth.

Taking statics as an example: We know it is insufficient to say only that a student should learn the text, solve the assigned problems, do the lab work, and pass the examinations. While such aims are more detailed than the broad goals cited above, we can break down the task even further. Unfortunately, we often do this the wrong way, or we do not go far enough. For example, we say that students must

Know how to apply $\Sigma F_v = 0 \quad \Sigma F_h = 0 \quad \Sigma M = 0$;
Understand the free body method;
Be familiar with solving composition of force problems;
Appreciate the time-saving potential of the digital computer;
Be aware of three-dimensional problems;
Be sensitive to the existence of insoluble problems (redundant members).

Now, let's pause and examine the language we have used to describe the student who successfully completes our course in statics. The student will

know ...
understand ...
be familiar with ...
appreciate ...
be aware of ...
be sensitive to ...

Although these are all common words and phrases used in writing course objectives, they all communicate poorly because they are *subjective*; they suggest something which is *in* the student, beyond direct observation. And here we come to our first principle:

First principle: Instructional objectives must be phrased in language that describes *observable* performance or a product of performance.

Reprinted with permission from *Engineering Education,* pp. 1–15.

What is the observable performance when a student demonstrates that he "knows," "understands," or "is aware of?" Instead of subjective expressions, you must be sure to use action verbs, such as *draw, solve, write,* and *list*:

The student shall be able to

- *Draw* a free-body diagram;
- *Solve* a three-force member problem;
- *Write* an outline of a chapter;
- *List* four problem-solving methods.

Although we are still some distance from a thoroughly defined instructional objective, we have removed a major vagueness by defining the goal in *terms of an observable act.* Or, we could define the *product* of an act, such as a project report. If students *know* how to solve a class of problems, we really mean that they have been given problems to solve and they *have solved them.* How else can the instructor "know" that students "know?" (And how else can students "know" that they, themselves, know?)

To demonstrate to you the vagueness of a subjectively written objective, consider this one:

The student must be familiar with applying the computer to statics problems.

You notice immediately that "be familiar with" is not an action verb. Say that students study a lesson; here are the six different learning outcomes telling what the six, respectively, can do:

Student 1: Can write a two-dimensional force problem in computer language, punch the IBM cards, order the program, cvaluate the results against a non-computer solution, and compare the two methods costwise.

Student 2: Can tell, verbally, the steps to be followed, but cannot actually do them.

Student 3: Can write a force problem in computer language and can interpret the printed results, but cannot evaluate the potential cost benefit.

Student 4: Can say, in principle, which classes of problems are more economically solved by computer, but is unable to program a problem himself.

Student 5: Can give a verbal account of the tour through the campus computer facility, but only that.

Student 6: Can recall watching someone demonstrate the procedure.

All these students are "familiar with" applying the computer to statics problems, but obviously the objective was too vague for us to know which, if any, of them have learned what the instructor may have had in mind. Or, the instructor simply did *not* have a specific objective in mind, and that is the difficulty!

Before taking up the qualities needed in writing complete objectives, let us first touch base with some pertinent rhetoric of educational psychology. We have stressed phrasing objectives in *action* terms: to the psychologist, action is *behavior.* Instructional objectives are therefore often called "behavioral objectives" for the very reason that they define the action, or behavior, that students demonstrate as they show what they have learned. For some nonpsychologists, behavior is unfortunately associated with "misbehavior." A problem child is a "behavior" problem.*

Conditions

Now we are ready to move on. Besides describing observable action, an objective must also prescribe *conditions* of the action. For example:

- Using a model XX calculator, the student shall be able to solve truss problems of the type illustrated on p. 99 in the text.

The student is to use a particular calculator and solve a particular type of problem. Another example:

- Using standard drawing instruments, and given three elevations, locations and depth-to-stratum, the student shall graphically determine the strike and dip of the stratum.

The student shall use graphics to solve the problem, and the conditions are given.

Second Principle: Instructional objectives must specify the conditions under which the student is to demonstrate his learning.

Performance Criteria

Clearly lacking from the illustrative objectives, so far, is any prescription of what performance is acceptable. Therefore:

- With a model XX calculator, the student shall be able to solve at least 9 out of 10 truss problems of the type illustrated on p. 99 in the text, in 50 minutes, to an accuracy of $\pm 3\%$.

Performance criteria are now included. The number of permissible wrong solutions is one out of ten, accuracy is $\pm 3\%$, and the time is 50 minutes.

Performance criteria are not always so simple. How, for instance, does one define acceptable quality of hand lettering, of organization and clarity in a lab report, or of an oral presentation? One answer is to save and display examples of previous students' work, accompanied by an instructor's written evaluation. Audio and video recordings of previous students' oral reports help clarify standards in this difficult area. By performing in front of a panel of judges (who give their comments) and by hearing instructors' comments when other students perform, students learn to recognize a high quality performance.

Third Principle: Instructional objectives must include performance criteria that enable the instructor, and student, to know if the objective has been met at the requisite standard.

* For readers of contemporary science, behavior is often associated with studies of animals. B. F. Skinner's manipulated pigeon behavior springs to mind. Understandably, we prefer not to equate students with pigeons. It is hoped you will rid your subconscious of prejudices and think of "student behavior" as no less dignified and no less desirable than "student performance," in the best sense.

Summary

Before pushing ahead with some generalizations about the writing and use of instructional objectives, let us pause to review the three main characteristics of an instructional objective. An objective:

1) Is written in action, or behavioral, terms, describing what students do, or produce, as they demonstrate their competence.
2) Specifies the conditions under which the action is to occur.
3) Includes a standard of performance that enables a qualified person to ascertain the degree to which the objective has been met.

Problems, Practicality and Philosophy

Any green thumb gardener tempers what the gardening book says with the wisdom of experience; here are some observations based on experience in writing and using instructional objectives.

1) *Objectives alter the instructor-student relationship.*

The old practice of specifying vague objectives can be compared to holding all the aces. It is like selling merchandise in an unlighted store. Students guess what the instructor expects and what they will be tested on. ''Psyching out'' the teacher is the name of the game. To write and distribute explicit objectives is to turn on the lights and lay your cards face-up on the table. This relinquishes teacher power, but if you use objectives successfully you have the compensating satisfaction of fair play and of seeing your students—spurred by explicit goals—study and achieve.

2) *Write objectives for ''higher-order'' learning, too.*

Do not become so hypnotized with objectives that you find yourself teaching only to the objectives that neatly satisfy all three ''principles.'' It is easy enough to write objectives for simple recall, but it is not easy to write an objective demanding, say, independence of thought and action. So what do you do? The answer is, do not give up! Write the objective as best you can. Don't worry if it takes several sentences; it probably will. Consult the Bibliography for further reading and for ideas.*

3) *Include subordinate objectives.*

How frequently do we discover students stumbling with the lesson because of a deficiency in prerequisite knowledge! An earlier illustration stated that the students ''... shall determine the strike and dip of a stratum,'' but suppose they do not know the terminology? A subordinate objective should be written stating that the student shall be able to define ''strike'' and ''dip.'' Write subordinate instructional objectives for each successively subordinate learning goal until—working backwards—you get to where the student is. Turned the other way, this says: write your objectives beginning with the students' existing knowledge, and go forward with a hierarchy of objectives until you arrive at the *terminal* objective for, say, the semester course. (In some writings a subordinate objective is referred to as an *enabling* objective.)

4) *Include objectives for attitude development.*

Goals of engineering education include development of proper attitudes in such areas as the profession, efficiency, ethics, natural resources, team efforts, and service to society.** So it is necessary and proper to include objectives in the *affective,* i.e., ''attitude,'' domain. For example, what does a student believe is the role of thermodynamics in the future progress of humanity? What do you want him to believe? How does he feel about the possibilities for human betterment in terms of integrated circuits or of operations research? And what do you want him to believe? What is a responsible attitude about natural resources? This is an area of teaching that is rarely attacked head-on. All of us do believe, deep down, that some attitudes are vastly to be preferred over others. Recognizing that if we desire certain outcomes in the affective domain, we should write objectives for them.

5) *Broad and lofty goals have their place.*

The thermodynamics student should have a ''... thorough understanding of the entropy principle.'' This has a profound ring, and if that is all that is said, we are up to our knees in vagueness. But the instructor can flesh out such a goal with many detailed objectives written in action terms, and then compare these detailed objectives, from time to time, with the broadly stated goals. A sailboat skipper sets aim on a prominent landmark across the lake and translates this goal into detailed actions with tiller, jib, and mainsheet; but from time to time, he compares the outcome of the detailed actions with the boat's progress towards the main goal. In the entropy example, detailed objectives refer to demonstrated ability to solve certain very specific types of thermo problems, but the instructor should still be asking himself, ''Am I satisfied that there *is* a 'thorough understanding'?''

6) *''Vest pocket'' objectives.*

I have emphasized that it helps to clarify one's thinking by setting down one's objectives in writing, but having done so, it may be the better part of wisdom to keep some objectives in the ''vest pocket.'' An explicit objective calling for more initiative, say, in using the engineering library—if it is handed out to the students—may only meet with rebuff. The artful instructor keeps the objective pocketed, while meantime devising tasks and providing hints, incentives, cajolery, or whatever, to get students to explore and use the library. He can use various means—questionnaires being one—to ascertain the degree of success in meeting the objective, which, by

* Systematically writing out objectives is useful because it enables you to see if your teaching is directed too much towards lower-order learning. If you collect all the quizzes and exams given in your course last semester and write corresponding objectives for each of the test items, you may be surprised (see Bibliography: Vargas; Bloom).

** The list of attitudes of concern to the engineering educator could go on. Further examples include confidence, a critical attitude, interest, tendency to be observant, curiosity and so forth. Once an instructor formalizes the attitude type of course objectives, he will begin to think of ways to develop, and even test for, these attitudes in students. Some ''attitudinal'' behaviors border on, or actually overlap into, the ''cognitive'' domain (see Bloom; Krathwohl; Ringness). In fact, as one begins to think about it, the divisions of learning into ''cognitive'' and ''affective'' (a third is ''psychomotor'') are seen as convenient, but arbitrary. On a map, meridians are convenient, but in our travels we do not actually encounter them.

the way, is more formally termed a *voluntary performance* objective. There is a voluntary performance objective for this guide; it does not appear in the introduction with the others. It is:

> At the end of this lesson, you will show, or soon show, evidence of plans and intention to use instructional objectives to improve your teaching in the near future.

To have displayed such an objective at the outset may have turned off the reader. If, on the other hand, these pages have done their job, this objective is met.

7) *Student participation in goal setting.*

Anyone who complains that objectives are originated only by the *teacher* is blind to the possibilities for student-selected and student-constructed objectives. Objectives are compatible with a philosophy of student participation in education. One well-known idea is "contract learning," in which a student—most often in a tutorial relationship with an instructor—agrees to a specified body of work that he will do for a specified grade. Although student goals and teacher requirements frequently change as a student gets into his "contract," an important aid to communication between student and tutor is a set of current goals, in the language of instructional objectives. One of my headaches with contract learning, before I became aware of instructional objectives, was the considerable vagueness of the contracts. Particularly for contract learning in new territory, flexibility is important, and it would be counter to the principle of educational freedom to hem the student in with objectives, even though they may be wholly or partly of his own making. On the other hand, the specificity of continually revised objectives would help student and tutor understand where each stood at any moment during the term of study.

One other illustration of student participation is *selective objectives.* Where individualization is practical, a student may choose objectives he wishes from a selection made available by the instructor.

8) *Do not confuse the objective with the pedagogical means.*

A team of sophomores receives an assignment to design, build and test a portable solar hot-dog cooker. Clearly, building the cooker is not the instructional objective; rather it is the pedagogical means to that end. The objective itself (and there are obviously several objectives associated with this project) is to be able to provide a solution to an open-ended design problem (with further conditions and qualifiers).

9) *Objectives help uncover gaps in teaching.*

"Input" is the instruction. "Outcome" is student learning. Regular use of objectives can help instructors in their "input-outcome" analysis. For example: A prevalent frustration for instructors is the seeming inability of students to apply their learning to problems that diverge from the ordinary. An algebra student is able to solve equations, but when first confronted with word problems is unable to succeed.

Two observations: a) People rarely make big mental leaps; they have to be taught a step at a time. Equation solving ability is a little different from word problem ability. If you expect students to solve word problems, then teach and drill with word problems. b) Instructors nevertheless wish that students would take bigger mental leaps. In fact, we repeatedly insert questions a little out of the ordinary into our exams, only to have our hopes dashed by the failure of our students to translate what they know to a slightly different context. If you expect students to discover the not-so-obvious, then teach techniques of translation and discovery. These techniques of teaching, and for teaching creativeness in general, are beyond the scope of this guide, but they *are* teachable. Where do objectives fit into this? When you think deeply about input-outcome relationships, you may discover that some goals (implicit in your examination questions) are not backed up by explicit teaching. As illustrated here, if you discover that you expect students to learn to be creative, then you will need to write objectives for creativeness and seek ways of developing in students this important capability.

10) *Learn to live with the gap between ultimate performance and tested performance.*

"Ultimate" is what you want the student to be able to do in real circumstances. "Tested" refers to the practicalities of academic testing. An ultimate objective might be:

- Given sudden notice of the failure of production-line components to meet performance specifications, the graduate shall be able to investigate and furnish the chief engineer with a verbal report within 30 minutes (with, obviously, further qualifiers and conditions).

Although instructors seldom or never have the opportunity to test their students under such realistic conditions, an instructor who fails to keep ultimate objectives in view will miss opportunities to test students under more realistic conditions. The ultimate objective is a continual reminder and challenge to figure out ways of testing as close as possible to the ultimate. Another example: Engineers are criticized for poor writing and verbal ability. An ultimate objective might be:

- The student shall be able to describe verbally and discuss with a critically inquiring referee three nontrivial illustrations of the second law of thermodynamics (and the objective would continue with qualifiers and conditions).

The objective suggests an oral examination. We use nonverbal problem-solving examinations, almost exclusively, because of the impracticality of oral examinations in large classes. As instructors, we rarely have the opportunity to observe the ultimate achievement—when the student has graduated and is on the job—but the deliberately constructed objective serves us as an excellent guiding principle in course and curriculum design.

11) *Unanticipated consequences of instruction.*

A criticism of objectives is that they obstruct the possibility of any other beneficial, but unanticipated, consequences of instruction. This criticism has little foundation. We all know that a chance remark or the morning news can trigger an unanticipated, stimulating exchange. Often the experience is accepted as a bonus, and later the class returns to its schedule. A sensitive instructor, even though using explicit objectives, may revise plans to capitalize on an unscheduled interest.

Another possibility: In studying the outcome of instruction, an instructor may observe that students finishing a course have made certain desirable, but unanticipated, gains. He therefore may write additional objectives for the next offering of the course intending to perpetuate a good thing. Those who complain that objectives are stultifying and obstruct the unanticipated outcomes of a free-form classroom are really setting up a smokescreen that permits outcomes to be totally *un*anticipated, for better or worse!

12) *Writing objectives is an iterative process.*

There is no such thing as a perfect set of objectives written before a course is given. Objectives become tempered by the process of instruction itself. Hindsight may reveal the objective to be too easy, too difficult, off target, unclear, and so forth. One cannot know one's course objectives until one wrestles with the problem of trying to reach them.

13) *Objectives and examination questions.*

Written in action terms, an objective leads very nicely into construction of tests and examinations. If the objective says what students must demonstrate they can do as a result of instruction, then the basic framework for the corresponding examination question is already accomplished. The instructor has the task of directing students' attention, via the objective, to the behavior he wants them to be able to demonstrate, and he has the companion task of designing examinations that measure that ability fairly. In the process he must employ skill in phrasing both the objective and the exam question so that the objective does not give away or subvert the measure of the essential ability for which the question is intended.

14) *Scientific evidence in education is elusive.*

Experiments show that behavioral objectives have succeeded in improving learning. This statement, however, must be regarded with great caution. There has been a rush to apply the successful model of physical research—experimental and controlled groups with statistical analysis of results—to classroom settings. With research on instruction, there are numerous pitfalls. Another tendency common in reporting such experiments is the error of generalizing from a particular experiment to a broad conclusion. What is true for one instructional situation is not necessarily true for another. Few will ever have the opportunity to prove scientifically to themselves whether or not objectives have been beneficial in a particular case or will be beneficial in future teaching. On the other hand, the philosophical and logical arguments for objectives are nigh unassailable. For any arguments against objectives, there are exceedingly strong arguments—logic and common sense—in favor. Good luck!

Annotated Bibliography

Anderson, Richard C. and Faust, Gerald W. Chapter 1, "Behavioral Objectives," in *Educational Psychology: The Science of Instruction and Learning*. Dodd, Mead and Company, 1974, pp. 13-56. A self-teaching text and a refreshing change from stodgy, confusing texts of earlier times.

Bloom, Benjamin S., Ed., *Taxonomy of Educational Objectives Handbook I: Cognitive Domain*. David McKay, 1956. Widely acknowledged early leader in the field. Makes important and useful distinctions among objectives, beginning at rote learning and proceeding into increasingly complex intellectual processes.

Cohen, Arthur M. *Objectives for College Courses*. Glencoe Press, 1970. A well-written, 140-page self-help paperback. Comprehensive grasp and meaty interpretation of the subject.

Combs, Arthur W. *Educational Accountability: Beyond Behavioral Objectives*. Association for Supervision and Curriculum Development, Washington, D. C., 1972. A group of thoughtful and philosophical essays (40 pp.) by the director of the Center for Humanistic Education, University of Florida-Gainesville. Recommended.

Ericksen, Stanford C. "Defining Instructional Objectives." *Memo*, No. 43, December 1970, Center for Research on Learning and Teaching, University of Michigan, Ann Arbor. One of a series of well-written, authoritative "Memos" on educational topics.

Gagné, Robert M. "Behavioral Objectives? Yes!" *Educational Leadership*, N.E.A., Vol. 29, No. 5, February 1972, pp. 394-396.

Gagné, R. M. *Essentials of Learning for Instruction*. The Dryden Press, 1975. See especially Chapters 3, "The Outcomes of Learning," and 4, "Conditions for Learning," for a readable, analytical discussion of objectives by a well-known author.

Houle, Cyril O. *The Design of Education*. Jossey-Bass, 1972.The book is directed to the adult learner, a refreshing shift from K-12, the object of most writings on education. See pp. 136-150 on objectives. Practical thoughts for applying objectives to the less rote aspects of education.

Instructional Objectives Exchange Catalog. Center for the Study of Evaluation, Graduate School of Education, University of California, Los Angeles 90024. Lists courses for which sets of objectives are available.

Kibler, Robert J. *et al. Behavioral Objectives and Instruction*. Allyn and Bacon, Inc., 1970. If for no other purpose, see this book for W. James Popham's "Probing the Validity of Arguments Against Behavioral Goals," gratuitously thrown in as a postscript. With authority and wit, Popham demolishes the critics.

Kibler, R. J. *et al. Objectives for Instruction and Evaluation*, Allyn and Bacon, 1974. A revision of Kibler *et al*'s 1970 *Behavioral Objectives and Instruction* but omitting, unfortunately, the Popham article.

Kneller, George F. "Behavioral Objectives? No!" *Educational Leadership, op cit.* (see Gagné above), pp. 397-400. A debate of sorts between Gagné and Kneller. Kneller, a noted writer and authority in educational philosophy, seems ambivalent, not wishing to join behaviorists but unwilling to throw out objectives altogether. His compromise, employing a substitute word, ("specified," rather than "behavioral") is a little confusing. On the positive side, Kneller is opposed to *mis*use of objectives, of whatever stripe.

Krathwohl, David R. *et al., Taxonomy of Educational Objectives Handbook II: Affective Domain*. David McKay, 1964.

Mager, Robert F. *Preparing Instructional Objectives*. Fearon, 1962. Widely acclaimed, short, sparkling rationale and "how-to" book in exemplary programmed style.

McAshan, H. H. *Writing Behavioral Objectives*. Harper and Row, 1970. Subtitled, "A New Approach." Author integrates institutional and teaching goals with behavioral objectives. A short, thoughtful discussion, including pros and cons, pp. 1-10.

Performance Objectives in Education. The Educational Technology Review Series, No. 7, 1973. Educational Technology Publications, Englewood Cliffs, New Jersey 07632. Thirty-five articles relating to objectives, reprinted from *Educational Technology*: "When Should You Lie to Students," "Behavior Control: The Matter of Ethics," and many more. Also, see subsequent issues of the magazine. Objectives is a recurring topic.

Ringness, Thomas A. *The Affective Domain in Education*. Little, Brown and Company, Boston, 1975. Good support reading as you set out to write objectives in the affective domain.

Vargas, Julie S. *Writing Worthwhile Behavioral Objectives*. Harper and Row, 1972. Although lacking sample objectives from engineering, the book is useful for illustrating how one converts simple objectives to those demanding higher levels of intellection.

What to Teach: Understanding, Designing, and Revising the Curriculum

MADHU S. GUPTA, SENIOR MEMBER, IEEE

***Abstract*—A systematic procedure for curriculum design is briefly summarized, and the crucial step of content selection is then discussed in detail. The content is described in terms of the domain, principles, and problems of the discipline, and some suggestions regarding the teaching of each are made. The various reasons for curriculum revision are listed, with particular emphasis on the ways of solving the dilemma posed by the fact that time devoted to learning cannot increase in proportion to knowledge itself.**

I. Introduction

BROADLY used, the term "curriculum" refers to a systematic group or sequence of courses or educational experiences that is either offered or prescribed by a school or under a program, or that is required for graduation, certification, or as a preparation in a field, trade, or profession. The task of designing a curriculum deserves a great deal of careful thought because the effectiveness and outcome of education are critically dependent on it. It is also a difficult and time-consuming task because it must deal with such basic issues as the goals of education, and with the conflicting requirements of perpetuation and change. Few discussions and case histories of curriculum design are available in the literature of engineering education [1]-[3].

Ideally, the design of a curriculum proceeds through the four major steps outlined in Fig. 1, leading to a curriculum which consists of a set of courses. A "course" is a useful basic unit with which to construct the curriculum because: 1) it deals with a single, or narrowly defined, subject matter; 2) it represents a learning effort of a few weeks (full-time equivalent); 3) it is under the control of a small number of, and usually only one, instructors; and 4) while the revision of an entire curriculum is undertaken only after long intervals, the content of a single course evolves more frequently. A course is itself defined in terms of its own "curriculum," or syllabus, and the same four steps of Fig. 1 can be used to design the curriculum of a single course.

This paper is concerned with curriculum design at a single course level, an activity in which every teacher must engage. The scope of this paper is further restricted to the content selection stage (step 3(a) in Fig. 1) in a discipline-based course. The primary purpose of this paper is to present one viewpoint on what are the essential elements of a discipline, and the alternative ways of responding to curricular pressures.

II. Analysis of Course Content

A. Elements of a Discipline

A discipline can be ascribed, and differs from other disciplines in, three basic elements: the *domain*, the *rules*, and the *history*. The *domain* of a discipline refers to the subject matter boundaries of the discipline. The boundaries are only a convenient artifact, erected because the totality of knowledge is too great for comprehension by an individual, and are invariably approximate, tentative, and ever evolving. The *rules*, sometimes also referred to as the fundamentals, concepts, or powerful ideas, are the established ways and tools for knowing

Manuscript received January 5, 1979; revised September 26, 1980.
The author is with the Department of Information Engineering, University of Illinois at Chicago Circle, Chicago, IL 60680.

Reprinted from *IEEE Trans. Educ.*, vol. E-24, no. 4, pp. 262–266, November 1981.

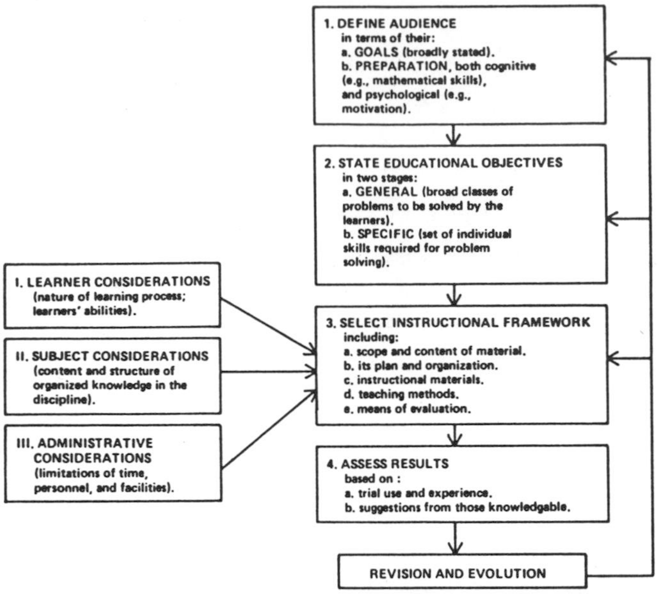
1. DEFINE AUDIENCE
in terms of their:
a. GOALS (broadly stated).
b. PREPARATION, both cognitive (e.g., mathematical skills), and psychological (e.g., motivation).
2. STATE EDUCATIONAL OBJECTIVES
in two stages:
a. GENERAL (broad classes of problems to be solved by the learners).
b. SPECIFIC (set of individual skills required for problem solving).
I. LEARNER CONSIDERATIONS
(nature of learning process; learners' abilities).
II. SUBJECT CONSIDERATIONS
(content and structure of organized knowledge in the discipline).
III. ADMINISTRATIVE CONSIDERATIONS
(limitations of time, personnel, and facilities).
3. SELECT INSTRUCTIONAL FRAMEWORK
including:
a. scope and content of material.
b. its plan and organization.
c. instructional materials.
d. teaching methods.
e. means of evaluation.
4. ASSESS RESULTS
based on :
a. trial use and experience.
b. suggestions from those knowledgable.
REVISION AND EVOLUTION

Fig. 1. Outline of steps in the design of a curriculum.

in the domain of that discipline. They consist of demonstrable and describable methods of discovering, generalizing, and validating the knowledge and hypotheses within a discipline. Finally, the *history* of a discipline is the accumulated set of answered (or recognized to be answerable) questions that the discipline has dealt with in the past. A discipline is fully described by these three elements. It follows that instruction in a discipline also requires these elements.

In engineering disciplines, the rules and the history may be liberally translated as "*principles*" and "*problems*," respectively. Each of these two terms, as used here, is to be interpreted broadly. By *principles* we mean two kinds of knowledge. The first may be labeled "axioms," and includes definitions, conventions, reference standards, and symbols; nomenclature, terminology, and classification schemes; units and typical values of parameters; and invariants, like material properties, specifications, and constants. The second may be called "key results," and includes physical laws and mathematical theorems; their corollaries, special cases, and alternative forms; their significance, consequences, and limitations; the established methods and techniques; the empirical rules, formulas, and relationships; and, for engineers, even analogies and solutions to broad, generic, or repeatedly occurring problems. The key results in an academic discipline can have one of two sources. In empirically based disciplines, the key results follow from practice and observation. In theoretically based disciplines, where the models constructed to represent physical objects work well, the key results take the form of a few simple theorems which can be conveniently applied.

Similarly, the term *problems* is used here in a broad sense as case studies in the application of principles, selected from the repertoire of problems which can be solved using the principles involved. It also consists of two parts, namely the formulation of a problem and the solution of a problem. Formulation of the problem includes the assumptions and idealizations, their justification or verification, and the reduction of the problem to a recognizable canonical form. Solution of the problem includes the search and selection of a suitable strategy or alternative, subdivision of the problem into smaller steps, the use of pattern recognition or shortcuts, and an examination of the results for self-consistency or agreement with expectation.

The distinction between principles and problems can be summarized as follows. Principles have a longer lasting and wider range of utility in a broad class of problems. By contrast, problems are more ephemeral, and of a narrow, local interest. Understandably, it is mostly the principles that are common among different courses or books devoted to the same subject. However, the labels "principle" or "problem" are not universally attached to a unit of curricular material; they are only applicable locally within the context of a course. For example, the equivalent circuit of a transistor is a "principle" in a course on transistor circuit design, because it is used for finding the response of a wide variety of circuits. In a course on the physics of semiconductor devices, obtaining the equivalent circuit is a "problem," exemplifying the use of the principles of carrier injection, collection, and transport.

B. Teaching of Domain

The domain of a discipline does not always receive explicit attention during instruction, possibly because teachers believe that it will be learned implicitly, by osmosis, or as a gestalt. In general, the domain of a subject can be described in terms of either its interior or its boundary. The interior of a course is best conveyed by identifying a unifying theme around which the course is structured. The boundary is often harder to delineate because it refers to the hazy periphery and frontiers of the discipline. However, there are sound reasons for teaching both.

Almost all courses have some central theme, usually implied by their titles. Most commonly, the theme is a small set of principles or techniques (example: in a course on electromagnetics, the problems solved may appear to vary a great deal, from electric motors to waveguides, but they all exemplify the use of Maxwell's equations). A particular application can also serve as a theme (example: in a course on solar energy systems, the important principles are many and come from a large number of disciplines, like thermodynamics, semiconductor physics, electrochemistry, heat and mass transfer, and thermal properties of materials, but the common theme is their application to a single problem, viz. obtaining useful energy from solar radiation). In either case, it is desirable that the theme be clear, pervasive, and significant. An unclear focus (or a complete lack of focus) in a course is not only confusing, it prevents seeing the conceptual connections between ideas. The more closely the content is tied to a single theme, the better the momentum, motivation, and attention can be maintained. An extreme example of a weak theme is a book on decibels [4] which covers the subjects of acoustics, noise, instrumentation, transmitters, receivers, television and microwave engineering, and still other topics, connected by the theme that they all express quantities in terms of dB.

The boundaries of a course may be specified by specifying the bounds imposed on the range of validity of principles involved, or on the class of problems or applications of interest. This is desirable both to warn against overgeneralization and to make links with other disciplines. Often, however, a mention of the scope is entirely neglected. For example, it is a rare undergraduate textbook on electromagnetism that points out the limitations of Maxwell's equations by mentioning the need for reinterpretation of field vectors at quantum level and the need for additional terms at very large fields. The purpose of such a mention need only be an awareness of the boundary rather than an understanding of the details of the generalizations.

The domain of a course is best discussed not only at the beginning of the course, but also at the end of the course because its understanding requires a familiarity with the subject matter of the course.

C. Teaching of Principles

Isolation of significant principles from the agglomeration of information is an intellectually demanding activity that all teachers must carry out in their own disciplines. The size of the set of principles is a matter of some concern. Some subjects are rich in specifics and details which are essential, and which cannot be (or have not yet been) replaced by a few general rules distilled from the specifics. As a mastery of copious detail is both difficult and inefficient, the instructor must attempt a clustering on the basis of some common features. Such packaging of details imposes a structure on them and proves helpful because patterns are easier to grasp than amorphous lumps. Learning is still further aided if the packages can be labeled with some (hopefully descriptive) names because of the ease with which words can be mentally processed.

It is clear that the sequential ordering of the individual parts of a course depends only on the principles; the problems normally follow the principles they are intended to exemplify. Often, the internal logic of the ideas determines a unique ordering; for example, in a circuit theory course, Thevenin's theorem must be preceded by the linear superposition theorem, which in turn must be preceded by Kirchhoff's voltage and current laws. Sometimes, the geometrical, hierarchical, temporal, or physical arrangement, rather than logical implication, dictates the order; for example, a radio transmitter may be described by starting at the microphone and proceeding towards the antenna. But there are many circumstances where the order appears to be discretionary. In such cases, an excellent basis for ordering is the richness and motivational value of the problems that can be employed.

D. Teaching of Problems

The problems serve several different but related purposes in a course: 1) illustration (they illustrate the technique or procedure for problem solving), 2) motivation (they show how the more abstract principles relate to real life situations and thus enliven and motivate discussions), 3) drill (they provide practice for reinforcing the already learned problem solving skills), 4) spacing (they act as separators between principles, allowing time for the principles to sink in, demarking the transition from one principle to the next, and accommodating learners with varying learning rates), 5) prototyping (they serve as prototypes and building blocks for the solution of other problems), and 6) scope delineation (they define the domain of utility of a technique of problem solving).

The choice of problems greatly influences the flavor of a course as perceived by the learners, particularly in well-established subjects like electromagnetic field theory, where the individuality of a course is determined almost entirely by the choice of problems. One of the most effective methods of collecting problems of classroom utility is through the professional literature [5]. A gradual substitution of problems dealing with older applications with problems pertaining to newer applications can "modernize" a course. Even a simple change of context, for example, a calculation of B-H power loss due to hysteresis, when transferred from power transformers to computer ferrite-core memory, may help increase student interest and motivation.

It is important that the instructor both solve as well as assign problems to be solved. The solution and assignment of problems serve two distinct purposes, each based in a different principle of learning theory. The inclusion of solved problems is important because imitation is one of the most effective modes of learning. Observing problems being solved is the classroom equivalent of apprenticeship. The assigned problems, on the other hand, require the learners to perform the very task for which they are being prepared. It is well known that a learner learns a task by doing it, and watching the problem being solved is not the same task as solving it.

III. Revision of Course Content

A. Forces Inducing Curricular Change

Changes in the curriculum of a discipline are caused by a variety of reasons, some internal to the discipline and others external to it. The "internal" reasons are those that can be traced back to block II in Fig. 1, i.e., they stem from changes in the discipline itself; all other changes are "external."

The external reasons for curricular change include such factors as changes in the preparation of entering students; the changing job market for graduates (and hence postgraduation goals of students); feedback from recent graduates concerning the success or utility of the existing curriculum; changes in the availability of instructional materials, personnel, or time; and changes in educational philosophy or interests, either prompted by a change in personnel or simply due to current fads. Finally, changes in the content of one course may influence the curricula of several other related or sequential courses, the so-called "ripple effect."

The internal reasons for curricular change can arise either from changes in the principles of the discipline or from changes in the class of problems of interest. A very striking example of each of these two in electrical engineering curricula occurred during the growth of the semiconductor device field in the 1950's and 1960's. First, the courses in physical electronics were modified to include a new set of principles dealing with transistors, concerning such processes as the depletion, injection, and collection of carriers. These gradually displaced the older principles relating to electron ballistics and electron emission from cathodes, ultimately leading to courses in semiconductor device theory. These courses themselves then shifted to a different and broader set of principles relating to such processes as carrier pair generation and transport theory, which describe not only the transistors but also LED's, solar cells, and microwave devices. Second, the curricula of courses in electronic circuits underwent a change in the class of problems of interest when vacuum tube circuits were replaced by transistor circuits as the vehicles serving to illustrate the same methods of analysis. In addition, the class of problems was enlarged, for example, due to the temperature sensitivity of biasing.

B. Expansion of Knowledge

Of the various factors causing changes in the curriculum, the enlargement of the set of principles in a discipline is undoubtedly the most profound and challenging to the curriculum designer. While the research in the physical sciences creates an ever smaller set of fundamental principles from which all other results can be deduced, the set of engineering principles grows with time because the engineering practice employs an ever widening range of hardware (i.e., physical phenomena, devices, and materials) as well as software (theorems, algorithms, models, etc.). In time, some new knowledge in a discipline may be considered sufficiently basic, or essential to the profession, to merit inclusion in the curriculum of that discipline, thereby creating new demands on facilities, teachers' time, and learners' time. For example, the teachers must invest a sizable fraction of their time to professional renewal so as not to compromise the interest of future students, and this reduces the time available for the current students.

Of interest to a curriculum designer is the demand that the expansion of knowledge creates on the learners' time. On the broader level of the entire engineering curriculum, this is an old problem with few proposed solutions, such as requiring a higher level of preparation for entrance into engineering (presumably from high schools), providing continuing education after graduation, and increasing the number of years required for graduation. At the level of a single subject, an instructor faced with the problem of knowledge expansion has even fewer choices. The duration of a course is usually an invariant due to practical considerations, such as the total length of a program and similar curricular pressures in other courses competing for the learners' time. The rate of presentation of new principles is limited by the comprehension abilities of the learners.[1] When neither of these two options is available, the only alternative is the revision of the curriculum.

C. Responses to the Expansion of Knowledge

A look at the past history of engineering curricula shows that the educators have responded to the expansion of knowledge in several different ways.

1) *Substitution:* The simplest method of incorporating new principles is to replace some of the existing principles with new ones. The main constraint to this method is that only those principles can be dropped which are not required in the subsequent work. As one example of this method in the electrical engineering curriculum, it appears that in the last decade many courses on circuit theory have added new principles, like Tellegen's theorem, and have dropped work with 3-phase circuits, which in turn has become the responsibility of power engineering courses, the principal users of the dropped principles.

2) *Idealization:* As pointed out in Section II-A, the "problem" part of the curriculum consists of "problem formulation" and "problem solution." A favorite method of saving classroom time is to present problems in an already formulated form, stripped of their real-life context, so that they are immediately amenable to solution without going through the exercise of describing the unfamiliar setting or background of the problem and reducing physical or practical situations to a solvable form via assumptions and approximations. This practice is responsible for the often-heard criticisms, for example, that a picture of a working hardware is becoming a rarity in engineering textbooks.

3) *Condensation:* As a next step, new principles can be added to the curriculum by displacing some problems entirely. Usually it is possible to relegate the examples and drill exercises to homework. The net effect is that more principles and fewer illustrative examples or problems of each remain. The process is obviously limited when the problems in the course are one-of-a-kind examples with little overlap and serve only as illustrations rather than as drill. This emphasis on principles is often criticized on the grounds of students having a lack of practice at problem solving, an age-old criticism of academics by many in industry. The criticism has been answered in various ways, for example, by pointing out that the emphasis on principles helps fight obsolescence, that education is only a preface and therefore should be concerned with the principles, and that it is easier to interest a broadly trained scientist in technology than a technologist in broader scientific principles.

4) *Abstraction:* Perhaps the most powerful way of solving the curricular problem due to knowledge expansion is by transferring the principles to a higher level of abstraction. This is done by selecting more general and powerful principles, and treating the older principles as special cases, derived results, or problems. This has the effect of reducing the number of principles. While the principles learned are more powerful, they are also one more step removed from actual applications, thus making the problem solving more involved. This method has also been criticized for decades, indirectly, by saying that engineering education is training "scientists" rather than "engineers." Any number of examples of this method in electrical engineering curricula can be found by comparing typical textbooks in a single field, written about two decades apart. For instance, the various theorems on "tee" and "pi" circuits, discussed in network analysis courses a generation ago [6], are now only examples of the theorems on two-port networks.

5) *Stabilization:* A course that has been reduced to highly abstract, condensed, and essential principles is essentially in a stable state. In electrical engineering, modern courses in some areas, such as electromagnetics, appear to have reached a simi-

[1] Operationally defined as the sense that an experienced instructor develops for what is an adequate rate in a given situation, such as the rule of thumb that no more than one "major new idea" or two to three "minor new ideas" can be accommodated in a class hour.

lar state. That such a steady-state is reached is to be expected and indeed justified. If the duration of a course is fixed, say at 30 weeks, the course can only attempt to provide the first 30 weeks of learning experience in the field regardless of the present state of the field. And if the field experiences a growth, it just means that after finishing the course a person will take longer to become a practitioner in the field.

6) *Specialization:* Finally, the ultimate response to the expansion of knowledge is narrower specialization. This is indeed how the discipline of electrical engineering was born from the "electrical option" in the curriculum of mechanical engineering early in this century. The birth of the various types of electrical engineering degrees in recent years (e.g., "computer and information engineering" and "instrumentation and control engineering") is a similar attempt at reducing pressures on the curriculum.

In some respects, the above six responses to the expansion of knowledge are increasingly more "drastic" in the order listed. It is natural to expect each to be employed, in that order, only after the previous one has been "used up." Courses in newly introduced subjects often are initially composed entirely of case studies, which are then gradually supplanted by principles as the subject matures and the fundamentals crystallize over the years. One may expect the natural evolution of the curriculum in a field to take place along this sequence of six stages, and many academic offerings can be identified which have indeed gone through these stages sequentially. The maturity of a subject is then indicated by which of the options the educators are presently exercising to deal with knowledge expansion.

IV. An Example

I have taught an introductory course in circuits and electronics for some time in which the curriculum was influenced by the viewpoint presented in this paper. Some of the discussed ideas apply to such a course, and their influence on the course content is briefly pointed out here.

I begin the course on the first day by describing what electronics is in terms of its applications in communication, computation, and instrumentation with which the students may already be familiar. On the last day of the course I come back to the same question, and define electronics in terms of the work done in the course with which the students are now familiar. In addition, I point out the limitations of their models at high frequencies due to the distributed nature of circuit elements (when a wire is no longer a short circuit), at very small sizes due to material inhomogeneity (where a junction transistor is no longer a three-distinct-layer structure), at very small signals due to fluctuations (where a voltage is no longer a steady reading on a voltmeter), and at very large signals due to heating (where charge flow and heat flow are coupled).

In an introductory course there is obviously little opportunity to isolate new principles. I do explicitly point out and label each principle, and make a sharp distinction between it and a problem. The ordering of principles was chosen to enhance the motivational value of the problems. Thus, operational amplifiers were introduced as physical manifestations of controlled sources very early in the course, before capacitors or diodes; as a result, the problems that could be used for illustrating later principles became more interesting and practical than the otherwise dry circuit analysis problems.

Interesting problems for use in the course were found through trade magazines and, although idealized, were stated in their real-life context; thus, a water-level indicator illustrated comparators, an automobile spark plug illustrated transient response, and a fire alarm illustrated the use of diode characteristics. Each solved problem was identified with a specific principle which it was meant to illustrate, and which justified its presence. The number of problems was thus tightly controlled, with more difficult principles illustrated by a larger number of problems, both solved and assigned.

Finally, the possible or proposed changes in course content were examined to classify them under the categories listed in Section III. Thus, the suggestion to drop digital circuits in favor of linear amplifiers was viewed as "specialization," while the proposal to drop hybrid-pi parameters in favor of h-parameters was viewed as "abstraction." Such classification helped in making more informed choices.

References

[1] L. P. Grayson, "On a methodology for curriculum design," *Eng. Educ.*, vol. 69, pp. 285–295, Dec. 1978.

[2] L. D. Feisel and R. J. Schmitz, "Systematic curriculum analysis," *Eng. Educ.*, vol. 69, pp. 409–413, Feb. 1979.

[3] *IEEE Trans. Educ.*, Special Issue on Curriculum Development in an Era of Rapid Change, vol. E-22, May 1979.

[4] V.V.L. Rao, *The Decibel Notation and Its Application to Radio Engineering and Acoustics*. New York: Asia Publishing House, 1966.

[5] J. D. Horgan, "Engineering reality in single-answer problems," *IEEE Trans. Educ.*, vol. E-21, pp. 65–68, May 1978.

[6] W. L. Everitt and G. E. Anner, *Communications Engineering*. Englewood Cliffs, NJ: Prentice-Hall, 1943.

Part IV
The Learning Process and Learner Characteristics

Scope and Purpose

LEARNING is a process resulting from the experience with some task and reflected in a relatively permanent change in performance under appropriate circumstances. Only in recent years have we come to understand the theoretical underpinnings of how learning takes place, and learning theory is now a major and active branch of the behavioral sciences. It seems reasonable to expect that a knowledge of how people learn would be of immense direct and practical value in designing and optimizing a teaching strategy in a given situation. Unfortunately, things are not quite that straightforward. First, there are several schools of thought on the details of the learning process, which differ from each other in fundamental ways. Indeed, it is unlikely that any single, completely general process governs all learning. As a result, it is not possible to state the basic principles of this theory in a universally applicable manner. Second, even if that was possible, the implications and consequences of the principles will not be immediately apparent in a form to be of direct use to the teacher. Roughly speaking, learning theory is to instructional design what thermodynamics is to the design of automobile engines: it provides a theoretical framework for the task, but a great deal of work is necessary to make its results useful at the operational level. Finally, some results of learning theory (such as those related to nonsense syllables) are not directly relevant to instructional design.

For these reasons, the likely subject of greatest interest to a beginning teacher of engineering is the implications of learning theory for a classroom teacher, to which Part IV of this volume is devoted. Just what exactly are the relevant principles from learning theory? The list of "learning principles" is not unique, and different authors construct their list of significant principles differently, although there are broad areas of agreement in all such lists. Some of them can be found in the papers reprinted in this part. Another such list of principles is contained in [9] listed below. Still another list of seven principles from learning theory, and their implications in teaching, is contained in the article by Paskusz and Stice, reprinted in Part V of this volume. In particular, that article examines a number of different methods of teaching, and shows how these principles are used in each method.

The Issues

The issues of interest to learning theorists are many and varied; a small sampling of these follows:

- Why does learning take place or not take place in a given circumstance?
- What are the effects of the various forms of positive and negative reinforcements on learning?
- Is learning a process or a product?
- How does one learn to generalize, discriminate, and differentiate?
- When and why does recall or forgetting occur?
- What are the various learning styles of individuals, and what type of learning are they suitable for?
- How is learning influenced by the mental state of the learner, e.g., by anxiety or motivation?

The Reprinted Papers

The following five papers are reprinted in this chapter:

1. "What Instructors Say to Students Makes a Difference!" A. A. Root, *Engineering Education,* vol. 60, no. 6, pp. 722–725, Mar. 1970.
2. "A Model of What Happens in Teaching and Learning," A. A. Root, *Engineering Education,* vol. 60, no. 6, pp. 726–731, Mar. 1970.
3. "Some New Views of Learning and Instruction," R. M. Gagne, *IEEE Trans. Education,* vol. E-14, no. 1, pp. 26–31, Feb. 1981.
4. "Improve Your Teaching Tomorrow with Teaching-Learning Psychology," C. E. Wales, *Engineering Education,* vol. 66, no. 5, pp. 390–393, Feb. 1976.
5. "Principles of Design and Analysis of Learning Systems," C. H. Durney, *Engineering Education,* vol. 63, no. 6, pp. 406–409, Mar. 1973.

One of the fundamental principles in the theory of human learning, somewhat akin to Newton's laws in classical mechanics, is Maslow's hierarchy of human needs. The first reprinted paper by Root is a highly condensed introduction to this principle. The central purpose of Root's paper, however, is to present the basic ideas of interaction analysis (a classification of conversational remarks), and to employ the hierarchy of needs to show how the different category of remarks address different needs.

The second reprinted paper, also by Root, is a very brief and non-technical description, from the viewpoint of a classroom teacher, of the classical stimulus-response type of learning theory, and the learning and forgetting curves, along with a listing of the various aspects or factors of the stimulus, the observer, the response, and the reinforcements, that affect learning.

The next paper reprinted here is authored by Gagne, one of the principal figures in the field of learning theory. It summarizes some experimental results which conflict with the classical stimulus-response-repetition based learning theory, develops from them an information processing theory of learning, and deduces four major recommendations for instruction: (a) determine the hierarchy of prerequisites to learning, and each individual learner's status within that hierarchy; (b) design learning to allow mastery of the set of prerequisites; (c) assist the process of mental coding of information to facilitate retrieval; and (d) use drill only for periodic and spaced review which requires the learner to retrieve previously learned strategies. An example of just how Gagne's theory may be employed by a teacher is available in the literature [7].

The next two papers are more pragmatic, and deal with day-to-day application of learning principles. The article by Wales states five principles of teaching—guide learning, provide practice, evaluate and feedback, motivate, and individualize. It then supplies suggestions on how to employ these principles to improve teaching, and summarizes the suggestions in a point-by-point manner. Most of his suggestions are practical, easily implemented, and have a rapid payoff. Durney's article lists eight principles of learning—five dealing with procedures and methods, adapted from Erickson, and three dealing with organization of content, adapted from Gagne [6]. It then evaluates a typical, hypothetical engineering lecture course taught in the traditional manner to determine if the course makes use of the eight principles in aiding learning, and finds a large scope for improvement. Next, it examines a hypothetical course taught by a personalized system of instruction and finds that it makes a better use of the eight principles.

The Reading Resources

[1] J. W. George Ivany, ''Resource Letter EP-1 on Educational Psychology,'' *American Jour. Physics,* vol. 37, no. 11, pp. 1091–1099, Nov. 1969. An annotated bibliography of books and articles introducing the field of educational psychology from the point of view of a teacher in a scientific field; now somewhat dated.

[2] C. Fincher, ''What Is Learning?'' *Engineering Education,* vol. 65, no. 5, pp. 420–423, Feb. 1978. Attempts to define learning; briefly describes how the viewpoint of learning theory has changed over the years; mentions the major factors influencing learning; and lists categories of learning.

[3] W. J. McKeachie, ''The Decline and Fall of the Laws of Learning, '' *Educational Researcher,* vol. 3, no. 3, pp. 7–11, Mar. 1974. Points out the problems in trying to apply the laws of learning to teaching, due to the difference between humans and other animals, and due to the numerous interacting variables which occur in a natural setting.

[4] S. C. Ericksen, ''Learning Theory and the Teacher, '' *Memo to the Faculty,* Center for Research on Learning and Teaching, University of Michigan, Ann Arbor, MI; in four parts:

Part I. Memory (Memo No. 33), Jan. 1969.

Part II. Transfer of Learning (Memo No. 34), Mar. 1969.

Part III. Defining Instructional Objectives (Memo No. 43), Dec. 1970.

Part IV. The Reinforcement Principle (memo No. 48), Apr. 1972.

[5] R. Good, E. K. Mellon, and R. A. Kromhout, ''The Work of Jean Piaget,'' *Jour. Chemical Education,* vol. 55, no. 11, pp. 688–693, Nov. 1978. A summary of the major contributions of Piaget, the leading developmental psychologist in limelight in the 1960s and 1970s. Although Piaget's work deals with children up to 15 or 16 years of age, there is increasing experimental evidence that many college students have not reached the ''Formal Operational'' stage described by Piaget, and therefore his ideas have relevance to college teaching. An annotated bibliography of studies based on Piaget's work is contained in the same author's ''Piaget's Work and Chemical Education,'' *Jour. Chemical Education,* vol. 56, no. 7, pp. 426–430, July 1979.

[6] R. M. Gagne, ''Instruction Based on Research in Learning,'' *Engineering Education,* vol. 61, no. 6, pp. 519–523, Mar. 1971. Classifies learning into three categories, and discusses conditions for each to occur; the categories are (i) verbal information, which directs learning, and mediates the transfer of learning, (ii) intellectual skills, which make possible the rule-governed behavior, and requires learning all essential prerequisites, and (iii) cognitive strategies, which are the highest level of learning, and make an individual creative.

[7] A. Ozsogomonyan, ''An Application of Gagne's Principles of Instructional Design,'' *Jour. Chemical Education,* vol. 56, no. 12, pp. 799–801, Dec. 1979. An actual example of instructional design based on Gagne's work, and an experimental evaluation of its effectiveness.

[8] M. H. McCaulley, ''Psychological Types in Engineering. Implications for Teaching,'' *Engineering Education,* vol. 66, no. 7, pp. 729–736, Apr. 1976; and M. H. McCaulley, ''Application of Psychological Types in Engineering Education,'' *Engineering Education,* vol. 73, no. 5, pp. 394–400, Feb. 1983. Classifies engineering students based on Jung's theory of psychological types; describes their characteristics; and the relevance of this classification to the training of, and communication with, the students.

[9] R. K. Irey, ''Four Principles of Effective Teaching,'' *Engineering Education,* vol. 72, no. 2, pp. 143–147, Nov. 1981. States four principles of learning: instruction should be based on objectives, absolute standards should be used for assessment of learning, learning strategies should be varied, and time available for learning should be variable; then makes suggestions on applying these principles in various kinds of situations.

[10] R. Battino, ''The Humanistic Psychology Movement and the Teaching of Chemistry,'' *Jour. Chemical Education,* vol. 60, no. 3, pp. 224–227, Mar. 1983. One teacher's suggestions on the practical, classroom use of the principles of humanistic (as opposed to the formal, academic) psychology, including confluent education and non-linguistic programming; discusses these and other miscellaneous tips for supplementing the classroom teaching in the affective domain.

[11] *Engineering Education, vol. 64, no.6, Mar. 1974. Special Issue on Motivation and the Management of Learning.* Articles on psychologist's view of motivation, motivation by removing obstacles to learning, enhancing motivation by case studies, and motivation in individualized methods of teaching.

[12] G. H. Flammer, ''Applied Motivation—A Missing Role in Teaching,'' *Engineering Education,* vol. 62, no. 6, pp. 519–522, Mar. 1972. Appeals to the teachers to give more attention to developing the students' level of motivation, points out the resulting advantages, and mentions some methods used by the author to achieve this end.

[13] W. J. McKeachie, "Student Anxiety, Learning, and Achievement," *Engineering Education,* vol. 73, no. 7, pp. 724–730, Apr. 1983. A survey of a lifetime of research by a distinguished researcher on questions like who is affected by test anxiety, and how can the test-anxious students be helped.

[14] N. Ellman, "Reducing Test Anxiety," *Clearing House,* vol. 55, no. 1, pp. 27–28, Sept. 1981. Compiles representative comments from test-anxious students, showing several sources of anxiety; provides suggestions to educators on what changes might be made to eliminate each of the sources.

[15] G. Root and D. Scott, "The Interpersonal Dimension of Teaching," *Engineering Education,* vol. 66, no. 2, pp. 184–188, Nov. 1975. Attempts to raise an instructor's awareness of the alternative ways in which his comments can be interpreted, by giving examples.

what instructors say to students makes a difference!

A. A. Root

Let us suppose that a sophomore student stopped to talk with his instructor after class, and in a halting, uncertain manner, said something like this:

"You know, I'm having an awful time with stuff we're supposed to be learning in this course—and in some of my other courses too. I don't think I'm stupid, but—well, a lot of people told me I should go into engineering because I was good at math in high school—but now I don't know. You're doing a good job with this material, but I wonder if I shouldn't drop this course."

Most instructors would like to be helpful to a student, at a time like this—to say the "right thing."

How would *you* respond? Below are seven responses that might be made to this student. Read the statements carefully and select the one which is closest to what *you* would be likely to say, in order to be *most* helpful. At the same time, mark an *x* beside the statement that you think would be *least* likely to be helpful. (Although you are reading this in *Engineering Education,* actually take the time to mark the responses—the *most* and the *least* likely to be helpful!)

1. "What do you think you should do?"
2. "I've watched what you have been doing and am convinced that you'll be able to make it if you keep trying. Engineering is a profession well worth working for."
3. "Now, in spite of the advice people have given you, you're not sure that engineering is the right thing for you."
4. "Hmm—it certainly would be a shame if you were to quit after all the time and effort you've put in. I hope you'll think seriously before you give it up."
5. "Uh-huh. Hmm."
6. "This has been bothering you for some time, and—now, you're really not sure what is best."
7. I recommend that you talk with a counselor who can help you decide what you are best suited for."

In almost any group of engineering instructors (i.e., anyone who teaches—graduate student or professor), there are some who choose each of these different responses. They give many different reasons *why* they selected the response they did, and almost always the motives behind their selections are laudable. Most instructors truly *want* to help a young man experiencing this kind of uncertainty.

Categories of Conversation

Before commenting on the merits and shortcomings of different answers, let us examine a scheme for categorizing the different kinds of things an instructor *can* say to a student. With a moderate amount of training, almost anyone can learn to listen to another person's conversation and classify each statement he makes, labeling it with one of the category numbers from figure 1.

We are going to be using these categories of conversation repeatedly in the material that follows, so it would be worthwhile for you to read them carefully. Let us pause a bit, so that you can become acquainted with them—right now.

Now, go back and re-read the different replies you might have made to that uncertain student, and next to each statement write down the number of the conversational category to which it belongs. (There's one statement for each category, so don't duplicate yourself.)

The "correct" categorizing of these seven statements is printed, upside down, below. When you've marked each statement, check yourself.[1]

When instructors find their responses labeled in this manner, they sometimes feel a little uncomfortable, because they may not have intended to appear in quite that way to the student. Unfortunately, the student has no way of knowing the instructor's intentions—he can only hear the words.

About this time, someone usually points out that any one of these statements can be completely changed by changing the tone of voice used to say the words—and that is absolutely right. The instructor could say, "You're uncertain," as a gentle category 1, to express genuine concern for the feelings being expressed—or could say, "*You're* un-*cert*-ain!" and have it come out as a strong category 7, with sarcastic criticism. For the moment, let us assume that an instructor's tone and facial expressions match his words—as they usually do. In repeated experiments using these categories to record classroom interactions, it is observed that an instructor's verbal behavior corresponds closely to his nonverbal performance—a kind of consistency most people develop.

Before going back to comment on the helpfulness of the different statements considered earlier, several other ideas will be presented to help put these statements into perspective. The next two basic ideas are: 1. a framework for analyzing the things that attract or disturb people, i.e., their "motives," and 2. a set of suggestions for how to respond to a person in order to

---

1

Statement 1: category 4 (question)
Statement 2: category 5 (opinion)
Statement 3: category 8 (reflect ideas)
Statement 4: category 7 (mild criticism)
Statement 5: category 2 (encourage)
Statement 6: category 1 (label feeling)
Statement 7: category 6 (give direction)

Reprinted with permission from *Engineering Education,* vol. 60, no. 6, pp. 722–725, March 1970.

The Instructor	Indirect (Inter-personal)	1. Identify and label feelings. Tell him how you think he feels. Sense his emotional state and put that into words.	"This really upsets you." "You wonder if --." "If you just knew --."
		2. Encourage and assure. Indicate that what he does is acceptable. Nod, say "Uh huh," or "I see."	"Hmm -- yeah." "Uh huh, go on." "I see. Tell me more."
		3. Identify and reflect ideas. In your own words, tell him what you heard him say. Demonstrate your understanding.	"As you see it, this --." "You're saying that --." "Your idea is --."
	Content	4. Ask question. With the intent that he answer, ask any kind of question which urges him to answer. Inquire.	"Where is --?" "What kind of --?" "Why do these --?"
		5. Give information.or opinion. Introduce new data or add something which the other has not said. Lecture. Present ideas.	"The solution is --." "I'm sure that --." "Another point is --."
	Direct (Organizational)	6. Give directions. Give instructions which the other is to follow. Tell him what to do. Inform him of procedures.	"Do these problems --." "Go talk with --." "After that, then --."
		7. Criticize or correct. Indicate an error with the intent that he change his behavior. Either mild criticism or harsh scolding.	"No, the right way --." "How could you --." "Damn, you didn't --."
The Student		8. Responds. He answers you or follows instructions or implied direction. He stays on the subject.	"Yes, I did --." "The answer is --." "Well, when I --."
		9. Initiates new idea. He goes beyond the previous information. Brings up new material. Asks or suggests novel item.	"I think that --." "But, how about --." "Another way would be --."
		10. Silence or confusion. There is no structure to the conversation. This may be constructive silence or bedlam.	

Figure 1. Categories of conversation. There is no value judgment implied in assigning statements to any of these categories; they are merely descriptions of different types of things that people say (adapted from Amidon and Flanders, 1967).

be most helpful. After this last set of suggestions, the meaning of each response will become more apparent.

Sensitivity and Responsibility

Instructors often feel caught in a bind between responsibilities to subject-and-schedule and to the individual student. The pressure to "cover" the material in the course is real and demanding. At the same time, instructors often want to be understanding and friendly with their students. One instructor phrased this conflict, "How can I be understanding and sensitive with students without giving up my responsibilities for the everyday business of life?" There is no simple answer to such a question. The best answer seems to come from an examination of the different types of motives that must be dealt with.

One way of representing the interests, motives and needs of man is shown in figure 2, where different classes of needs are shown in a "hierarchical" structure. The "stair-step" form indicates that there is a kind of progression in which the higher levels build upon the lower ones. The lower left represents the needs of man for physical survival. When these are threatened, they generally take precedence over any other need (for example, try to stay under water for ten minutes while thinking "beautiful thoughts"; the demand for air will trigger most people to violent exertions). When survival is reasonably well assured, one next becomes concerned for the security of a home, good health, adequate pay, life insurance, a regular job, and the regularity of a reliable environment. When these physical and organizational needs are moderately satisfied, affiliation and social needs begin to become of major importance, i.e., the desire to be accepted by others and to have a special place in their regard. These four, lower level needs are sometimes called deficiency or maintenance needs as they seem to be the necessary human requirements a man must receive before he can be productive and give—rather like inhaling before one can exhale. An interesting feature of these needs is that they grow stronger when *denied* by one's environment. This is in contrast to the higher level growth needs (intellectual and aesthetic) which grow stronger when *fulfilled.*

The achievement and intellectual needs are concerned with an individual's desire to know enough about this environment to be able to influence it and, eventually, to understand how significant decisions can be made to achieve important goals. At the highest levels of this hierarchy are the aesthetic needs—the need to appreciate the order and balance of life and to find a unique, yet consistent, style of relating to the environment out of a sense of belonging to all of life.

Principles Based on Man's Needs

Out of an understanding of this general structure of man's needs and motives come three related principles which are applicable to education, business, family life, and other activities where men work together to achieve important goals.

PRINCIPLE 1: Human motives are ordered in a hierarchical structure such that when lower level motives are excited, they are stronger and take precedence over higher ones. (Examples of self-sacrifice by persons very high on this hierarchy are rare and do not affect the general principle.) This principle would suggest that a student worried about money or failing grades ("security" deficiencies), and who is socially ostracized or lonely ("social" deficiencies), would have a hard time functioning at the

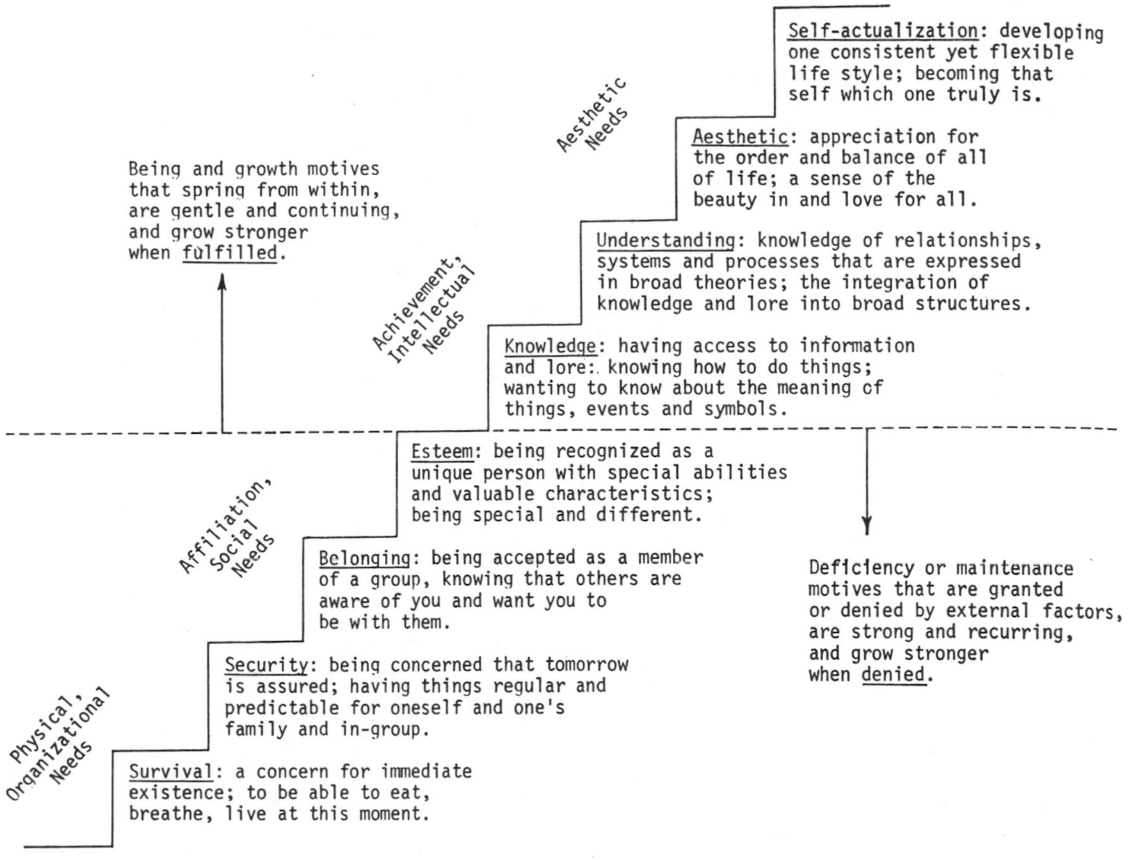

Self-actualization: developing one consistent yet flexible life style; becoming that self which one truly is.
Aesthetic Needs
Aesthetic: appreciation for the order and balance of all of life; a sense of the beauty in and love for all.
Being and growth motives that spring from within, are gentle and continuing, and grow stronger when fulfilled.
Understanding: knowledge of relationships, systems and processes that are expressed in broad theories; the integration of knowledge and lore into broad structures.
Achievement, Intellectual Needs
Knowledge: having access to information and lore: knowing how to do things; wanting to know about the meaning of things, events and symbols.
Esteem: being recognized as a unique person with special abilities and valuable characteristics; being special and different.
Affiliation, Social Needs
Belonging: being accepted as a member of a group, knowing that others are aware of you and want you to be with them.
Deficiency or maintenance motives that are granted or denied by external factors, are strong and recurring, and grow stronger when denied.
Security: being concerned that tomorrow is assured; having things regular and predictable for oneself and one's family and in-group.
Physical, Organizational Needs
Survival: a concern for immediate existence; to be able to eat, breathe, live at this moment.

Figure 2. Actions, motives, and incentives of man (adapted from Maslow, Harvey, McClelland, and others).

intellectual or achievement level. A similar thing is expressed in principle 2.

PRINCIPLE 2: Strong emotions may produce "internal noise" which prevents the reception of other stimuli and, therefore, the functioning of higher level reasoning. Some persons appear to generate their own emotional "noise," such as the young man at the beginning of this article. Efforts to help such a person lead to a third principle.

PRINCIPLE 3: If one can demonstrate that he hears, understands, and accepts what the other person says and does, the other person will be most likely to hear and understand him, and become able to make constructive decisions.

Based on these principles, with the accompanying hierarchy of motives, it seems best to provide someone who is worried about grades with an environment which, at least temporarily, provides for his deficiency needs. It is important for an instructor to learn to control his own behavior (both verbal and nonverbal) so as to reduce the intensity of these lower level drives. He can do this by following a sequence such as:

1. Whenever you sense that the student is reacting to strong internal emotion, identify and reflect your understanding and acceptance of these feelings. Continue doing this until the intensity of these feelings has decreased, saying something like, "You're really not sure that this is right for you." This kind of a statement can do much to meet both his security and social needs—because a significant person in the organization has thought enough of him to listen and respond, sensitively, to his immediate problem.

2. When the student's emotion has decreased in intensity, identify and reflect the content or *ideas* he has expressed, without adding anything new. Such a statement can show further evidence of your acceptance of him as a person-of-worth (positive social relations) and begins to lead up to the intellectual domain, starting with his own ideas as suggested by principle 3.

3. When the student knows you've heard him, then you can begin to add some *new material* that he has not mentioned. The first new ideas you add should be as noncontroversial as possible, e.g., "Something you didn't bring up was the fact that ....." Add controversial topics only when he is able to consider new ideas without getting emotionally involved again, e.g., "On the other hand, the way I see it is...."

4. Arrange for some specific *actions* both for the student and for yourself, including who will do what, by when, e.g., "Why don't you do ...., I'll do ...., and We'll get back together again on ....." Only as a last resort issue ultimatums about what *must* be done, as this drops back to exert force in the organizational-security level of the motivation hierarchy.

Now, look back at figure 1 and identify the category of conversation you would be using at each of the above four steps. The correct classification is printed, below and upside-down.[2] After analyzing each of these four types of statements, check yourself.

Now we have a basis for recommending which of the first seven possible responses to the uncertain student would be most likely to be helpful. As you can see, almost every statement has *something* to recommend it. What is most likely to help the student come to that solution which is best for him is a sequence of things you could say to him. Following the four steps recommended above, you'd probably be most helpful responding as follows:

...... 6. "This has been bothering you for some time."
...... 3. "In spite of advice, you're not sure engineering is for you."

2
First, category 1 (identify feelings)
Second, category 3 (reflect ideas)
Third, category 5 (add ideas and opinions)
Last, category 6 (directions for action)

......2. I've watched you, and think you'll make it. What do you think?"
......7. "Let me help you get in touch with a counselor who can help you."

And you notice that this arrangement has avoided using any category 7. Also, during any such conversation, there should be a liberal sprinkling of category 2. (All of us use this on the telephone, when we want the person on the other end to know we're still there, listening).

Application to Teaching

But, are these categories of conversation applicable to what students and instructors say in classrooms? Absolutely! They were originated in order to study the kinds of things said in educational settings. Some of the best data on good vs. poor teaching has come from research studies based on this kind of analysis of verbal interaction in classrooms. In the discussion that follows, one basic diagram summarizes the conclusions of a number of research studies. Finally, the research data are considered in relation to the ideas of motivation discussed above.

The data in figure 3 were obtained from a study of many mathematics teachers. Their classes were arranged in rank order according to their average scores on a standardized test of mathematics achievement. This list was divided into thirds, so that the researchers could talk about the "high," "moderate," and "low" achievers. At the same time, observers visited all of these classrooms and recorded the category of conversation *every three seconds*, for extended periods of time. These observers had been carefully trained in the use of these categories, and were shown to be "reliable" observers (multiple observers would record the same categories, while observing the same activities), but they did not know which classes were the high, moderate, or low achievers.

When observers' records were separated according to levels of achievement, as shown in figure 3, there were significant differences. The instructors of the high achievers used significantly more "indirect" behavior and less "direct." The reverse was observed for instructors of the low achieving classes, where there was significantly less "indirect" and more "direct" teacher behavior. It is interesting to note that all of the teachers used approximately the same amount of categories 4 and 5, asking questions and presenting information (lecturing).

In a parallel study, almost identical results were obtained when classes were arranged in rank order according to attitudes toward school and the specific course. Evidently, students are aware of and respond to differences in the patterns of instructors' verbal behavior in the classroom. What instructors say to students makes a difference!

Success of Indirect Behavior

There are at least two reasons why instructors who use more indirect types of behavior have students who perform better and have more positive attitudes: 1. as described above, students whose lower level organizational and social needs are met, rather than excited, are free to function at the intellectual level, to be curious and to strive for the achievement of significant goals, and 2. as will be described in the next article, instructors who learned to provide reinforcement and feedback to students for their efforts tend to have students who try harder and like the classroom environment. Thus, both reasons tend to be "motivational" explanations.

It is common for instructors to say that they want to "motivate" their students, yet it is uncommon for instructors to know how to provide the kind of "indirect" environment that achieves the result they want. Many programs to develop "effective teaching" have concentrated on the analysis of what instructors say to students. Teachers can learn to control their own verbal behavior and, when they do, students can tell the difference. For those who wish to learn more about "interaction analysis" and who may want to work with others to modify their own classroom behavior, the references listed below should provide an introduction to a challenging and valuable new experience.

References

1. Amidon, E. J., and Flanders, N. A., *The Role of the Teacher in the Classroom,* Association for Productive Teaching, 1040 Plymouth Bldg., Minneapolis, Minn., 1967.
2. Amidon, E. J., and Hough, J. B., *Interaction Analysis: Theory, Research and Application,* Addison-Wesley, Reading, Mass., 1967. △

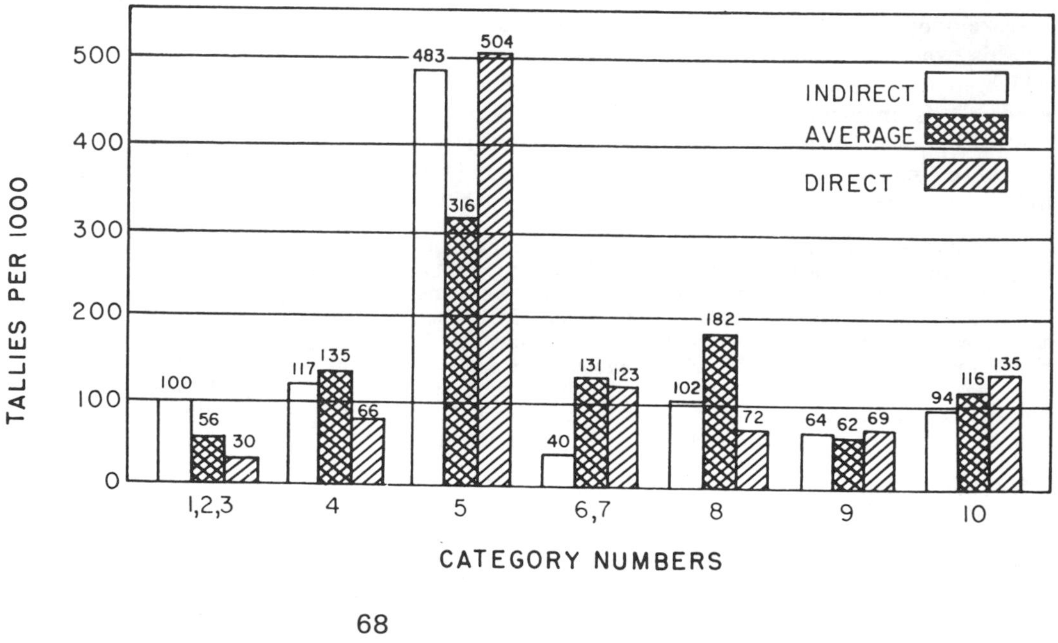

Figure 3. Influence patterns of most indirect, average, and most direct mathematics teachers.

a model of what happens in teaching and learning

A. A. Root

In trying to understand, predict, and eventually control any phenomena, the language and models used become more complicated. The phenomena of teaching and learning, as they occur in engineering education, are extraordinarily complicated and, therefore, a model that is adequate to express even a few of the major factors which affect university education must become intricate and involved. The discussion which follows gradually develops some major elements in one model of teaching and learning.

An S-O-R-C Model

In the early days of psychology, it was observed that there was some functional relationship between a person's situation and his response to that situation. Those who began to study these relationships used a simple *S-R* model. In its rudimentary form, the model suggested that an individual learns to associate certain responses (*R*) with particular characteristics of his stimulus situation (*S*). Some critical characteristics of the situational stimuli (*S*-factors) and the responses made during learning (*R*-factors) which helped or hindered the learning of the association were identified. Many of these functional relationships have been refined and are still very useful.

Others became concerned that this basic *S-R* model left out any recognition of the unique characteristics of the person doing the learning. Often, the learner modifies the stimulus because of his own previous experiences. For example, observers had no trouble identifying the color of a small, irregular shaped red spot when flashed on a screen; but when the spot was shaped like the symbol for spades on playing cards, adults tended to see the color of the spot as gray. Thus, it was said that the observer's *perceptions* modified the stimulus received. On the other hand, *S-R* psychologists responded that the observer merely had two competing sets of "internal" responses and that his associations to the *shape* of the stimulus were stronger than his response to the color. Either way, it was apparent that an *O* must be added to the model to account for what goes on inside the observer, giving the basic *S-O-R* model of learning and performance.

Another factor that has always been of concern to both of the above kinds of psychologists has been the rewards, reinforcement, or pay-off that the learner has received as a result of his efforts—the "consequences" (*C*) of his behavior. Without doing violence to either set of theories, it is easy to add this symbol to the basic model, giving *S-O-R-C*, which emphasizes that "nobody keeps doing anything unless it pays off." Again, a number of characteristics of the consequences (*C*) of the observer's behavior (*R*) have been identified that make certain kinds of learning (within the *O*) more likely to occur. Examples of critical *S*, *O*, *R*, and *C* factors are considered below.

S-factors

There are many characteristics of lectures, films, printed texts, case studies, and other stimulus materials which make them either easy or hard to perceive, respond to, and learn from. Let us look at a few critical *S*-factors and show how these can be controlled through effective teaching to optimize learning.

1. *Stimulus overload.* People have a very small "short term memory" (STM) storage capacity. This is very easy to demonstrate. When a person looks up a seven-digit telephone number, he often has to look back at the book while dialing the number. It is as though there were STM capacity for storing about seven organized but unrelated "chunks" of information, and if some of this memory is used for the subroutine of dialing and thinking of the person being called, then some of the telephone number will drop out. Almost everyone can handle five "chunks" and almost no one can store and do anything with nine or more.

In giving an engineering lecture, most people *vastly* exceed seven "chunks" of new information when talking for more than just a few minutes. How can teachers avoid this natural limit to the student's information processing? One suggestion is to pass out semi-complete lecture notes, so that students can listen without having to make their own notes on everything. Blanks can be left where they have to fill in material that ties together the separate ideas presented so that they have to organize the separate pieces of information into *one larger* chunk, which then takes less storage space. Simple problems can be put into the incomplete notes, therefore requiring students to organize the pieces of information and use them. The stimuli must be adjusted to prevent repeated overload of STM capacity.

2. *Contiguity*, or the principle of togetherness. When two or more objects or events appear together, they come to be associated with each other in such a way that the later presentation of one makes the observer think of the other. This principle seems simple and only a matter of common sense, yet how often it is

Reprinted with permission from *Engineering Education*, vol. 60, no. 6, pp. 726–731, March 1970.

ignored. Let us consider how often most teachers follow a sequence something like this: (a) lecture and write the equations for an R-L-C circuit, (b) draw a set of diagrams representing the phenomena, (c) assign a homework problem accompanied by a circuit diagram drawn in a neat rectangular pattern, and (d) send the student to the lab to connect the components and observe the phenomena on a scope. The principle of contiguity would suggest that the student hear the names, see the symbols for, and hold in his hand each of the components at the same time. He would see the diagram and the physical circuit at the same time so that one comes to mean the same thing as the other. The further apart things are in time, space or context, the more difficult it is to associate them.

Other S-factors which help or hinder learning have been identified, such as: repetition (and the timing of this is tricky), figure-ground relationships (sort of a signal-to-noise phenomena such that the critical aspects of an idea stand out against all of the other things students are exposed to), information source credibility, information organization, use of multiple input channels (visual, audio, tactile), etc.

O-factors

It is popular to talk about adjusting one's teaching to the individual student, yet this is extremely difficult to do. Most teachers cannot or do not do it. When talking about student characteristics which should be considered in designing instructional sequences, teachers use terms like "what he knows," "his level of motivation," and "what he can do." What is proposed here is that an individual student has learned many associative responses which are activated by internal or external stimuli (the arguments as to whether they should be called perceptions, motives, or responses seems relatively irrelevant). A 4 x 2 matrix, as shown in figure 1, represents one simple way of looking at different types of associations stored in long-term memory.

Four classes of associations of major concern to educators are:

1. *Cognitive-symbolic*, or the intellectual meaning of words, mathematical symbols, musical notation, engineering symbols, or any other special set of notations, in which the shape of a symbol is unrelated to the object or event it represents. Such symbols may represent the *static* attributes of the object or event (data), or the *dynamic* instructions for an action to be performed (subroutines, operations, or transformations). For example, the teacher can test himself by thinking or writing out the first ten words that come to mind when seeing each of these words: *house, integration, viscosity, red, reading.*

2. *Pictorial-imagery*, or the "mental image" of an object, event, scene, map, activity or operation, etc. As in the cognitive domain, images may be either static or dynamic, such as the *attributes* of a bicycle or machine, in contrast to the image of a person *riding* a bicycle or *operating* the machine. For example, when someone asks me to describe the pyramids of Egypt, my first response is a vivid picture of a scene with one towering pyramid, several small ones off to the left and a camel-riding Bedouin in the foreground. Quite a different type of image is produced in response to the stimulus words, "dislocation slippage," where I see contour-like lines moving through a lattice and clustering at impurities.

3. *Affective-emotional*, or the glandular and smooth-muscle responses associated with interest in or feelings about something. These may be accompanied by such "internal" reactions as coronary acceleration or deceleration, galvanic skin response (GSR), pupillary contraction or dilation, and other types of static associations. Dynamic or operational associations of this type would include such responses as focusing one's attention, approach-avoidance tendencies, committing energy to some activity, etc.

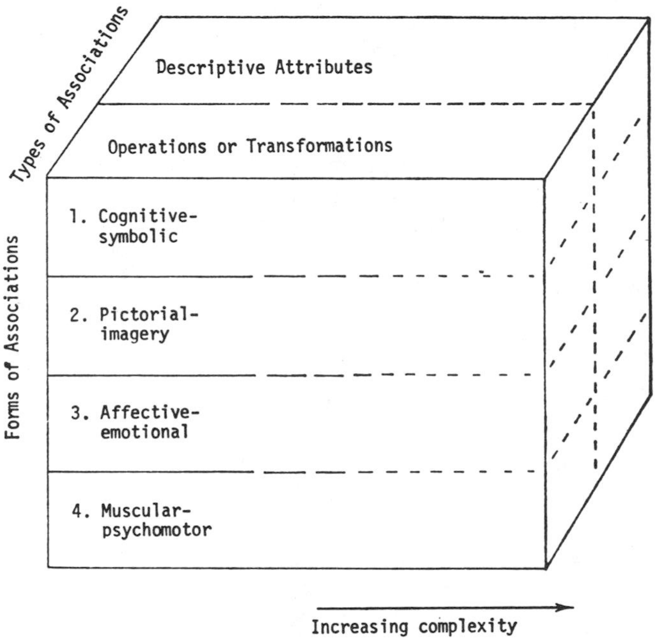

Figure 1. An organization of different types of associations. Simple types may exist independently; complex associations may be interrelated.

4. *Muscular-psychomotor*, or the production of a set of physical responses without any mediating verbal or pictorial associations. Static associations would consist of the tactile or kinesthetic sense of how something feels, and dynamic associations would be the actual performance of an activity. For example, simple reactions such as finger withdrawal from a hot stove or the writing of notes during a lecture, and complex performances such as the playing of a violin sonata or the skillful operation of complex electronic instruments are all psychomotor associations.

An important characteristic of a student which determines the ease with which he learns new materials is the degree to which his associations are rich and interconnected. When one engineering student was asked to freely associate to the expression,

$$i(t) = I_o e^{-at} \cos (bt+c)$$

he produced many descriptive and operational associations, and then commented, "There's no end to it, and I don't want to tell you about my wife who made it possible for me to study engineering." He was a good student and could tie together all that he had learned.

The result of learning is a change in these associations, but this can only be inferred from observation of the performances (the *R's*) as they are the basis for action.

R-factors

The old expression, "students only know what they undergo," has considerable truth to it. While learning-by-doing is an important goal, a student cannot practice *all* of the different competencies he is expected to acquire. Thus, he must practice skills which have broad general application, i.e., skills which he can transfer from one situation to another. In the following discussion, two points will be emphasized: 1. when and how the student should *practice,* considering such questions as the concentration vs. distribution of practice, and the problem of immediate vs. delayed practice, and 2. the different *kinds* of things he should practice.

1. *Practice.* The question of *how* the student should practice what the teacher wants him to learn is actually a complicated question and deserves a more sophisticated answer than will be given here. For present purposes, this discussion will be based on the general characteristics of some curves of learning and forgetting which were derived from studies of relatively simple kinds of learning, but which seem broadly applicable to many other types. Figure 2a indicates that the rapidity with which learning takes place is a function of both practice and the relevance of prior learning. What is not shown is the necessity for massive doses of practice when new learning appears to contradict prior learning (e.g., technicians need a great deal of help in substituting an engineering approach for their customary methods of attack). Figure 2b suggests that forgetting is rapid for information that is only partially learned, while well-learned material is retained for long periods. Figure 2c implies that forgetting proceeds more rapidly after massed practice than when the practice is distributed over time. Another major factor not shown in these simple diagrams is the influence of activities that take place between periods of practice or between practice and testing for the amount learned. As suggested by Figure 2a, when the "intervening" activities are relevant they can help, whereas activities that contradict the earlier learning will cause rapid forgetting. The conclusion is that distributed practice has far-reaching benefits, except (a) when students have considerable "relearning" to do or (b) when their between-practice activities contradict or interfere with the desired learning. This supports the efforts of those who try to have mathematics courses, mechanical engineering laboratories, and electrical engineering lectures all considering related topics at the same time.

2. *Types of Behavior.* Most efforts to improve the quality of teaching have emphasized the importance of describing in detail the performance desired in competent students when they *finish* the course, i.e., objectives stated in terms of post-course student behavior. When instructors do this, student performance on these objectives almost always improves. But it is also important to specify the different types of performances students should be able to demonstrate. Several different "taxonomies" have been proposed, to

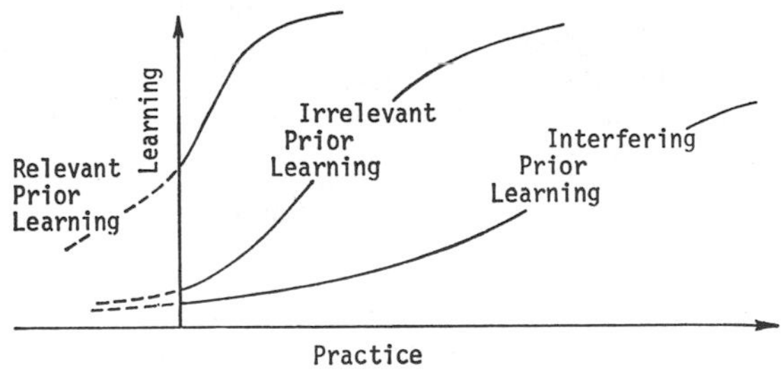

Figure 2a. Learning as a function of prior learning.

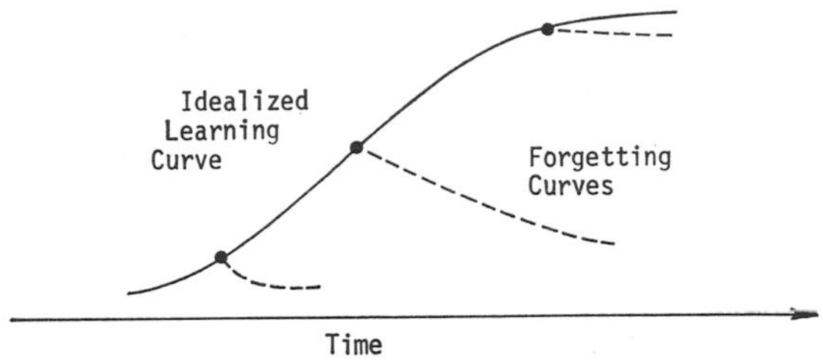

Figure 2b. Forgetting, or the decrease in amount retained, as a function of the level of learning at the time when practice is stopped.

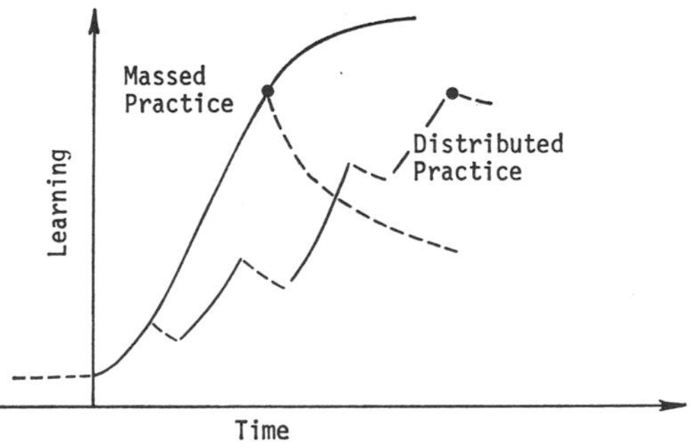

Figure 2c. Forgetting as a function of the distribution of practice.

organize objectives in some meaningful way. The framework described below and illustrated in figure 3 has been drawn from a number of different taxonomies and shows a kind of logical progression from the simplest behaviors to the most complex performances. It is important to ask whether engineering students are taught to perform the tasks required at each level of this framework.

a) *Associating,* or the process of calling material from memory, including all of the different types of associations described above. The recalled materials may be descriptive attributes or operations (data and subroutines); cognitive, pictorial, affective or motoric; and any combination of simple or complex structures of associations (such as structures commonly called concepts, principles, values, etc.).

b) *Discriminating,* or the process of comparing two or more attributes and concluding either, "Yes, they are the same," or "No, they are not the same." Such discriminations are important to engineers in reading instruments, setting controls, reading diagrams, interpreting equations, and reading case descriptions.

c) *Conceptualizing,* or the process of making discriminations and grouping together into one organized pattern those objects or events which have certain related attributes. Instructors often talk about "defining the problem" as the first step in problem-solving, and this is the process of conceptualizing an organized pattern of the information given. What most instructors do *not* realize is that the subroutines necessary to conceptualize are complex and that they need to be taught and learned. It is instructive to prepare the flow diagram of a computer-like program that humans could follow in performing the operation of conceptualizing. Several programs have been written that permit computers to "learn" certain concepts.

d) *Valuing,* or the process of comparing the attributes of an object, event or concept, with the attributes of a "value" concept where many of these attributes have high affective loading, to obtain a set of value attributes which can now be attached to the original object, event, or concept. This is merely finding out whether x is really worthwhile for a person. This can be done simply by seeing whether one likes one or two of x's attributes or by a very complex process of examining all x's attributes, placing a value on each, and combining the different values through some weighting process.

1. Associating, or recalling attributes and operations from memory
2. Discriminating, or comparing two or more attributes
3. Conceptualizing, or grouping according to common attributes in a pattern or organization
4. Valuing, or comparing the attributes of an event with those of a "value" concept
5. Predicting, or forecasting the results of performing an operation on an object or event
6. Deciding, or selecting an object or event on the basis of the relative worth of several alternatives

Increasing Complexity within each type

Figure 3. A framework for classifying types of human processes according to the complexity of the basic operation involved in each type.

e) *Predicting,* or the process of forecasting the result of performing an operation upon the attributes of an object, event, or concept. The resultant would be a new, modified or transformed concept, with many of its attributes being uncertain or probabilistically defined. Predictions that extend over a period of time, and include many different operations and concepts, are very complex processes and usually require the assistance of external information processing and storage, as the amount of information involved vastly exceeds the limit of one man's capacity.

f) *Deciding,* or the process of comparing the relative worth of two or more objects, events or concepts, and selecting the one with the highest worth (which is some complex function of gains, losses, costs, risks, etc.). Although simple decisions (choices) may be made on the basis of a single attribute and value, complex decisions represent one of the most complex performances of which man is capable. The algorithms and heuristics of decision-making need to be taught and learned by engineers; once learned, they will consist of some of the most complex operational associations.

Referring to figure 2 of the preceding article ("What Instructors Say to Students Makes a Difference"), one notes the similarity between this six-point framework and the simpler two-step levels of the "achievement, intellectual needs" of man. They are both intended to represent the same phenomena, with this more specific treatment being more useful as instructors work toward the detailed specification of students' tasks and their desired outcomes.

C-factors

The manner in which teachers, business managers, parents, and others use rewards and punishments has been of concern for many years. With the very best of intentions, individuals have acted and argued as though they held diametrically opposed positions on the management of the consequences to be provided to students, employees, and children. In the discussion that follows, an attempt will be made to show how different points of view can be related and how each position may be appropriate under particular situations.

Effect of Positive Consequences. Building on the hierarchy of motives and incentives shown in figure 2 of the preceding article, some of the different kinds of rewards (positive consequences) which instructors and educators can use are listed in figure 4. Readers may be able to think of others which *could* be used and of how few *are* actually, deliberately employed to foster the achievement of the goals of engineering education.

Levels of Need	Consequences Which Could Be Provided To Students
Self-Actualization	Resources for independent exploration Encourage and assist in publishing, giving editorial and typing assistance A "sabbatical semester" with full academic credit, giving resources and facilities, introductions to others in areas of interest
Aesthetic	Seminars for reading and discussion with experts in diverse fields which could be related to areas of interest to the student Field trips to places of unusual interest Outside specialists brought in for talks and visits (artists, social anthropoligists, psychologists)
Understanding	Provide him with all data and results of his own work, access to reports, plans, budgets, etc. Review with him rationale and decision criteria for curriculum content and methods Give him key to laboratory and facilities to perform related work on his own schedule
Knowledge	Provide access to data on courses and instructors Make diverse library resources available Provide tutoring on request Arrange for student to see practical application of course contents
Esteem	Refer others to him for help in his area of competence Publicize his performances in bulletin boards, newspapers, notices, meetings, classes Give him significant responsibility when faculty or administration out-of-town Have him represent school at professional meetings
Belonging	Call him by name, provide him personal space in the building with his name plate Visit him and have him visit you, on schedule Express publicly and privately your pleasure in working with students, and with him in particular
Security	Provide grades that permit continued study in areas of his personal interest and abilities Prepare and publish policies on all administrative matters and follow them consistently Act consistently, avoiding punishment and sarcasm Schedule time for relaxation and fun
Survival	Provide strong, consistent safety measures Have trained first aid personnel in the areas, ambulance service, hospitals Provide for health and sanitation Assure adequate diet

Note. Lower level consequences can deactivate lower level needs, freeing the student for activity in the growth domain.

Figure 4. Consequences which instructors and educational administrators can provide students, contingent upon the demonstration of appropriate professional performances.

Particular attention must be called to one aspect of these consequences which is often overlooked, frequently misunderstood, and sometimes rejected as a significant factor. These consequences are effective for the improvement of student performance only when they are *contingent* upon the student's demonstration of some desired performance. Recalling that the four basic domains of figure 4 are physical, social, intellectual and aesthetic, it may be appropriate to provide the simplest kinds of consequences in each domain to all students merely as the consequences of their *attending* school—the very simplest type of performance expected of students. For humanitarian and cultural reasons, most teachers would provide all possible consequences at the physical-organizational levels, to assure that these factors do not detract from attention to the more important functions of a university (those in the social, intellectual, and aesthetic domains). In fact, there is growing pressure to abolish grades (which function primarily at the security level), as they are not seen as functioning to promote learning (or functioning at the achievement level). This would be an interesting argument to explore at another time.

The function of feedback in any control or servo-system is well known to engineers, as is the effect of time delay in the feedback loop of such a system. Therefore, it should not be surprising to engineers that the consequences illustrated in figure 4 serve effectively to improve function only if they are directly related to and contingent upon specific performances of the student and provided with a minimum of time delay. There are several things an instructor can do to provide immediate and relevant feedback to students through learning to control his own behavior in the manner suggested in the preceding article. Before considering how this *S-O-R-C* model can be used to improve the effectiveness of instruction, it is well to look at another aspect of *C*-factors.

Effect of Negative Consequences. The above discussion of C-factors focused on the "rewards" instructors can use to enhance learning. It is also useful to look at the influence of the negative side of these issues—reprimands, correction, or punishment. At best, the effect of even mild forms of punishment is uncertain and, at worst, punishment is quite likely to produce the opposite effect from that intended or hoped for by the one who punishes.

Generally, there are three kinds of reaction to punishment:

1. *Avoidance*, or staying away from the source of the punishment (people tend to avoid what is unpleasant). Most school drop-outs have experienced massive amounts of failure, frustration and punishment, and they quite naturally leave as soon as the law allows.

2. *More aggression*, or more frequent acting in the way that was punished. Punishment often rewards the punish*ee* in two ways: (a) it focuses the attention and energy of important people on the punish*ee*, and that can be important, and (b) it puts the behavior of significant persons under the control of the punishee, so that he forces them to pay attention and act in response to his behavior. Both of these rewards can be sufficiently reinforcing that the punished activity, or a similar behavior, will be repeated more frequently in the future. This makes punishment and aggression a vicious spiral.

3. *Behavior re-adjustment*, or the changing of action in order to obtain more desirable consequences. Unfortunately, this type of behavior is only possible for those who are able to function at a relatively high level of motivation—well up in the achievement-intellectual domain. To be able to process punishment as information feedback, rather than social rejection or a threat at the security level, requires a remarkably well-adjusted individual. It is probably only possible for an individual to use punishment as information feedback if he has other, adequate sources of need-fulfillment at the security and social levels.

Behavior Modification. If punishment is uncertain and usually undesirable, there should be at least one alternative—and there is. The following approach has grown out of studies in the area of "behavioral modification" or "contingency management." There are two aspects to this alternative and both must be practiced together, as neither is complete by itself.

a) *Ignore* the undesired misbehavior. (If severe property or personal damage makes this impractical, use only that amount of force required to stop the behavior without pain or punishment to the actor.)

b) *Reward* successive approximations to desired or acceptable behavior. The reward must be some thing or activity the individual or group wants which can be controlled by the instructor, manager, parent, etc. The first rewards may be at the very lowest level of the motivational hierarchy that is appropriate to the individual (e.g., money or food), and then the contingency manager may move slowly up the hierarchy with the kinds of "motivators" that work and are more desirable.

There are usually a surprising number of arguments brought out to refute or attempt to discredit this type of approach. The first concerns the notion of "bribing" or manipulating individuals to act the way one wants them to, and the second concerns the moral issue of the justice of punishment for wrongdoing. The counter-arguments are basically pragmatic; punishment is very seldom effective in changing another's behavior while positive consequences are.

The S-O-R-C Model and Instructor Behavior

I sat in on a class, recently, where the instructor walked in at the appointed hour, opened his notebook, picked up a piece of chalk and began, "Last time, we discussed equation 17 on page 129. Now let's continue with . . ." He presented a well-organized lecture on some detailed characteristics of the equations which model the behavior of fluids under various conditions. He gave several good examples from practical experience. His lecture was well timed, as it rounded off nicely just before the end of the hour, giving time for the assignment of a problem for next time and the final phrase, "Are there any questions?" There were none, and students picked up their corrected homework on the way out. Almost anyone would have agreed that his was a well-organized, carefully prepared class session with significant material presented and illustrated. Now, let us look at it from the points of view presented in this and the preceding article.

First, using the *S-O-R-C* framework for analysis, the instructor spent 100% of his time in class controlling some *S*-factors, with no use of *R*-factors or

C-factors. His intent appeared to be the simultaneous use of words and chalkboard equations to induce verbal *associations* to the mathematical symbols, to make complicated and very fine *discriminations* between some verbal descriptions of the problem situation and the mathematical model, and to lead students to *conceptualize* a method for attacking other problems similar to those presented. He made the assumption that all students in the class (the *O*-factor) had adequate and equivalent associations to the words and symbols he presented—an assumption which he knew was invalid, as not all students were performing equally well. In marking homework, he used a red pencil to indicate negative consequences (pointing out errors) and very general evaluative feedback through a letter grade (which functions at the security level of motivation).

Second, using the categories of conversation to analyze the session, about 95% of the time was spent in category 5 (giving information and opinions), 3% in category 6 (giving directions), and 1% in category 4 (asking questions). There was no use of categories 1, 2, or 3, which have been shown to be positively related to higher achievement and more favorable attitudes toward the subject and the class. There were no category 8 or 9 statements (students did not respond or initiate any conversation).

Let us look at some of the arguments used to support this kind of class session and, then, at some alternatives open to the instructor. The ever-present need to cover as much material as possible tends to make instructors feel that their function is to be technically competent and to keep up-to-date in their fields, and this leaves them no time to consider any "educational frills." Also, few instructors have been trained in pedagogy, most don't think they have the time to learn educational psychology, and many have so little respect for the competence of teacher-educators that they would be unwilling to learn from them even if they had the time. And besides, many fine engineers have been produced by this "lecture method," so what proof is there that it is so bad, after all? Now, that is a formidable array of defensive arguments.

What the Instructor Could Have Done

What could the above instructor have done differently?

1. *S-Factors.* He could have started with a brief overview of the problem, even using slides of a real situation in conjunction with (contiguous with) a simplified explanation of the problem characteristics and solution methods (the attributes and operational associations to be learned reduced to 5-7 "chunks" of information). His lecture notes could have been partially duplicated (even in hand-written form), including problems which the students could have worked on in class.

2. *R-factors.* Student work to complete the lecture notes (adding only the integrative or most significant elements) and to partially solve some problems presented in the notes, would have assured that all students had the chance to respond, no matter how large the class. Time could have been allowed for students to ask questions (category 9), or to answer those raised by the instructor (category 8).

3. *C-factors.* Social reinforcement could have been provided in several ways. Greeting students as they arrived could reward them merely for coming to class, in the same way as smiling at the class and sometimes telling a humorous story. Calling students by name as they responded or asked questions could be a form of social reward for active participation. Restating and using students' ideas (category 3) could provide both social reinforcement and achievement-oriented feedback. Even a little thing like using red pencil marks on a homework paper could be avoided, because of the number of students who have learned to associate "red marks" with punishment; a green nylon-tip pen produces remarkably different associations. And letter grades provided feedback at only the security level; they could have been accompanied by comments at the social and intellectual level, such as "Good, Charley. Your initial equations and diagrams were correct, except that you left out . . ."

Design of Instruction

There are a number of instructional design principles which can be applied to engineering education and which have the likelihood of increasing (a) the effectiveness of an instructor's performance as he interacts with students, (b) the amount of learning which students achieve, and (c) students' attitudes toward the learning of engineering. The simple models of learning and teaching presented in these two articles can provide guidance to those interested in becoming more sophisticated about the design of instruction.

References

1. Bloom, B. S., *Taxonomy of Educational Objectives I: Cognitive Domain*, Longmans Green, New York, 1956.

2. Bourne, L. E., Jr., *Human Conceptual Behavior*, Allyn & Bacon, Boston, 1966.

3. Deese, J., *The Structure of Associations in Language and Thought*, Johns Hopkins Press, Baltimore, 1965.

4. Gagne, R. M., *The Conditions of Learning*, Holt, Rinehart and Winston, New York, 1965.

5. Jakobowits, L. A., and Miron, M. S., *Readings in the Psychology of Language*, Prentice-Hall, Englewood Cliffs, New Jersey, 1967.

6. Klausmeier, H. J., and Harris, C. W., *Analysis of Concept Learning*, Academic Press, New York, 1966.

7. Krathwohl, D. R., *Taxonomy of Educational Objectives II: Affective Domain*, McKay, New York, 1964.

8. Mandler, G., Mussen, P., Kogan, N., and Wallach, M. A., *New Directions in Psychology III*, Holt, Rinehart and Winston, New York, 1967.

9. Reitman, W. R., *Cognition and Thought*, Wiley, New York, 1965.

10. Schroder, H. M., Driver, M. J, and Streufert, S., *Human Information Processing*, Holt, Rinehart and Winston, New York, 1967. △

Some New Views of Learning and Instruction

ROBERT M. GAGNÉ

***Abstract*—Does learning require repetition? What factors affect recall? How important is diagnostic testing?**

A generation ago learning psychologists based answers to such questions on Thorndikean formulations. In this paper a leading contemporary educational psychologist discusses the new theories and explains some of the provocative experiments on which they are based.

DURING recent years there has been an increased recognition of, and even emphasis on, the importance of principles of learning in the design of instruction for the schools. This recognition of the central role of learning in school-centered education seems to be accorded whether one thinks of the instruction as being designed by a teacher, by a textbook writer, or by a group of scholars developing a curriculum.

Manuscript received July 30, 1970. This paper is reprinted from the May 1970 issue of the *Phi Delta Kappan*, and is based on a paper delivered before a seminar on recent scientific developments, sponsored by the American Association of School Administrators, Washington, D. C., October 26–28, 1969.

The author is with Florida State University, Tallahassee, Fla.

When the findings of research studies of learning are taken into account, one usually finds questions about instruction to be concerned with such matters as these.

1) For student learning to be most effective, how should the learning task be presented? That is, how should it be communicated to the student?

2) When the student undertakes a learning task, what kinds of activity on his part should be required or encouraged?

3) What provisions must be made to insure that what is learned is remembered and is usable in further learning and problem solving?

Questions such as these are persistent in education. The answers given today are not exactly the same as those given yesterday, and they are likely to be altered again tomorrow. The major reason for these changes is our continually deepening knowledge of human behavior and of the factors which determine it. One should not, I believe, shun such changes nor adopt a point of view which makes difficult the application of new knowledge to the design of novel procedures for instruction.

Reprinted from *IEEE Trans. Educ.*, vol. E-14, no. 1, pp. 26–31, February 1981.

The opportunities for improvement seem great and the risks small.

Status of Learning Research

As a field of endeavor, research on how human beings learn and remember is in a state of great ferment today. Many changes have taken place, and are still taking place, in the conception of what human learning is and how it occurs. Perhaps the most general description that can be made of these changes is that investigators are shifting from what may be called a *connectionist* view of learning to an *information processing* view. From an older view which held that learning is a matter of establishing *connections* between stimuli and responses, we are moving rapidly to acceptance of a view that stimuli are *processed* in quite a number of different ways by the human central nervous system, and that understanding learning is a matter of figuring out how these various processes operate. Connecting one neural event with another may still be the most basic component of these processes, but their varied nature makes connection itself too simple a model for learning and remembering.

My purpose here is to outline some of these changes in the conception of human learning and memory, and to show what implications they may have for the design and practice of instruction. I emphasize that I am not proposing a new theory; I am simply speculating on what seems to me to be the direction in which learning theory is heading.

The Older Conception

The older conception of learning was that it was always basically the same process, whether the learner was learning to say a new word, to tie a shoelace, to multiply fractions, to recount the facts of history, or to solve a problem concerning rotary motion. Thorndike [1] held essentially this view. He stated that he had observed people performing learning tasks of varied degrees of complexity and had concluded that learning was invariably subject to the same influences and the same laws. What was this model of learning that was considered to have such broad generalizability?

One prototype is the conditioned response, in which there is a pairing of stimuli, repeated over a series of trials. The two stimuli must be presented together, or nearly together, in time. They are typically associated with an "emotional" response of the human being, such as an eyeblink or a change in the amount of electrical resistance of the skin (the galvanic skin reflex). The size of the conditioned response begins at a low base-line level, and progressively increases as more and more repetitions of the two stimuli are given. Such results have been taken to indicate that repetition brings about an increasingly "strong" learned connection, with an increase in strength that is rapid at first and then more slow.

Learning curves with similar characteristics have been obtained from various other kinds of learned activities, such as simple motor skills like dart-throwing and memorization of lists of words or sets of word pairs.

Remembering

What about the remembering of such learned activities? Is learning retained better as a result of repetition? Is something that is repeated over and over at the time of learning better recalled after the passage of several weeks or months? The curve which describes forgetting is perhaps equally familiar. Forgetting of such things as lists of nonsense syllables is quite rapid in the beginning, and after several weeks descends to a point at which only about 20 percent is remembered. A motor task is usually retained a great deal better, and after the same amount of time its retention may be as much as 80 percent.

These are the basic facts about remembering. But how is it affected by repetition? Is retention better if the original learning situation has been repeated many times? Evidence is often cited that this is so. Increasing the number of trials of repetition during original learning has the effect of slowing down the "curve of forgetting," i.e., of improving the amount of retention measured at any particular time. Underwood [2], for example, has stated that "degree of learning" of the task to be recalled is one of the two major factors which influence forgetting in a substantial manner. The second factor is interfering associations, whose strength is also determined by their degree of learning. It should be pointed out that when Underwood uses the phrase "degree of learning" he refers to amount of practice—in other words, to amount of repetition.

Here let me summarize what I believe are the important implications for instruction of what I call the "older" conceptions of learning and memory. The designer of instruction, or the teacher, had to do two major things. First, he had to arrange external conditions of presentation so that the stimulus and response had the proper timing—in other words, so that there was *contiguity* between the presentation of the stimulus and the occurrence of the response. Second, he had to insure that sufficient *repetition* occurred. Such repetition was necessary for two reasons. It would increase the strength of the learned connections; the more the repetition, within limits, the better the learning. Also, repetition was needed to insure remembering; the greater the number of repetitions, the better the retention. Presumably, whole generations of instructional materials and teacher procedures have been influenced in a variety of ways by application of these conceptions of learning to the process of instruction.

Questioning Older Conceptions

During recent years, a number of significant experimental studies of learning and memory have been carried out which call into question some of these older conceptions. (Of course, there have always been a certain number of individuals—voices in the wilderness—who doubted that these principles had the general applicability claimed for them.) I shall describe only a few of the crucial new studies here, to illustrate the perennial questions and their possible answers.

Does Learning Require Repetition?

A most provocative study on this question was carried out by Rock [3] as long ago as 1957. It has stimulated many other studies since that time, some pointing out its methodological defects, others supporting its conclusions [4]. The finding of interest is that in learning sets of verbal paired associates, practice does not increase the strength of each learned item; each one is either learned or not learned. To be sure, some are learned on the first practice trial, some on the second, some on the third, and so on; but an item once learned is fully learned.

As far as school subjects are concerned, a number of studies have failed to find evidence of the effectiveness of repetition for learning and remembering. This was true in an investigation by Gagné, Mayor, Garstens, and Paradise [5], in which seventh-graders were learning about the addition of integers. One group of children was given four or five times as many practice problems on each of ten subordinate skills as were given to another group, and no difference appeared in their final performance. A further test of this question was made in a study by Gibson [6], who set out to teach third- and fourth-graders to read decimals from a number line. First, she made sure that subordinate skills (reading a number in decimal form, writing a number in decimal form, locating a decimal number on a number line) were learned thoroughly by each child. One group of students was then given a total of ten practice examples for each subordinate skill, a second group 25 for each, and a third none at all. The study thus contrasted the effects of no repetition of learned skills, an intermediate amount of repetition, and a large amount of repetition. This variable was not found to have an effect on performance, both when tested immediately after learning and five weeks later. Those students who practiced repeated examples were not shown to do better, or to remember better, than those who practiced not at all.

Still another study of fairly recent origin is by Reynolds and Glaser [7], who used an instructional program to teach ten topics in biology. They inserted frames containing half as many repetitions in one case, and one-and-a-half times as many repetitions in another, as those in a standard program. The repetitions involved definitions of technical terms. When retention of these terms was measured after an interval of three weeks, the investigators were unable to find any difference in recall related to the amount of repetition.

I must insert a caveat here. All of the studies I have mentioned are concerned with the effects of repetition immediately after learning. They do not, however, test the effect of repetition in the form of *spaced reviews*. Other evidence suggests the importance of such reviews; in fact, this kind of treatment was found to exert a significant effect in the Reynolds and Glaser study. Note, though, that this result may have quite a different explanation than that of "strengthening learned connections."

Modern Conceptions of Learning

Many modern learning theorists seem to have come to the conclusion that conceiving learning as a matter of strengthening connections is entirely too simple. Modern conceptions of learning tend to be highly analytical about the events that take place in learning, both *outside* the learner and also *inside*. The modern point of view about learning tends to view it as a complex of processes taking place in the learner's nervous system. This view is often called an "information-processing" conception.

One example of an information processing theory is that of Atkinson and Shiffrin [8]. According to this theory, information is first registered by the senses and remains in an essentially unaltered form for a short period of time. It then enters what is called the short-term store, where it can be retained for 30 seconds or so. This short-term store has a limited capacity, so that new information coming into it simply pushes aside what may already be stored there. But an important process takes place in this short-term memory, according to Atkinson and Shiffrin. There is a kind of internal reviewing mechanism (a "rehearsal buffer") which organizes and rehearses the material even within this short period of time. Then it is ready to be transferred to long-term store. But when this happens it is first subjected to a process called *coding*. In other words, it is not transferred in raw form, but is transformed in some way which will make it easier to remember at a later time. Still another process is *retrieval*, which comes into play at the time the individual attempts to remember what he has learned.

It is easy to see that a much more sophisticated theory of learning and memory is implied here. It goes far beyond the notion of gradually increasing the strength of a single connection.

Prerequisite for Learning

If repetition or practice is not the major factor in learning, what is? The answer I am inclined to give is that the most dependable condition for the insurance of learning is the prior learning of prerequisite capabilities. Some people would call these "specific readinesses" for learning; others would call them "enabling conditions." If one wants to insure that a student can learn some specific new activity, the very best guarantee is to be sure he has previously learned the prerequisite capabilities. When this in fact has been accomplished, it seems to me quite likely that he will learn the new skill without repetition.

Let me illustrate this point by reference to a study carried out by Wiegand [9]. She attempted to identify all the prerequisite capabilities needed for sixth-grade students to learn to formulate a general expression relating the variables in an inclined plane. Without using the exact terminology of physics, let us note that the task was to formulate an expression relating the *height* of the plane, the *weight* of the body traversing downwards, and the *amount of push* imparted to an object at the end of the plane. (Wiegand was not trying to teach physics, but to see if the children could learn to formulate a physical relationship which was quite novel to them.) The expression aimed for was, "Distance pushed times a constant block weight equals height of plane times weight of cart."

Initially, what was wanted was explained carefully to the students; the plane and the cart were demonstrated. Thirty students (out of 31) were found who could not accomplish the task, that is, they did not know how to solve the problem. What was it they didn't know? According to the hypothesis being investigated, they didn't know some *prerequisite* things. Fig. 1 shows what these missing intellectual skills were thought to be.

What Wiegand did was to find out which of these prerequisite skills were present in each student and which were not present. She did this by starting at the top of the hierarchy and working downwards, testing at each point whether the student could do the designated task or not. In some students, only two or three skills were missing; in others, seven or eight. When she had worked down to the point where these subordinate capabilities *were* present, Wiegand turned around and went the other way. She now made sure that all the prerequisite skills were present, right up to, but not including, the final inclined plane problem.

The question being asked in this study was: If all the prerequisite skills are present, can the students now solve this physical problem which they were unable to solve previously? Wiegand's results are quite clear-cut. Having learned the prerequisites, nine out of ten students were able to solve the problem which they were initially unable to solve. They now solved the problem without hesitation and with no practice on the problem itself. On the other hand, for students who did not have a chance to learn the prerequisites, only three of ten solved the problem (and these were students who had no "missing" skills). This is the kind of evidence that makes me emphasize the critical importance of prerequisite intellectual skills. Any particular learning is not at all difficult if one is truly prepared for it.

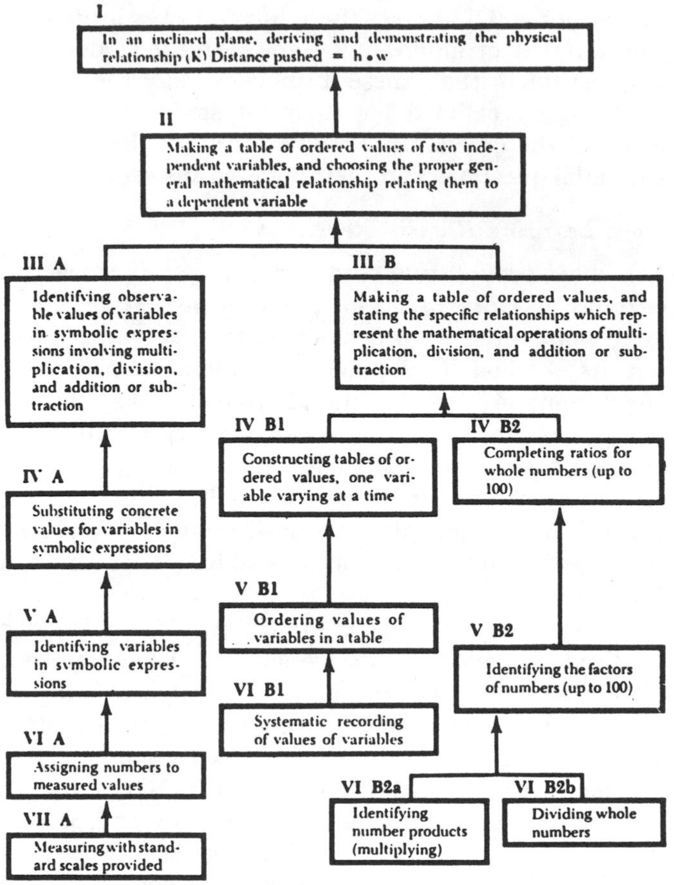

Fig. 1. A hierarchy of subordinate intellectual skills applicable to the problem of deriving a general expression relating variables in an inclined plane (Wiegand, 1969).

Coding and Remembering

Quite a number of studies appear in the experimental literature pertaining to the effects of coding of information on its retention. I choose as an illustration a study by Bower and Clark [10]. These investigators studied the recall by college students of 12 lists of 10 nouns apiece. In learning each list, each student was encouraged to make up a story connecting the nouns. For each student there was a yoked control who was not encouraged to make up a story, but who was permitted the same amount of time to learn each list of nouns.

Here is an example of a story which one of the subjects constructed for the words *vegetable*, *instrument*, *college*, *nail*, *fence*, *basin*, *merchant*, *queen*, *scale*, and *goat*: "A vegetable can be a useful instrument for a college student. A carrot can be a nail for your fence or basin. But a merchant of the queen would scale that fence and feed the carrot to a goat."

The subjects were asked to recall each list immediately after their study of it. They recalled 99 percent under both conditions. The subjects were later asked to recall all of the lists, after they had learned all 12. In this case there was an enormous difference: the recall of the narrative group averaged 93 percent, that of the nonnarrative group only 13 percent. In other words, deliberate coding had increased recall by seven times.

Retrieval and Remembering

Suppose that learning has indeed occurred. What will insure that whatever has been learned will be remembered? There seems to be at least some absence of evidence that simply practicing or repeating things after they have been learned has the effect of improving retention. What the individual does when he is asked to remember something is to *retrieve* it; that is, he brings to bear a process of searching and finding, in his memory, something he is looking for. This process is probably very little understood at present, but there is increasing evidence that it does occur and that it plays a crucial role in remembering.

Some interesting work has been done on the subject of retrieval. In one experiment, Tulving and Pearlstone [11] had groups of high school students learn lists of words of various lengths: 12 words, 24 words, or 48 words. The words themselves were instances of categories, such as four-footed animals (cow, rat), weapons (bomb, cannon), forms of entertainment (radio, music), professions (lawyer, engineer), etc. The words were presented one at a time in mixed-up order. Different lists also used one, two, or four words in each category.

Once the lists of words had been learned, recall was measured under two different conditions. In the first, the learners were simply told to write down all the words they could remember. In the second, the category names were used as cues to recall; that is, the learners were asked to write down all the words they remembered which were "forms of entertainment," all which were "four-footed animals," and so on. These extra cues worked wonders on recall. The effect was more marked the greater the number of words that had to be recalled. The differences among those learning 48 words was striking, amounting to a twofold increase.

These results show in a rather clear way how powerful is the effect of such extra cues on retrieval of information that has been learned. In this study, the words themselves can be said to have been "equally well learned" in all the groups. What was different between the groups was the aid they were given in retrieving what they had learned. This is only one of the accumulating pieces of evidence that remembering is markedly affected by retrieval at the time of recall, more than it is, perhaps, by events taking place at the time of learning.

Implications for Instruction

The contrasts between older and newer conceptions of learning and memory seem to me quite remarkable. What implications do they have for instruction? If there are indeed newly discovered ways to affect learning and remembering, how might they be put to use in the classroom and in materials of the curriculum?

First, there is the very fundamental point that each learner approaches each new learning task with a different collection of previously learned prerequisite skills. To be effective, therefore, a learning program for each child must take fully into account what he does and what he does not know how to do already. One must find out what prerequisites he has already mastered—not in a general sense, but in a very precise sense for each learner. Does this mean one must do "diagnostic testing"? Yes, that's exactly what it means. To do so, of course, one must first develop the requisite diagnostic tests. By and large, we do not have them.

Second, the most important guide to the learning that needs to be accomplished is the set of prerequisites that the student has not yet mastered. Remember here Wiegand's experiment. When she systematically saw to it that students climbed the hierarchy, skill by skill, this was what was specifically needed to get them to engage in the problem solving they were originally unable to do.

Third, do students need additional practice to insure retention? If by this is meant, "Should they be given many additional examples so that what they have learned will be 'strengthened'?," I think the evidence says it probably won't work this way. Periodic and spaced reviews, however, are another matter, and it seems likely that these have an important role to play in retention. Notice that when a review is given, the student has to exercise his strategies of retrieval.

This brings me to the final point, which concerns the processes of coding and retrieval. Probably what should be aimed for here is the learning by students of strategies of coding. These are by no means the same as what are called "mnemonic systems," although it is possible that such systems have a contribution to make in teaching us how coding might be done. For meaningful learning, it appears even more likely that notions like "advance organizers" and "anchoring ideas," as studied by Ausubel [12], may be particularly powerful.

Similarly, retrieval strategies are also a class of objective that might be valued for instruction. From the evidence we have, I should say that retrieval strategies might very well consist of networks of superordinate categories into which newly learned specific information, or specific intellectual skills, can be placed. Having students learn to retrieve information by a process of search which first locates such superordinate networks may be a major way of providing them with the capability of good retention.

Even these two or three aspects of modern learning conceptions, it seems to me, lead to a very different view of what instruction is all about. In the most general sense, instruction becomes not primarily a matter of communicating something that is to be stored. Instead, it is a matter of stimulating the use of capabilities the learner already has at his disposal, and of making sure he has the requisite capabilities for the present learning task, as well as for many more to come.

References

[1] E. L. Thorndike, *Human Learning*. New York: Appleton-Century, 1931, p. 160.

[2] B. J. Underwood, "Laboratory studies of verbal learning," in *Theories of Learning and Instruction. Sixty-Third Yearbook, Part I*, E. R. Hilgard, Ed. Chicago: National Society for the Study of Education, 1964, p. 148.

[3] I. Rock, "The role of repetition in associative learning," *Am. J. Psych.*, pp. 186–193, June 1957.

[4] W. K. Estes, B. L. Hopkins, and E. J. Crothers, "All-or-none and conservation effects in the learning and retention of paired associates," *J. Experimental Psych.*, pp. 329–339, December 1960.

[5] R. M. Gagné, J. R. Mayor, H. L. Garstens, and N. E. Paradise, "Factors in acquiring knowledge of a mathematical task," *Psych. Monographs*, no. 7, 1962 (whole no. 526).

[6] J. R. Gibson, "Transfer effects of practice variety in principle learning," Ph. D. dissertation, University of California, Berkeley, 1964.

[7] J. H. Reynolds and R. Glaser, "Effects of repetition and spaced review upon retention of a complex learning task," *J. Educ. Psych.*, pp. 297–308, October 1964.

[8] R. C. Atkinson and R. M. Shiffrin, "Human memory: A proposed system and its control processes," in *The Psychology of Learning and Motivation: Advances in Research and Theory*, vol. 2, K. W. Spence and J. T. Spence, Eds. New York: Academic Press, 1968, pp. 89–195.

[9] V. K. Wiegand, "A study of subordinate skills in science problem solving," Ph. D. dissertation, University of California, Berkeley, 1969.

[10] G. H. Bower and M. C. Clark, "Narrative stories as mediators for serial learning," *Psychonomic Sci.*, pp. 181–182, April 1969.

[11] E. Tulving and Z. Pearlstone, "Availability versus accessibility of information in memory for words," *J. Verbal Learning and Verbal Behavior*, pp. 381–391, August 1966.

[12] D. P. Ausubel, *Educational Psychology: A Cognitive View*. New York: Holt, Rinehart and Winston, 1968.

Charles E. Wales
West Virginia University

Improve Your Teaching Tomorrow with Teaching-Learning Psychology

You can be a better teacher tomorrow through the common sense application of a little teaching/learning psychology. Five basic "psychological principles"[1] every teacher should attempt to use in the organization and presentation of each course are:

1. *Guide* the student's learning
2. Provide for *practice*
3. *Evaluate* and *give feedback*
4. *Motivate*
5. *Individualize*

Simply put, the effective teacher is likely to be someone who guides students while they learn, allows them to practice, gives students the feedback needed to know whether what they have learned is correct, who reinforces correct behaviors, motivates the students so they want to learn, and recognizes that people are different, that they have different backgrounds and interests and learn at different rates, and who attempts to meet this variation by individualizing work.

Guide the Student

Some teachers translate *guidance* into a syllabus for their course, the choice of textbook, homework assignments and lectures. That may be a good start, but effective guidance involves more. You can do more tomorrow by going to class prepared to pass out a set of content-performance objectives describing what each student should be able to do after the current unit is completed. Also tonight you might prepare a set of notes to be duplicated and handed out, which supplement or complement the text by organizing new concepts from simple to complex and which provide organizers, such as lists, tables, or graphs, to help students learn.

While you might prepare these objectives and notes tonight, you should also consider the more involved task of preparing objectives for your whole course. More effective teaching usually springs directly from preparing objectives which specify both the subject matter involved and the intellectual operations expected of the student. A sample set of objectives and the intellectual operations appropriate for work with single-answer problems or ideas is shown in table 1. If the goals of the course involve more sophisticated intellectual operations, such as those related to the decision making skills required for professional development, then objectives similar to those shown in table 2 should be prepared. Of course, these objectives should be given to the student to guide the learning experience, whether it takes place in or out of class.

Both the teacher and the student must have clearly stated objectives if the teaching/learning process is to be most effective. The student also needs to know how the course relates to the real world: the many and varied applications of the subject matter may be obvious to the teacher, but not to the student, who can learn more effectively if this relationship is made explicit. To accomplish this end the teacher may organize course work so the student uses what is learned to solve both single-answer and open-ended problems.

Applying new concepts by solving single-answer problems is the key to acquisition and assimilation of new knowledge. Engineering faculty have followed that practice for many years; other faculty are just now learning the importance of such work. Using both single-answer problems and values to solve open-ended problems is the final step in a student's development—the application of what has been learned to complex problems, using simulations, games, case studies and finally, if possible, some sort of internship.

The teacher who makes the effort to develop objectives and provide both single-answer and open-ended problems will almost automatically find ways to organize the subject matter sequentially from simple to complex and to help guide students by providing the organizers mentioned earlier. This teacher will find ways to help the student evaluate his or her own performance so that a great deal can be learned independently, outside of class. This change frees class time for guidance on either the objectives in table 1 or those in table 2, or both.

The way you plan to use tomorrow's class time also should be carefully considered. If you have prepared objectives and notes to guide your students, there is no reason to give a lecture; students should be able to learn on their own outside class. You might instead plan to model what you expect them to do, by solving problems at the board, for example. A steady diet of modeling is not what guidance is all about,

Reprinted with permission from *Engineering Education,* vol. 66, no. 5, pp. 390–393, February 1976.

either, so you should look beyond tomorrow in your thinking to consider how you might arrange your class for students to experiment with new ideas and discover the desired behavior. Whatever your choice, students should *use these new ideas in class,* so that you can supervise their initial trials and prompt them until they can do the work correctly on their own. Psychologically, this is a far better use of class time than is the presentation of subject matter.

All these ideas on guidance can be summarized in the following way.

How To Guide Learning

- Prepare content-performance objectives which describe the values, subject matter and the intellectual modes and abilities involved in the student's work and relate these objectives to specific activities through single-answer and open-ended problems.
- Organize concepts and principles sequentially from simple to complicated and provide organizers: verbal, visual, lists, mnemonics, tables, graphs.
- Demonstrate (or model) or provide a situation in which the student can experiment with the desired behavior or discover it.
- Supervise the initial trials.
- Use the necessary prompts. Withdraw this support gradually as ability develops.
- Help each student learn to evaluate his own performance.

Table 1. Example of Content-Performance Objectives for Intellectual Abilities.

At the end of a period of study each student should be able to solve *single-answer problems* and:

Intellectual Ability	**Action**
Recall:	*Write* concept X.
Manipulate:	*Restate* concept X in a new form.
Translate:	*Convert* concept X from verbal to graphical or symbolic form.
Interpret:	*State* the results derived from the use of concept X.
Predict:	*State* the expected effect of concept X.
Choose:	Independently *select* concept X and use it to solve a single-answer problem.

Table 2. Example of Content-Performance Objectives for Intellectual Modes.

At the end of a period of study each student should be able to use the *decision-making process* including steps such as:

Intellectual Mode	**Action** (with an open-ended problem)
Analysis:	*Break down* a problem into its constituent parts.
Synthesis:	*Combine* elements from many sources into a pattern not previously known to the student.
Evaluation:	*Make* purposeful *judgments* about the value of ideas, methods, or designs.

Provide for Practice

Effective learning requires activity on the part of the student. To accomplish the objectives listed in table 1 students should, for example, study a textbook, a programmed text, or work through an audio-tutorial lesson and have an opportunity to clarify fuzzy concepts by talking to other students, the teacher or assistants. Then they should do homework, which might involve solving single-answer problems, preparing an outline or writing a report. If the high-level objectives of table 2 are involved, students should be working on the solution of an open-ended problem, gathering information from the library or the laboratory, and developing decision-making skills under the teacher's guidance.

The teacher's job is to plan and manage student work and provide what it takes to ensure that students get needed *practice.* Part of the teacher's job is to pace work so that students can manage several classes simultaneously. The teacher should realize that inexperienced students often do not know how to plan their work, and many of them are totally inept at managing self-study. Thus, the teacher must find ways to help students plan their work. Furthermore, varying the setting of the assignments will help stimulate interest by demonstrating the relevance of the work to a variety of real situations. To summarize, the following elements should be considered.

How To Provide for Practice

- Offer both supervised and independent practice with single-answer and open-ended problems.
- Ensure that the student is active.
- Pace the work; spaced practice is best.
- Vary the context.

The first principle we presented—guidance—plays an important role in practice because students are more likely to do work if they have an effective model to work from, including supervised in-class practice, the guidance provided by a set of objectives, a study guide, notes or whatever else the teacher can devise. While some of this planning can be done for tomorrow's class, the teacher might also be thinking in terms of a programmed text or an audio-tutorial presentation to provide much greater individualized support for out-of-class practice with single-answer problems. In fact, it should be noted that these two methods of presentation can satisfy all five of the "psychological principles," and therefore they are likely to be the best type of self-study material.

Evaluation and Feedback

A "psychological principle" the teacher can put into effect immediately is that of giving *feedback.* An obvious place for this is the classroom, where the teacher is supervising the students' initial trials. This feedback can also be supplied by other students, if the teacher organizes teams to work and learn in class. Feedback might also be provided in writing, through answers to homework problems or solutions used by students in class to check their work. One simple technique which seems to have value is to disguise the correct answer by giving three answers to each homework problem, one right and two wrong. When students finish each problem they can tell if their work is right or wrong. If the result does not match one of the three given answers, the student knows the work should be continued. This is much more valuable than the common practice of working backwards from a single correct answer to find an error as, later, there will be no answers.

Feedback is one of our more obvious needs—we all look for it. (That's why we prefer elevator buttons that light up.) The student needs it, too, to be confident that the practice of specific behaviors is adequate. Feedback all too often occurs only on a test, where errors count against a student's record and no further learning is involved—what's missed is often lost. Of course, there is an appropriate time for a test, where students demonstrate learning. But before that takes place students should have practice with guidance, either directly by the teacher, as mentioned earlier, or through some technique such as programmed instruction or audio-tutorial instruction, which provides feedback on the correct response so that comprehension and performance can be checked by the student. These ideas are summarized below.

Evaluate and Give Feedback

- Correct inadequate responses.
  1) Immediately during initial learning.
  2) Frequently thereafter.
- Provide the student with diagnostic-progress information about performance: formative evaluation.
- Determine if the student has mastered the stated objectives and is ready to move on: summative evaluation.

The teacher who wants to implement the three ideas of guidance, practice, and feedback can also do so through a competency-based testing program which allows students to restudy and repeat any quiz or exam not up to some minimum level. Competency-based testing provides guidance by letting a student discover where a test performance is weak. It gives practice with the concepts in a testing situation, practice which is likely to reduce fear or anxiety. And two or more chances to be successful can do a lot for a student's self-respect, resulting in a more confident graduate. This pattern also gives the needed feedback to correct learning errors, which is desirable any time, but particularly important where concepts build through a course or from one course to the next.

When a complex or sophisticated performance, such as that described in table 2 is expected, the student should have the more immediate feedback provided by direct interaction with other students, a teaching assistant, or the teacher. Since the teacher may not have the help available to work directly with 30 or more inexperienced students on the sophisticated tasks involved in decision-making, an approach such as guided design[1,2] should be considered. With the printed "instruction-feedback" material characteristic of guided design the teacher can model the decision-making process in slow-motion and provide the guidance, practice, and feedback needed to help students learn.

Motivation

Another "psychological principle" you can apply in tomorrow's class is *reinforcement,* which has been shown to have an important effect on people's motivation. To begin, the teacher might couple reinforcement with the feedback given during class practice by not only confirming a correct response, but also giving an open, enthusiastic "that's right." With more sophisticated tasks the reinforcement might involve an elaborate congratulation for a task properly accomplished. Whatever the form of reinforcement, it should be clear that a student is more likely to learn the correct response, continue learning, and enjoy learning if practice includes appropriate guidance, feedback and reinforcement.

> "All too often education is based on the promise that what is learned will be useful later in life, a promise that fails to be fulfilled within a meaningful time span."

This point is so important that it is worth a few more words. The type of feedback used can be critical to the learning process. Appropriate feedback has two components: informative and affective. It's not enough to supply feedback if it is presented with the wrong attitude. The response must be one of genuine pleasure that the student is learning and developing appropriate skills. The teacher should realize that one of the most effective types of feedback is to help the student learn how to correct errors so they do not occur again.[3]

Another way in which the teacher can help motivate students is to relate what they are learning to meaningful work. If the goals of the course include objectives from both table 1 and table 2, so that students use what they learn to make decisions, this motivation may be automatic. If the course is focused on single-answer problems and ideas, the teacher must look for ways to make the real-world relationship meaningful. It is not motivating to tell a student that three years of course work must precede the use of the material in some type of "capstone" course. Every course in the curriculum should attempt to justify what is being learned through the application of ideas studied.

The teacher can also motivate students by helping them be successful. There is surely an important difference in the attitude of a student who gets an A on the first exam in a course and tries to maintain that level and the student who gets a C or D grade and knows it is next to impossible to raise that grade. The teacher can build this success and motivation into the course by initiating the competency-based testing system described earlier, which provides success if the student is willing to keep working and learning.

In summary, the teacher who wants to increase the student's motivation should consider the following factors.

How To Motivate

- Reinforce, encourage—don't discourage.
- Show the value of learning; show that concepts and principles to be learned are relevant to meaningful work.
- Help the student achieve success.

Reinforcement properly used is likely to be motivating, and therefore this addition to the concept of feedback provides an important part of what the student needs to be an effective learner. Guided practice can also be highly motivating, as can the success which is likely to follow if activities are organized to suit abilities, or if retesting is allowed. Further motivation can be provided by the teacher's encouragement, by acceptance of the student's ideas and viewpoints, and by helping the student see the relevance of the subject matter. All too often education is based on the promise that what is learned will be useful later in life, a promise that fails to be fulfilled within a meaningful time span. Much better motivation can be achieved by making the learning process relevant to real-world activities that the student can see as valuable.

Individualize

Whatever else can be said about students, one thing is sure: as individuals, they are vastly different from each other. They differ in aptitude, in intelligence, in motivation, in background, in their ability to learn, in the rate at which they learn, in learning style, in the time of day when they are at their best, in the personal problems they face, in their interests and in the goals they seek. To accommodate this vast array of differences the effective teacher should consider the factors given below.

How To Individualize

- Provide for students who learn at different rates.
- Provide enrichment for the fast learner.
- Give extra help to the slow learner.
- Consider the goals of the learner.

The range of options available to a teacher are limited by many constraints, in particular the time and learning materials available. Time is limited for both the teacher and the student; the student has several classes to prepare for and the teacher has several classes to teach, as well as other duties. This may mean the teacher must limit the use of a competency-based testing approach to one a week for repeats in early classes, or two tests as the student gains both experience and confidence. This approach still provides more individualization than one "all or nothing" event which is bound to hit some people at the wrong time.

Another way the teacher can individualize course work is to *offer different learning materials.* Sophisticated learners know that one textbook may not be as easy to understand as another—isn't that why faculty keep writing them? Therefore, when a problem occurs, this learner looks for another book. The teacher can help by suggesting students try other sources and recommending one or two alternate textbooks or a programmed text, if one is available. The teacher can also supplement confusing sections in the text with a set of printed notes or example problem solutions—in other words, provide the guidance we considered earlier.

Group work in class can also help individualize learning by providing peer tutors for students who have a problem. Since teaching is often a way of learning, some students in a group will reap this benefit, while the others get the help they need. If a team of students are working together on the solution of an open-ended problem, this may increase the motivation of those who are having trouble. The teacher should also carefully consider that slow learning often reflects poor prerequisite skills, which may be the fault of the student or a previous teacher. If algebra, for example, is a necessary tool, poor skills may handicap a student in many other classes. If appropriate materials or tutorial help are available, a properly motivated student can often overcome this type of handicap quite rapidly.

Finally, the teacher should consider the *goals of the learner.* Is this a course in the student's major, is it a required non-major course, or is the class a freely chosen elective (and if it is an elective, on what basis do students usually elect it—as a cake course or ?). A chemistry course for non-majors which dwells on sophisticated atomic theory important only to a Ph.D. chemist is not likely to meet the needs of education students or political science majors who need to know how important chemistry is to the operation of our society.

The same concern for a student's goals should be demonstrated in a competency-based testing program. A course in a student's major which is focused on fundamental concepts may demand a high level of performance and a passing grade of 85. A required non-major course which some students would be pleased to pass with a C grade may require the same quality work and an 85 on examinations, but the teacher can set a minimum number of units which must be passed to earn a C and allow those students who have an interest to cover more ground and earn a better grade.

Summary

The teacher who wants to improve tomorrow's class can do so through the use of the five "psychological principles" described in this paper. These principles can be applied to the pattern of class work, the student's homework, to a testing program, or to the planning of any part of the education process. In fact, it appears that those faculty who are identified as effective teachers use all or most of these principles—it's just a matter of common sense to them. But any teacher who wants to improve the teaching-learning process in his or her class can be equally effective by organizing course activities around these concepts. We hope we have helped those teachers who would like to improve their performance with this description of how each principle might be applied.

References

1. Wales, C.E., and R.A. Stager, *Educational Systems Design,* 1970, 1974, available from C. Wales at West Virginia University.
2. A series of eight articles on guided design was published in *Engineering Education* between February and May, 1972, vol. 62, nos. 5, 6, 7 and 8.
3. McKeachie, W.J., "The Decline and Fall of the Laws of Learning" *Educational Researcher,* vol. 3, no. 3, March 1974.

Carl H. Durney

A Review:

Principles of Design and Analysis of Learning Systems

Engineers pride themselves on being able to design devices, systems, and processes. They define the problem and establish the specifications, search out the relevant fundamental principles, apply these principles to obtain a design, build a prototype, test it, and modify it until it is acceptable or discarded. Because of this background, it would seem that engineering educators are in a good position to lead the way in designing better learning systems. Yet it appears that the percentage of engineering educators who attempt to follow this engineering procedure in systematically designing learning systems[1] is not significantly greater than the very small percentage of other educators who make such attempts. Why have more engineering educators not followed engineering design procedures in designing learning systems?

Many new approaches to learning have been proposed and tried by educators in various fields. The reader of the literature on the subject will find new methods described by such phrases as Proctorial System of Instruction (PSI), Personalized System of Instruction (PSI), Self-Paced Instruction (SPI), Individually Prescribed Instruction (IPI), Keller Method, Contract Grading, Audio-Visual Tutorial (AVT), Guided Design, Programmed Instruction (PI), Learning Machines, Computer Aided Instruction (CAI), Behavioral Learning System, and so on. Some new methods appear to hold promise for significant improvement in learning; others appear to be little or no better than traditional methods. On what basis can these methods be analyzed, compared, and evaluated? Just as engineers analyze devices and systems, educators should be able to analyze and evaluate learning systems.

In either case, in designing learning systems or in analyzing and comparing learning systems, educators need to know the fundamental principles which describe these kinds of systems. They need to know the laws for learning systems which are analogous to Newton's laws for mechanical systems. This article attempts to formulate a set of principles which can be used as the basis for designing and analyzing learning systems. Educators may benefit by using it as a checklist for evaluation. These principles are not purported to be all inclusive, but they have been found to be workable, and they seem to be a useful first step.

A Set of Fundamental Learning Principles

The set of principles described below has been formulated in connection with my efforts to develop better learning systems for the courses I have taught over the past several years. They are my version of principles which seem to be widely accepted, but seldom applied in designing learning systems.

The principles are listed below in two categories. The first five are based in part on Erickson,[2] and the last three are based on Gagné.[3, 4]

Principles of the Management of Learning

1) *The learner must be active.* Telling a student something does not necessarily mean that he has learned it. As Galileo said, "You cannot teach a man anything, you can only help him to find it for himself." Significant learning occurs only when the learner is actively trying and working; very little learning occurs through inactive listening.

2) *Feedback and a second try are essential.* The learner must be able to find out what he did poorly, try to improve it, and try again. He must also feel rewarded (as opposed to penalized) in some sense for doing well.

3) *The learner must know the learning objectives;* i.e., just what it is that he is trying to learn. If the learner is confused about just what it is that he is trying to learn (his learning objectives), his learning will proceed very slowly, if at all, and he will be discouraged. If he knows what his objectives are, he can focus his energy on learning them, his progress will

Reprinted with permission from *Engineering Education,* vol. 63, no. 6, pp. 406–409, March 1973.

be much faster, and he will tend to be happy and enthusiastic as he masters the objectives.

4) *The learner must be committed and he must be motivated to learn.* If the learner is not committed, he will not put forth the effort required to learn, and, consequently, he will not learn. One powerful motivator is the situation in which the learner commits himself to a task and agrees to give an accounting of himself to someone else at a specified time, and then receives some kind of recognition for his accomplishments. In themselves, success and recognition for accomplishment are great motivators.

5) *Each person learns at a different rate and in his own way.* No two individuals are exactly alike, and no two learn in exactly the same way, even though there is much that is common to everyone's learning. The learner's progress will be much more rapid if he is encouraged to learn in ways which are consistent with his individual characteristics. It is particularly important that the learner be able to find out what the stumbling blocks are in his own method of solving a problem or looking at something, so that he can modify his own thinking. His progress is impeded if he is forced into another way of thinking rather than being helped to improve his own way.

Principles of Learning Activities

6) Gagné [4] lists eight types of learning. Of these eight, the three types of learning most important in the design of engineering learning systems are:

a) *Learning concepts.* A concept is a response to an abstract stimulus. For example, "middle" and "voltage" are concepts.

b) *Learning to apply principles.* A principle is a chain of concepts. "Hot things burn" and $V = IR$ are principles.

c) *Learning to solve problems.* Solving problems is defined as combining known principles to get new ones. Working out a class schedule at registration time is solving a problem according to this broad definition. The definition also includes the more familiar kinds of engineering problems.

7) Learning to solve problems depends critically on two factors:

a) The learner's acquisition of an organized set of intellectual skills which constitute what is ordinarily called "understanding" of a subject area. The intellectual skills include being able to use concepts and to apply principles.

b) The learner's "practice" in solving typical problems of the kind he desires to be able to solve.

8) The set of intellectual skills is most efficiently acquired by the learner progressing through a structured set of learning experiences and demonstrating mastery of all the learning objectives in the set. The learning experiences are organized according to the prerequisite skills required for each one. This structured set of learning experiences is called a *learning hierarchy.*

The first five principles have to do mostly with the procedures and methods to be used—when and how exams are given, whether lectures are given, how the student receives help, whether group discussions are held, and the general format to be followed. The last three principles have to do mostly with the organization of the material to be studied and the tasks given to the student—what it is that the student should be able to do as a result of having taken the course (i.e., the learning objectives), what sequence he learns the skills in, what the exams consist of, the source material he is given, etc. Much more could be said about the seventh principle—methods used in problem solving, constructing models, and other ramifications, but space does not permit that here. The application of the first set of principles might result in a general procedure which could be used in several different courses. On the other hand, the application of the second set of principles would result in a set of materials applicable only to one course, because the materials contain the subject matter of the course.

The set of principles given above is not very startling. Surprisingly, although the principles are simple, and probably familiar to most educators through their own experience, there has been little effort to apply them in designing learning systems. Many courses could probably be improved significantly without an unrealistic amount of effort by direct application of these principles.

Just as in engineering design the application of the basic principles does not result in a unique device or system, the application of the fundamental learning principles does not result in a unique learning system. There could be many versions of a learning system which would be satisfactory for a given course.

As an example of how to apply these principles in analysis, consider a hypothetical, typical, traditional engineering course. Suppose the course is a four-hour course taught as follows:

- Four lectures per week are given in which the instructor goes through derivations and works examples. A good textbook is used.

- Homework problems are handed in weekly and graded by a graduate student and handed back to the students sometime later.

- A midterm exam and a final exam are given.

Evaluating a Traditional Course

Now let's evaluate this hypothetical course on the basis of the fundamental principles listed above. The comments below are numbered to correspond to the principles already given.

1) If the homework is well designed, the student could be moderately active. However, during the lecture most will not be actively learning most of the time because they will be trying to copy down what the instructor is doing, and they will be unable to understand or think critically about what he is saying. Usually only a very small percentage of the students really comprehend much of what is being done in the classroom, because they have not prepared well before they go to class, and there is so little opportunity to stop the instructor, reverse him, and have him repeat that critical missed thought or word. Almost all of the student's learning occurs through

There are basic principles of learning that almost everyone can agree on. How do traditional and innovative courses fare when analyzed in terms of these principles?

his work outside the lecture room. (If you would like to test whether or not students learn much during a lecture, give one of your best lectures and then give an unannounced quiz at the end of the class hour to see what the students have gained during the hour.)

The students will be most active just before the two exams, typically trying to learn most of what they should have been learning earlier. In summary, the students are probably not nearly as active as they should be, and since their learning is directly proportional to their activity, they probably are not learning nearly as much as they should be.

2) The feedback which the average student gets is very poor. He probably gets the homework back in two or three weeks (in extreme cases, sometimes not until the end of the quarter) with some red crosses distributed through his work and a score like six out of ten at the top. He probably mutters something like, "I wonder why that's wrong" or "That dumb reader," throws his paper in his briefcase, and never finds out what he did wrong nor tries to correct his errors. In a few cases, a student will go to an instructor to find out what he did wrong, but usually the system discourages him from doing so. Rarely does a student get a second chance to do a problem. Yet the process of trying something, correcting mistakes, and trying again, is at the very heart of almost any kind of learning.

3) Most students have only a hazy idea of what they are trying to learn. A typical plea before an exam is, "Would you please tell us what is important in the material we have covered?" The usual response amounts to saying that everything is important, which is very little help to the student.

When a teacher first reads about specifying learning objectives, his reaction is usually that specifying the learning objectives amounts to "spoon feeding." Properly written objectives do not amount to spoon feeding, and they can greatly increase a student's efficiency. Space does not permit further discussion of learning objectives here, but much can be found in the literature.

4) Unfortunately, the motivation for many students in this hypothetical course too often is to pass the exams and get a grade, rather than really to learn. Perhaps one main reason is that the student has little opportunity to get feedback on his work and to get some sense of accomplishment. Too often he feels frustrated and discouraged because there are so many things he does not understand, and since he is not sure of what it is that he is trying to learn anyhow, it is difficult for him to ascertain how much he does know.

5) In this hypothetical course, there is probably very little allowance for individual differences in the way students learn, and very little encouragement is given for novel or creative ways of solving problems. Many students probably try to memorize enough to get by, rather than gaining skill in applying basic principles.

6), 7), 8) The instructor probably follows the textbook, with some editing, and he has probably not made much further effort to develop a learning hierarchy or even to write a detailed list of learning objectives. Since only two exams are given, the students are not tested for mastery of each learning objective. There is no systematic method for the student to obtain the set of intellectual skills he needs to work problems, and he may or may not acquire enough skills to be proficient.

In the comments given above, the hypothetical course has been assumed to be a poor one in order to emphasize how much improvement could be made if the course were designed on the basis of the learning principles. Even so, the description is probably not far from many courses taught in engineering today, and significant improvements could be made in many courses by straightforward application of the learning principles.

Evaluating the Innovative Course

Now, in comparison to the traditional kind of course, how does an innovative system such as the Keller method, or the Personalized System of Instruction fare? Let's consider a hypothetical system that has the following characteristics:

- The student is given a set of learning objectives and appropriate resource material to help him reach the learning objectives. The resource material might include a good text, some written study guides which include examples, and access to audio and/or visual aids.
- The student is tested for mastery of a set of tasks based on the learning objectives. He is pro-

vided with an immediate and individual evaluation of his performance (often by a student proctor), and he has the opportunity to get help in correcting his mistakes. The tasks which the student is tested on are structured in a learning hierarchy, and the student must demonstrate mastery of the tasks in sequence. He is not allowed to progress to a task until he has demonstrated mastery of prerequisite ones.

- Lectures are used for motivation. The instructor does not rely on the lectures to transmit information and methods to the student. That is accomplished by the resource materials mentioned above. The lecturer gives demonstrations, describes applications, invites guest lecturers to describe pertinent real projects, etc.

This description is necessarily brief, and probably not very meaningful to those who have not heard or read about this kind of system. Descriptions of some systems of this type can be found in *Engineering Education*, March 1971.

A few brief comments will illustrate the differences between a typical traditional system and a typical innovative system. Again, the comments are numbered to correspond to the learning principles listed above.

1) The student is much more active in the innovative system because he must demonstrate mastery of a whole set of learning tasks. Since he cannot progress until he demonstrates mastery, he is strongly encouraged to regular activity.

2) He receives a great deal more feedback because he discusses his work individually with someone (usually a student proctor).

3) The learning objectives are clear to the student because they are specifically stated to him.

4) The motivation appears often to be higher in some of these courses because it is easier for the student to see what to do to learn, and because he can see his own regular success.

5) The learning is highly individualized and the student is often encouraged to develop his own methods of thinking.

6), 7), 8) The student does progress through a carefully designed learning hierarchy. However, it appears that in many innovative courses, the emphasis is on the acquisition of the concepts and application of principles, and the practice in solving problems is neglected or omitted. This is perhaps a significant weakness of many innovative courses.

Conclusion

Even though these comments are not specific and detailed, they do help to point out the differences between the two hypothetical courses. It should also be pointed out that in some innovative courses, one or two of the above principles are emphasized and most of the others neglected. For example, self-pacing is only one aspect of the above principles, and if the designer of a learning system emphasizes self-pacing in his system and neglects other important features, such as feedback, the results will likely be disastrous. On the other hand, in a well-designed learning system, the learning can be individualized to a satisfactory extent even though there are some deadlines for work to be accomplished and the course is therefore not completely self-paced. It is important for the designer to use all the above principles in designing his learning system, and it goes almost without saying that an important part of the design process is to evaluate the performance of the initial design and make appropriate modifications until a satisfactory system is obtained. And, of course, the design of a new system always involves trade-offs and compromises. Many of the innovative systems tend to be more costly, and some involve controversial methods of grading. The designer's task is to get the best performance that he can within his constraints.

How good is your course or learning system when analyzed in terms of the learning principles above? The challenge is clear: either formulate your own set of learning principles or use the ones given here, and analyze your present learning system; then design a better one. The work is hard, but the rewards are great.

Acknowledgments

I am grateful to many friends and colleagues for stimulation in this work, and in particular to Curtis C. Johnson and L. Dale Harris for helpful comments on the manuscript.

References

1. Koontz, Jesse L., "An Engineering Approach to Designing Instructional Systems," *Engineering Education*, vol. 61, no. 6, March 1971, pp. 528-531.
2. Erickson, S. C., "Learning Theory and Educational Engineering," *ERM*, vol. 1, 1969, pp. 17-18.
3. Gagne, R. M., *The Conditions of Learning*, second edition, Holt, Rinehart and Winston, New York, 1970.
4. Gagne, R. M., "Instruction Based on Research in Learning," *Engineering Education*, vol. 61, no. 6, March 1971, pp. 519-523. ▲

Part V
Teaching Methods and Strategies

Scope and Purpose

WHEN one hears of college teaching, the most likely mental image that is evoked is that of a professor delivering a lecture in front of a room full of students. There are, however, numerous other teaching methods that are effective in various circumstances, including discussions, problem sessions, seminars, group projects, and a number of types of individualized or self-paced programs. It is unlikely that a beginning teacher will be equally conversant and comfortable with all of these methods. Since it is essential that a teacher be comfortable with a method of teaching before that method can become an effective one for him, many teachers give up, or never even try, a new method of teaching. This is unfortunate, since some of the methods are particularly suitable for meeting some of the educational goals. It is the purpose of this part to make the reader aware of the various teaching methods, with the hope that familiarity will lead to experimentation with them. A second purpose of this part is to provide advice and suggestions for improving the effectiveness of the various methods of teaching. Some of these suggestions deal with procedural matters that can enhance the efficiency or utility of a teaching method. Others deal with style, and here a word of caution is needed. Perhaps the single most important thing for a beginning teacher to remember is that there are as many styles of teaching as there are teachers, and there is no one "best" method.

The Issues

Some of the issues of interest to beginning teachers, and discussed in the literature of this subject, are as follows:

- What are the various methods of teaching and what are their domains of effectiveness?
- What educational objectives are best met by each of the methods of teaching?
- Does a particular teaching method make the best use of the known principles of learning?
- Is there one "best" way of teaching a given subject?
- Why are certain methods of teaching favored, or disliked, by the teachers or the students?
- What classroom style or manner is conducive to most effective learning?
- How can the lecture, or seminar, or group discussion be improved?
- What lecturing skills need particular attention?
- What style of teaching will encourage the students to ask questions in class?
- How should the instructor answer the questions from students?

The Reprinted Papers

The following four papers are reprinted in this chapter:

1. "A Pedagogical Palimpsest: Retracing Some Teaching Methods," G. F. Paskusz and J. E. Stice, *Proc. IEEE,* vol. 66, no. 8, pp. 902–911, Aug. 1978.
2. "The Art of Teaching," A. B. Giordano, *IEEE Trans. Education,* vol. E-13, no. 4, pp. 196–199, Nov. 1970.
3. "How to Improve Classroom Lectures," L. D. Reid, *American Association of University Professors Bulletin,* vol. 34, no. 3, pp. 576–584, Autumn 1948.
4. "Increasing Engineering Teaching Effectiveness," W. L. Cooley, *IEEE Trans. Education,* vol. E-18, no. 2, pp. 94–100, May 1975.

The first reprinted paper by Paskusz and Stice is a comprehensive and systematic examination of a large number of the common methods of teaching, including formal lectures, discussions, laboratories, self-paced instruction, mastery learning, programmed instruction, audio-tutorials, and computer-assisted instruction. Each of these methods is analyzed with respect to six measures of effectiveness: active learning, feedback to learner, positive reinforcement, distributed learning, specification of objectives, and pacing. In addition, this article provides remarks on the strengths and weaknesses of each teaching method, and the kind of situations where each is most effectively used.

The next paper by Giordano is a collection of tips and advice on increasing the effectiveness of two of the most common teaching methods used in engineering: lectures and discussions. His suggestions cover a very broad range of matters concerning the methods, including procedural matters (e.g., preparation for a class), attitudinal matters (e.g., understanding a student making an audacious statement), and mechanical details (e.g., the suitable rate of oral delivery), such as can only come from a very experienced teacher. In particular, Giordano provides, for an instructor planning a lecture, a checklist that should be thought provoking.

Lecturing, the oldest and the most common method of classroom teaching at the college level, has been the subject of more advice and admonitions than any other method. The reader must, however, be careful since (a) some of the advice is nothing more than personal bias, or tradition, or folklore, and (b) a great deal of advice on speaking is addressed to debators, salesmen, after-dinner speakers, public leaders, scientists presenting a conference paper, and others who speak to group audiences, and some of this advice is inapplicable to teaching. The next reprinted paper, by Reid, on the subject of classroom lectures, is the oldest paper reprinted in this volume, and well worth preserving and recycling. It is based on a survey of university students, and it confines its attention to just five pieces of advice, dealing with lecturer's personality, use of examples, oral delivery, use of notes, and use of student comments for improvement.

The last reprinted paper by Cooley does not deal with any single method of teaching, but makes suggestions on increasing the effectiveness of teaching in general. Although written as a ''guide'' for the teacher on how to improve his ratings in respect of each of the questions that normally appear on teacher evaluation forms, the paper contains many ideas that beginning teachers can use.

The Reading Resources

[1] W. J. McKeachie, *Teaching Tips: A Guidebook For The Beginning Teacher.* D. C. Heath, Lexington, MA, 1978. After seven editions in the last three decades, this is perhaps the most well-known guide on ''how to teach.''

[2] H. H. Skilling, *Do You Teach? Views On College Teaching.* Holt, Rinehart, and Winston, New York, 1969. Written in a light, conversational language, and presenting the views of some of the highly regarded teachers at Stanford University, on college teaching.

[3] T. C. Taveggia and R. A. Hedley, ''Teaching Really Matters, or Does It?,'' *Engineering Education,* vol. 62, no. 6, pp. 546–549, Mar. 1972. A comprehensive review of some 170 empirical studies published between 1924 and 1967, comparing different teaching media, or different methods of instruction, at the college level; the principal conclusion is that there is no significant difference between various methods and media used to convey the subject matter content.

[4] S. C. Ericksen, ''The Lecture,'' *Memo to the Faculty,* No. 60. The Center for Learning and Teaching, University of Michigan, Ann Arbor, MI. Makes and discusses some suggestions, such as talking with credibility and enthusiasm; proposes that three requirements for learning (motivation, understanding, and remembering) be used as guide in lecturing; recommends guidelines for lectures based on these requirements.

[5] A. W. Engin and A. E. Engin, ''The Lecture: Greater Effectiveness from a Familiar Method,'' *Engineering Education,* vol. 67, no. 5, pp. 358–362, Feb. 1977. Discusses capabilities and limitations of the lecturing format of teaching; makes ten practical suggestions for improving lectures; and describes the operation of a system in which two or three faculty members monitored each other's performance.

[6] D. S. Shupe, ''Improving Learning Efficiency by the Lecture Method,'' *Engineering Education,* vol. 69, no. 5, pp. 406–408, Feb. 1979. Identifies the shortcomings of the lecture method of teaching, especially as compared to individualized instruction; makes five suggestions, dealing with the mechanics and organization, for mitigating those shortcomings.

[7] R. L. Weaver, ''Effective Lecturing Techniques,'' *Clearing House,* vol. 55, no. 1, pp. 20–23, Sept. 1981. Or, ''Effective Lecturing Techniques: Alternatives to Classroom Boredom,'' *New Directions in Teaching,* vol. 7, no. 1, pp. 31–39, Winter 1982. Makes several suggestions for capturing and holding students' attention, keeping alive their interest, and using personal mannerisms that convey sincerity and dynamism.

[8] S. E. Palmer, ''The Art of Lecturing: A Few Simple Ideas,'' *The Chronicle of Higher Education,* vol. 26, no. 7, pp. 19–20, Apr. 1983. An itemized compilation of a number of ideas and suggestions for improving lectures, each elaborated by a sentence or two.

[9] P. M. Berthouex, ''Oral Communications: Some Guidelines on Answering Questions,'' *Engineering Education,* vol. 72, no. 3, pp. 243–244, Dec. 1981. Presents ten rules on answering questions in an oral presentation; not all rules are applicable in the teaching situation where a question from the students can be used as a motivational, diagnostic, or enrichment device.

[10] H. E. Stanton, ''How Might the Seminar be Improved,'' *Improving College and University Teaching,* vol. 28, no. 1, pp. 37–39, Winter 1980. Makes six suggestions for improving the effectiveness of seminar courses in which one student at a time presents a paper to his fellow students, who may be bored, uninterested, and passively listening; also describes a particular scheme used by the author, which may have a limited applicability in engineering disciplines.

[11] J. V. Jensen, ''Oral Skills Enhance Learning,'' *Improving College and University Teaching,* vol. 28, no. 2, pp. 78–80, Spring 1980. Suggests a number of ways of getting the students involved in active classroom participation, while at the same time helping them improve their oral communication skills.

[12] M. Benaim, ''P.S.I. versus Traditional Method. Criteria for Comparison and Results (1969–1979),'' *IEEE Trans. Education,* vol. E-27, no. 1, pp. 41–46, Feb. 1984. A review of a large number of published studies comparing the traditional classroom teaching with personalized system of instruction (P.S.I.), using different criteria of effectiveness; the results favor P.S.I.

[13] A. P. Castino, ''OK, So How Do We Teach It?'' *Engineering Education,* vol. 66, no. 5, pp. 399–401, Feb. 1976. Contains a chart for selecting an instructional method, based on the instructional objective and the pace of instruction, for each phase of instruction: introduction, practice, and testing.

A Pedagogical Palimpsest: Retracing Some Teaching Methods

G. F. PASKUSZ, SENIOR MEMBER, IEEE, AND JAMES E. STICE

Invited Paper

Abstract—Curriculum and course-design theory suggests that there may exist a scientific method for the development of effective courses. Some of the important steps in their design include writing of instructional objectives, selection of instruction method, and others.

The paper will deal primarily with one step in this sequence: selection of an instruction method. First seven currently deemed important concepts from learning theory, the "psychological building blocks" in the construction of a course, are indentified and analyzed. Next we discuss identifiable educational processes which may be used in a course. We then describe some of the popular teaching methods and show how the building blocks and processes are used in them. The methods discussed include the traditional lecture, lecture discussion, discussion group, laboratory teaching, self-paced instruction, mastery learning, programmed instruction, audiotutorial instruction, and the *open university*. The paper culminates with the development of two matrices. The first of these defines the interactions between these teaching methods and their usual building blocks. The second, the "utility matrix," attempts to summarize the effect of choice of method on learning, student acceptance, required faculty effort both in deveopment and in long-term use, need for special equipment or personnel, and ultimately cost.

I. Introduction

THE PRINCIPLES of curriculum design suggest that there are better ways to develop effective courses than are often used. To help systematize the process, there exists a variety of checklists. One of these gives the following steps for instruction development [21]:

1) analyze role that course will play in the curriculum
2) consider the backgrounds of the prospective students in the course
3) write instructional objectives
4) determine course prerequisites
5) develop testing materials
6) select method(s) of instruction
7) sequence instructional materials
8) develop lesson plan
9) teach course and obtain feedback
10) revise course as indicated.

The step with which this article will concern itself is number 6—select method(s) of instruction. First the major psychological building blocks underlying most learning are presented. Then some "process" blocks are added. Next a variety of teaching methods are described, the building blocks they depend on are noted, and some comments are made about the method's strong and weak points. Finally, a utility matrix is presented which summarizes all this information in a form which we hope will be useful for quick reference.

Manuscript received September 29, 1977; revised December 21, 1977.
G. F. Paskusz is with the Department of Electrical Engineering, University of Houston, Houston, TX 77004.
J. E. Stice is with the Center for Teaching Effectiveness, University of Texas, Austin, TX 78712.

II. Psychological Building Blocks for Learning

In this section seven current concepts from learning theory are presented briefly. These concepts influence the learning behavior of most people positively most of the time and a knowledge of them can help in the analysis of teaching problems or in the choice of a teaching method to be used, since they are not present uniformly in all of them. Not discussed are those concepts, such as motivation, which pervade the whole spectrum of methods of instruction.

A. Active Learning

In order to learn, a student should be actively involved with the material. It is not enough that he may hear material or see it; he must also think and talk about it, or do homework on the topic. In short, he must pay attention to it, and therefore, any procedure that helps direct the student's attention to the course material will aid his learning. When the student is paying attention to instruction, he is said to be "active." Many students pretend to be attentive. They may look at the lecturer or speaker, take notes, nod or smile, and behave in other ways which make it appear that they are interested and involved in what is going on (these "survival techniques" were well learned before the students ever got to college). If the student is not motivated (or obliged) to participate from time to time however, and is spending a sizeable fraction of his time thinking about something else, then he is more passive than active, and the degree of his learning will suffer commensurately.

B. Feedback to the Learner

Feedback is the provision of information about whether or not he has achieved the intended objective. Learning can occur without any overt feedback, but its direction is unpredictable and under little control. With feedback, the learner can see whether he is learning what he is supposed to learn and whether he is learning it correctly. The more immediate the feedback, the more efficient is the learning; the learner is positively reinforced for the things he has learned correctly and adequately, and errors may be corrected before it becomes necessary to unlearn any misconceptions or before these misconceptions cause further errors. The more feedback is delayed the less effective it is in motivating the learner and in controlling the learning process.

Reprinted from *Proc. IEEE,* vol. 66, no. 8, pp. 902–911, August 1978.

C. Positive Reinforcement

Positive reinforcement occurs when behavior is followed by a positive event such as a word of praise or appreciation, a feeling of success or satisfaction, or some other reward. All students are not reinforced by the same rewards, but all have reinforcers to which they respond. Positive reinforcement increases motivation and keeps students working (attending). The reward must follow the behavior fairly promptly however, for delay causes the strength of reinforcement to fall off rapidly with time.[1]

D. Distributed Versus Massed Learning

Research has shown repeatedly that studying on some sort of regular schedule results in more learning than the same amount of studying done all at once. Thus if a student is going to study 80 hours in a course, he will learn more and will remember it longer if he studies eight hours per week for ten weeks than if he attempts to study 80 hours during the week before the final exam.

E. Learning in Small Steps[2]

Learning proceeds by a sequence of more or less logical steps, and from simple to more and more complex ideas. If any of these steps is omitted, it is possible that no great harm will be done because the learner may fill in that blank later when he realizes that something important is missing. On the other hand, the learner may remain unaware that he has missed something and he may fail to learn other seemingly minor things as time goes on. Thus his ignorance accumulates and imperfect understanding of many small points ultimately results in his inability to grasp major concepts.

A good plan to follow is to keep the steps in the logical development of a course of study small enough that the students can follow them, and provide enough feedback that they know whether or not they have learned correctly what they were supposed to learn.

F. Knowledge of Course Objectives

Learning is more efficient when students have some sort of guide to let them know what they are expected to learn. It is unreasonable to expect students to learn *everything* in a textbook or a course, and most instructors have no such expectations. One of the most important steps, therefore, in the design of a course is the specification of the instructional objectives, which give detailed information about what the student is expected to learn and how well he is expected to learn it. One desirable result of preparing objectives is that the instructor also develops a very clear idea of what his course is about, which can be of help in the design of course sequences. Another is that the students, knowing the objectives, can use their study time more efficiently, using it to learn what the instructor really wants them to learn, rather than trying to learn everything in the book.

Although research results are mixed on the question of whether or not the use of objectives results in superior exam performance, no one has ever concluded that the use of objectives has had detrimental effects. On a subjective level, most instructors who use objectives are solidly in favor of them, and students overwhelmingly endorse them. They give the student an idea of the structure of the course and a yardstick against which to measure his own achievement. This can reduce his anxieties and give him feelings of security. He may then relax and begin to enjoy the course, and may even develop some self-confidence. These outcomes do not seem altogether undesirable.

G. Pacing

People learn at different rates. This has been demonstrated with learners at all levels and is no longer a subject for argument. If a fairly broad spectrum of learners takes a course which covers a given amount of material in a fixed period of time, the variety of learning rates will result in variable amounts of learning. On the other hand, if the slower learners are allowed to take more time, it is theoretically possible for all the learners to learn all the material.

These considerations have led to the development of several self-paced teaching strategies during the past ten to fifteen years. Most allow a student to take as much time as is required to master the material, at least during the semester or quarter the student is enrolled in the course. A voluminous literature has developed in this area, and most investigators claim superior learning results. The major drawback to this method of teaching is procrastination on the part of a few students [27]; given the opportunity to schedule their own studying, they do little or nothing. Their lack of motivation may stem from poor prior preparation, inability to discipline themselves, unwillingness to accept the increased responsibility, or lack of interest in the subject.

III. Defining Some Educational Procedures

There are some educational procedures with unique names that are used in several teaching methods or systems. Again, we are not attempting to present an exhaustive list of these, but prefer to introduce only those which are prominent in the teaching methods we discuss.

A. Modularization

Modularization means breaking a course down into smaller chunks, or modules. The modules are self-contained packages which deal with some part of the subject matter to be presented. The number of modules in a course is somewhat arbitrary, but usually ranges from about five to perhaps twenty for a one-semester course. Dividing a course into modules is analogous to breaking up a textbook into chapters. The purpose in each case is to break the material into readily digestible and logically distinct parts.

Most modules contain instructional objectives (either stated explicitly or implied by a set of study questions), instructions, suggested activities, and references. A module normally includes some sort of examination to assure that the learner undergoes a performance evaluation and receives feedback on attaining the objectives.

Modules come in a variety of forms. They may be a written study guide, an audiovisual package, may be presented by computer, etc. Modules are also known by other names, in-

[1] However, rewards do not inevitably seem to help improve learning. For a more detailed discussion of this and other psychological building blocks see [23].

[2] This is somewhat controversial. There is some evidence that our concept of sequential learning is not optimal for every learner, but that there are some "global learners" who learn better from the whole picture than by small steps. Ideally, teaching materials should be matched to the learning style of the learner. For a more detailed discussion of "holist" versus "serialist" learning styles, as they are also known, see [8], [26].

cluding "unit" (as in a Keller Plan course) or "learning activity package."

B. Learning for Mastery

Mastery learning is a relatively new arrival on the educational scene. Under this scheme the student is given detailed information of the course objectives and is expected to demonstrate superior performance on examinations over the material. In the Personalized System of Instruction (PSI, to be described later) the student must make 100 percent on each unit test. In Bloom's Mastery Learning Model (also to be described later) the student must make 85 percent on a single final exam over the course. In conventional courses, grading is usually done "on the curve," that is, students are graded with respect to whether or not they have attained a previously determined criterion; their grade is not contingent upon the performance of other members of thc class. If the criterion for passing the course is set at a high enough level so that students must have a superior grasp of the material in order to pass, then the course is a mastery course.

Mastery courses are nearly always modularized to break the material to be learned into manageable parts. These courses require detailed objectives, frequent testing with attendant feedback, and distributed learning. A student taking a mastery course does not have to learn all there is to know about a subject, but he does have to learn most of what the instructor considers important in the course as stated in the objectives. The students know at the outset that they cannot get by with sloppy work and that there will be no "gentleman's C." Thus they work harder and are kept on track by feedback from the many examinations. Adherents of mastery methods claim that both short-term and long-term retention are improved over conventional teaching methods.

C. Computer Management of Learning

The digital computer is more and more being used as a tool in education. It is used as a "teaching machine" in the sense that it can present instruction, tutor the student, and evaluate progress and give feedback. This use we shall discuss later, under the heading "Computer Assisted Instruction". It is also used to help *manage* instruction, but without being used to present that instruction. One situation in which the computer can be used to help the instructor in managing instruction is in generating and evaluating tests. If a pool of test questions is prepared, the computer can prepare individualized examinations by random selection from the item pool.

Course records can also be kept by computer. Grades on homework assignments and examinations can be entered into the machine and totals to date can be printed out on demand. This is particularly useful for large classes. It is also useful for PSI courses, where students may take several forms of the readiness test on each unit; the computer may be consulted to determine which forms of the readiness test a student has already attempted, and it can randomly select the form to give him for the next try.

In cases where testing is done on-line by computer, some very sophisticated programs have been developed. In addition to keeping scores for all students, it is also possible to provide students with references for further study for questions they have missed, to retrieve their answers to all missed questions to aid in analyzing their problems, to plot grade distributions for classes, to do item analyses on test items, etc.

D. Use of Proctors

This is one of the five basic ingredients of PSI. Keller [15] states, "The use of proctors . . . permits repeated testing, immediate scoring, almost unavoidable tutoring, and a marked enhancement of the personal–social aspect of the educational process." The proctor's job basically is to grade the readiness exam in the presence of the student and to apprise the student immediately of the results (feedback). This usually leads to some form of tutoring although proctors generally are cautioned not to replace the discontinued instructor's lecture with one of their own. Proctor-student ratios vary with course type, instructor preference, and certainly with availability of funds. A 1:10 ratio is probably a good average but ratios of 1:5 and 1:20 are not unusual.

Not all proctors are paid. Especially in the social sciences, proctors are often rewarded with course credit rather than money. In the usually stiff engineering curriculum such a solution to tight funds is less likely to be viable. The use of proctors often—but regrettably not always—is preceded by some form of proctor training. D. Born's *"Proctor Manual"* [4] and a self-paced training manual [32] from the Naval Post Graduate School can be useful adjuncts to this training but should not be substituted for it.

E. Instructional Television

Like the computer, television is used in a variety of ways in education. It can be used to present an entire course to a second, distant or dispersed audience, as it is in the Open University which we will discuss later.

Closed circuit TV with strategically placed monitors is used to allow students to see experiments and demonstrations which are either too dangerous for direct exposure, or which would otherwise require many repetitions to small groups of the class. And then there are of course the TV cassettes which are gaining in popularity. But these two uses, however interesting they may be, are not particularly related to the teaching methods we discuss and will, therefore, not be treated any further.

IV. Description of Some Teaching Methods

Following are descriptions of several teaching methods, together with a listing of the psychological "building blocks" employed by each. Finally, some strengths and weaknesses of each method are outlined. The reader should understand that it is difficult to compare one style of teaching with another in any meaningful way, since classroom attendance is not the only learning activity students engage in. Also, some kinds of material are learned better by one method than by another [5], but a student's outside activity will tend to decrease much of the difference occasioned by what the instructor does in class. Indeed, Dubin and Taveggia [10] and Taveggia and Hedley [30], after comparing over 40 years of comparative research, concluded that "there is no measurable difference between distinctive methods and media of college teaching *when evaluated by students' perfomances on course content examinations."* (Emphasis added.) Others have arrived at similar conclusions [28], and it seems reasonable to assume that their conclusions are valid if the *time* the learners spend on a course outside of class is not controlled [6]. But changes in student attitudes (affective outcomes)—for example, how they feel about a given subject or discipline—may be more important than much of the cognitive learning prized by instructors.

In the following discussions of methods, then, most of the emphasis is upon what happens in class, but comments are made about out-of-class activities from time to time.

A. Formal Lecture

The lecturer tells the students what he wants them to know. No questions are entertained by the lecturer, and so information flows in one direction only. The chalkboard is the only medium commonly used, although other visual aids may be employed. The lecture may be given to any number of students, provided they can hear the lecturer's voice and can see clearly any visual aids used.

1) Learner activity is very difficult to determine. Students may be paying attention, or they may be drifting in and out of contact with what the lecturer is saying. In a formal lecture situation it is not possible to evaluate accurately the degree of activity of the audience.

2) The student has little chance for feedback during class, because there is no opportunity to ask questions, and feedback generally is provided only on homework and exams, if these are returned.

3) It is difficult for the teacher to provide reinforcement directly, but motivated students can obtain satisfaction from feelings of interest in, and understanding of, the material.

4) It is difficult to assure that all students are working regularly in a lecture course. One way to encourage this is to give frequent tests.

5) Step size depends entirely upon the organization of the lecture material by the lecturer.

6) Course objectives often, but not always, are defined in the introductory lecture.

7) Lecture courses are nearly always completely instructor-paced.

From the standpoint of Bloom's taxonomy [3] of cognitive objectives,[3] the formal lecture is probably best used to encourage learning at the first and second levels, although it is not a very efficient way to transfer information. However, it is progressively less effective at levels three through six; the lecturer can model these activities, and this certainly has value, but the students do not have an opportunity to practice these intellectual skills in class. Homework assignments generally are used to help students acquire those skills.

The lecture is a particularly good technique for motivating students by creating interest in a topic; for covering a sizeable body of information to show the interconnections between subdivisions of the material, or for purposes of review; for providing students with a model of a practitioner of a discipline; and for reporting advances in a discipline before they have appeared in the literature.

B. Lecture–Discussion

In this technique the instructor lectures, but he also responds to questions from the class and can ask questions of his own. This technique works best with classes of around 15 to 40 students, although a skilled instructor can work with up to 60 or 70 students. As class size increases, however, students become less and less willing to ask questions and the instructor is less likely to know the students' names. Thus rapport between the instructor and the members of the class diminishes. In classes larger than about 70, lecture–discussion generally gives way to the formal lecture.

1) Learner activity *can* be high if the teacher is open to questions and also asks questions of his own.

2) Students can obtain feedback from their own participation, from the participation of their classmates, and from homework assignments and tests.

3) Students can obtain direct positive reinforcement from the teacher.

4) As with the formal lecture, a student can elect to cram before major tests, rather than to work regularly throughout the course. Frequent tests which may be ungraded are one method of encouraging most students to study regularly.

5) Course material can be developed in small steps.

6) Instructor can prepare detailed course objectives and make them available to students.

7) Such courses are almost always instructor-paced.

The lecture–discussion method is generally a better teaching technique than the formal lecture for classes of small to moderate size. The students have an opportunity to participate, and their participation causes a change of pace of the focus of attention and helps keep the attention of the class from wandering. This improves learning at the knowledge comprehension and application levels, and causes some gains at higher levels (if the lecturer attempts to work at these levels). However, the higher intellectual skills (analysis, systhesis and evaluation) are better taught using other methods.

C. Discussion Group

Group processes have been studied rather intensively over the past ten or 15 years. There are different kinds of groups for different purposes. Davis [9], for example, lists 17 different types of formal groups. The groups most commonly used in college teaching include discussion groups, task groups, seminars, case study groups and problem solving groups. In this paper these will be lumped under the general heading of "discussion groups," and only those having a leader (the instructor) will be dealt with here.

Discussion groups examine topics by talking about them among the members of the group. The group members prepare themselves ahead of time by reading and thinking about the topic. The instructor generally does not take a very active part in the discussion; his main functions are selecting the topics for discussion, keeping the discussion generally on target, and encouraging participation by *all* class members. He should avoid embarrassing (punishing) the members for their contributions and should allow them to carry the ball with minimum intervention on his part. However, it is difficult to handle more than 12 to 15 persons in a discussion group.

1) Learner activity is generally very high in a good discussion group. Even if a student is not speaking at a given time, he likely is listening pretty closely to what is being said.

2) There is a great deal of feedback in a good discussion group. Participation results in both verbal and nonverbal responses from other members of the group, and this feedback spotlights areas of misunderstanding, lack of information and conflict for further discussion.

3) Many opportunities exist for positive reinforcement, either by class members or by the instructor. There is also some danger that hostility may develop, but a skillful instructor can

[3] See Appendix.

avoid this, or at least minimize it. In a healthy group, disagreements can occur without causing bad feelings.

4) Learning is very likely to be distributed in discussion groups. At the start students may come unprepared, but if the instructor encourages participation by all, they will begin to be embarrassed by their lack of ability to add to the discussions, and will begin to prepare themselves.

5) The development of ideas in a group does not follow a logical pattern. The instructor must be alert to what is happening, and if he sees that some students have missed a point he can clarify.

6) Instructional objectives can be prepared for group discussion, but it is harder to adhere to them than with, say, a lecture class. This is because discussion groups may develop interests along lines somewhat different from those the instructor had in mind when he prepared the objectives. (This is not to say that the things the class develops an interest in are not desirable objectives; they are just different from those the instructor originally envisioned.) The objectives for a discussion group should not be made too rigid; if the class members are made aware of the instructor's objective, they will tend to keep their discussions closer to his goals.

7) Discussion groups are generally group-paced. The instructor must keep some pressure on to make sure the group moves along in the desired direction.

From the instructor's point of view, discussion groups give him a chance to find out much about both a student's general and specific level of knowledge of the subject. It also gives the instructor a chance to provide honest reinforcement. Bonuses for the student (some of which were mentioned by McKeachie [24], but which we wish to emphasize) include the following: the opportunity to develop skills in the oral communication of ideas (an opportunity which is lacking in the education of most engineering students); a chance to listen to several points of view and consider ideas that may not have occurred to them; the chance to become better acquainted with other students, which may result in his obtaining peer approval from his classmates and esteem from his instructor [24] (but may also work the other way if hostilities develop); and the motivation to learn more about a subject or discipline as a result of participation in a discussion group.

A discussion group is one of the best ways to change an individual's opinions, attitudes and beliefs. Readings, lectures, and most other learning activities are largely ineffective in this regard, because opinions, attitudes, and beliefs are not cognitive but affective, and are rooted in group behavior. On the cognitive level discussion groups are not effective for presenting new information; knowledge and comprehension elements should be acquired by the student outside of class. Groups are very effective for developing skills in analysis, synthesis and evaluation. They are, in fact, one of the best ways to teach conceptual thinking, since the student gets to practice the skills, rather than just seeing them modeled by a lecturer.

A potential drawback to the method is an almost certain decrease in coverage of material. A considerable amount of time may be required for ten or 15 students to read about a problem, to think about it and discuss it in some depth. A lecturer probably could cover the material in less time.

D. Laboratory Teaching

"Laboratory teaching" as used here refers to those courses in engineering and the natural sciences in which students carry out "experiments" to demonstrate natural laws, to learn how to obtain and reduce data, to develop their powers of observation, to learn experimenal procedures, to become familiar with various kinds of equipment, to obtain practice in formulating and testing generalizations, and to develop their technical report-writing skills. Probably 90 percent of laboratory courses have some or all of the above goals, but a growing number of educators is begining to speak out in favor of more "open-ended" laboratory courses, where the students are more involved in planning and designing the experimental work and where the results cannot be predicted in advance [11], [25].

1) Learner activity can be very high in a laboratory course. There is great potential for motivating students because they are experiencing a phenomenon first-hand, rather than merely reading or hearing about it. At the same time group size must be kept small and other controls must be instituted to keep unmotivated students from doing little or nothing themselves.

2) In simpler work, the learner gets adequate feedback from comparing his results with those predicted by theory. In more complicated situations it may take several hours of calculation before results are apparent, and there may be no opportunity to redo the work to obtain further data. In these cases the instructor often will accept the results if the reasons for the aberrations are discussed intelligently in the laboratory report. Here it is most important for the instructor to return the graded reports promptly.

3) Students can obtain satisfactory reinforcement from the work itself, and can acquire more from the instructor, if his contribution is not delayed too long.

4) Learning is distributed in the sense that most lab assignments adhere to a fairly strict schedule. Many experiments depend upon material that was supposed to have been learned previously.

5) Lab courses generally do not use the "learning in small steps" concept. They depend on the learning that occurred in previous courses, and the experiments often deal with relatively large chunks of material and/or knowledge drawn from a number of fields.

6) Lab objectives should be communicated clearly to students in the early courses of a discipline so that they will learn fundamentals and procedures well. In advanced courses it is desirable to make assignments more open-ended and to challenge speculation beyond the immediate experiment.

7) Except in project-lab courses, most instructors control the pacing because of equipment and time limitations.

Considerable care must be exercised in the preparation for, and conduct of, laboratory classes in order to avoid a number of potential "turnoffs". The students should be prepared for the laboratory work so that they can comprehend the implications of what they perceive. The necessary equipment and materials must be available and must function properly to avoid student frustration from inability to perform the experiment through no fault of their own. And if laboratory reports are required, they should be graded and returned promptly.

Learning can occur at all levels of the cognitive taxonomy, and psychomotor skills are often improved also. In practice the open-ended projects are superior for developing synthesis and evaluation skills.

E. Self-Paced Instruction

A number of self-paced teaching strategies have developed over the past ten or 15 years. Among them are Self-Paced Instruction (SPI), Individually Prescribed Instruction (IPI), and PSI, also known as the Keller Plan, after Fred S. Keller, its developer). PSI, the oldest and best-known of these strategies, will be discussed here [15], [16].

PSI is a teaching method which was developed and pioneered by Fred S. Keller, J. Gilmour Sherman, Caroline Martuscelli Bori, and Rodolfo Azzi at the University of Brasilia, Brasilia, Brazil, in 1964. Briefly, the method requires a teacher to make a careful analysis of what his students are to learn in his course. Having established the terminal and intermediate objectives, he then divides the course material into units, each containing a reading assignment, study questions, collateral references, study problems, and any necessary introductory or explanatory material. The student studies the units sequentially at the rate, time and place he prefers. When he feels he has completely mastered the material for a given unit, a proctor gives him a "readiness test" to see if he may proceed to the next unit. This proctor is a student who has been chosen for his mastery of the course material.

On the readiness test the student must make a grade of 100, but if he misses only a few questions the proctor can probe to see if the questions are ambiguous, and can reword the questions if necessary. If the student does not complete the test successfully, he is told to restudy the unit more thoroughly and return later for another test. He receives a different test form each time he comes to be tested, and no matter how many times a student is required to retake a readiness test, he is never penalized; the method requires only that he ultimately demonstrate proficiency. All students who demonstrate mastery of all course units and make a satisfactorily high grade on a final examination receive a grade of A.

The lecture is greatly deemphasized as a medium for information transfer. Lectures may be given at stated times during the course, but only to those students who have completed a specified number of units and can therefore understand the material to be discussed. The students who qualify for a lecture are not required to attend it, nor is the material discussed in the lecture covered on any examination. Thus the lecture is used as a reward. The professor gives his lecture on a topic he is personally interested in, and about which he feels he has something worthwhile to say. Those who attend come because they want to.

In various articles and presented papers, Keller has listed the following five essential features of the PSI method:

1) The go-at-your-own-pace feature, which permits a student to move through the course at a speed commensurate with his ability and other demands upon his time.

2) The unit-perfection requirement for advance, which lets the student go ahead to new material only after demonstrating mastery of that which has preceded.

3) The use of lectures as vehicles of motivation, rather than as sources of critical information.

4) The related stress upon the written word in teacher-student communication.

5) The use of proctors, which permits repeated testing, immediate scoring, almost unavoidable tutoring, and a marked enhancement of the personal-social aspect of the educational process.

We now refer again to the Psychological Building Blocks.

1) Learner activity is quite high in PSI courses if the learner is progressing. If he is procrastinating, however, he may do nothing for weeks at a time.

2) The learner gets a great deal of feedback, both from results of the unit tests and, if sought, from the study-hall proctor or the instructor. Further, the feedback is immediate. The method utilizes as much feedback as any educational system, and more than most.

3) Positive reinforcement is built into the system and is high. Success is rewarded, but inability to pass a unit examination at the mastery level is not punished.

4) If a student makes satisfactory progress, the learning is distributed. A student may "cram" for the unit tests (of which there may be 15 or 20), but this is not the same as cramming for one or two tests per semester.

5) The system is designed so that learning is by small steps.

6) The method requires the generation of course objectives by the instructor, and these objectives are made known to the student, either explicitly or implied by the questions in the study guides.

7) The courses are self-paced.

Proponents of PSI claim that students learn more under the method and that their long-term retention is better in comparison with conventional teaching methods. It is also claimed that, by modeling good self-questioning skills and encouraging good study habits, PSI helps students learn how to study. Many articles attest that students prefer the method over more conventional teaching styles.

Critics of PSI point to the very real procrastination problem [27]. Although the majority of students make satisfactory progress, a few are unable to do so, whether from a lack of self-discipline or a lack of motivation. That the majority of students make A grades is also viewed with suspicion by many in these days of "grade inflation." It is also claimed that PSI courses have higher than normal drop rates, although Stice [29] reports no significant difference in drop rate for a study involving mainly engineering courses.

After looking at the pros and cons, Kulik [17], Kulik, Kulik and Smith [18], and Kulik and Kulik [19] have concluded that PSI is more effective than conventional methods of college teaching from the standpoints of end-of-course performance, retention, transfer and student attitudes. This is in contrast to the conclusions drawn in the previously cited work of Dubin and Taveggia, but the latter study was done in 1968, before PSI had become known.

Under PSI learning can occur at all cognitive levels, although most PSI courses (and others as well) concentrate on learning at the knowledge, comprehension and application levels. Higher level objectives can be achieved if the necessary materials are prepared and if performance evaluations can be devised to test for the achievement of these objectives. However, the method is not very effective in achieving affective objectives, that is, in changing beliefs, opinions and attitudes. This is not so surprising, since these affects are rooted in group behavior, as noted previously.

F. Bloom's Mastery Learning Model

Benjamin S. Bloom has developed an educational system called "Mastery Learning" [2] derived from J. B. Carroll's Model of School Learning [7]. In Bloom's method, the instructor prepares detailed instructional objectives for the course, breaks these objectives down into "teaching-learning units," and teaches his course as he normally would by lecture, discussion group, or other method. Students take a short (15-minute), ungraded, formative test in class over each unit, and immediately after taking the test they are provided with the answers. The question sheet for the test has references for each question, and the student can refer to these references for further study of the questions missed on the test. These references are not the same as those provided when the students first studied the material; Bloom supposes that a student would not have missed the question if the original reference had been understood, so alternative references are provided as "unit correctives." At the end of the course students take a final (summative) test which is graded. Students must make a grade of 85 percent or more on this examination to receive a passing grade in the course. If they do not achieve mastery on the examination, they fail the course.

1) Learning would reflect the teaching method used by the instructor, although it should be superior to that obtained with the conventional method alone.

2) The learner receives more feedback than with most teaching techniques, and it is immediate.

3) Positive reinforcement received is contingent to a considerable extent upon the teaching method used. It can be augmented by feelings of satisfaction caused by good performance on the many formative tests taken, however.

4) Learning is distributed if the students keep up with the course and study properly for the frequent formative tests.

5) Learning is in small steps.

6) The method requires the careful detailing of instructional objectives.

7) Pacing depends upon the teaching method used; in most cases the instructor controls pacing.

Block [1] reports impressive results with Bloom's Mastery Learning Method; in Korea, students learning seventh and eighth grade physics and mathematics were instructed using both Mastery Learning and conventional methods. On the order of ten times as many students using Bloom's method achieved mastery as did the students in conventional classes. Bloom's method, which holds considerable promise for the improvement of cognitive learning, requires the delineation of objectives, the breaking down of these objectives into teaching-learning units, the preparation of unit tests on those units, and the provision of alternate study references for all test questions. This is nothing like the amount of work required to develop materials for a programmed course, a course taught by computer, or a PSI course. The only test graded by the instructor is the final (summative) examination, and the instructor can continue to use the teaching style with which he is most comfortable.

Bloom's method should be very effective for learning cognitive material at the knowledge, comprehension and application levels. Learning at higher levels tends to be dependent upon the teaching method adopted by the instructor. A major problem is encountered with the students who take the unit tests but never use the unit corrective references to restudy the points not learned or misunderstood. These students, very much like the procrastinators in PSI courses, would not profit from the advantages of Bloom's method.

G. *Programmed Instruction*

This is probably one of the oldest "innovative" methods of instruction. It generally is thought to have started with Rodney Pressly's rudimentary teaching machine, but flourished only as a result of B. F. Skinner's work on operant conditioning in the 1950's [31]. The heart of the system is the program, which may be linear, branched, or multitrack.

The linear program presents the same material to all students in the same manner and in the same order regardless of the individual student's background or proficiency. Conversely, in the branched program the student's answers determine the next material to which he will be exposed. The multitrack program actually consists of a number of linear (or branched) programs. A pretest is used to identify the track a particular student is to follow.

Programmed instruction often is associated with the use of teaching machines, but other media such as programmed texts, computer programs, etc., may be used.

1) Learner activity is extremely high, since every step in the program requires the student's active response.

2) Feedback is immediate at every step in both linear and brached programs since the student's response is followed either by a discussion of that response and its validity or by a simple display of the correct response.

3) Programs are generally so designed that student responses tend to be correct most of the time. Successful progress through the program thus provides continuing positive reinforcement.

4) Learning is very likely to be distributed although there is nothing in the method forcing it to be so.

5) The principle of small steps is the heart of programmed instruction. Step size varies in some—particularly branched—programs but is always small in comparison with other methods.

6) It would be difficult if not impossible to write a learning program without first considering the program's objectives. The author then generally is keenly aware of the program's objectives. However, individual learning programs do not necessarily emphasize or even identify these objectives.

7) One of the salient characteristics of programmed instruction is that it is necessarily self-paced.

Programmed instruction using programmed texts or teaching machines has never become very popular. This is due partly to the comparatively large investment of time and effort needed to generate programmed instructional materials, and partly to students' opposition to the lack of social interaction characteristic of these modes.

PSI, which is closely related to programmed instruction, has added the social aspects and is finding wide student acceptance. Computer-based programs which are dynamic are also continuing to gain popularity.

H. *Audio-Tutorial Instruction*

This instructional technique uses an audio tape recording as the primary, but usually not the only, teaching tool. Although much thought must necessarily go into the production of the audio tape, two factors make the method attractive as a first step away form the classroom lecture: the equipment is relatively inexpensive, and the lecturer still delivers a lecture. Techniques for preparing audio-tutorial tapes are described in the literature [20]. The method is particularly useful in the teaching of manual skills and the operation of equipment, since the student can listen to the instructions while manipulating the equipment. He can stop the tape any time or replay a portion for a missed point.

1) Learner activity is almost assured in a properly designed tape.

2) In manual skill and equipment operation situations, immediate feedback is naturally present.

3) Positive reinforcement also is provided by the eventual success of the student.

4) Distributed learning may or may not be present. It can be assured by requiring students to adhere to a regular schedule of instruction.

5) Step size depends largely on the design of the tape. Small steps can be assured in a properly designed tape.

6) As in programmed instruction, the instructor must have a clear set of objectives in mind before recording an audio-tutorial tape. Objectives may be described in the taped lecture but are more often provided in a printed handout accompanying the tape.

7) This method lends itself best to student self-pacing and almost always is used that way.

I. Computer-Assisted Instruction

The interface between the computer and education is extremely broad (for example see [12]). Here we will concentrate only on the use of the computer to "teach," i.e., to present information, to test for competencies, to diagnose knowledge gaps, and possibly to keep score. In the normal operation of a computer assisted instruction (CAI) system the student converses with the computer. The computer presents some course material and the student must respond. Depending on that response, the computer may display a question, a drill problem, new course material, or it may provide remedial action.

1) In this mode, there is essentially no learning without learner activity because the student response is needed to trigger the next step.

2) Immediate feedback is provided by the program.

3) The program normally guides the student to successful completion of a learning task; therefore, positive feedback is a prominent feature of this system.

4) It is relatively easy to keep track of a student's interaction with the program so the instructor can make sure that students work regularly in the course.

5) Programs are almost always written to present the material in small steps.

6) Objectives must be well formed prior to design of the program, but they are not necessarily communicated to the student.

7) The method is necessarily self-paced.

The activity in CAI started about 1960 with a few pilot projects. Now, more than 15 years later, the number of courses taught by this method is still very small in spite of the method's many proven and well documented advantages. There seem to be two main reasons for this lack of progress: the inertia of the educational establishment, and the economic factors.

For a more thorough discussion of the present state of CAI see [14].

J. The Open University[4]

The Open University took shape in the 1960's in Great Britain as an alternative to the rigidities of campus life. Its purpose is to make a college education available to those who could not attend college as regular students because of distance from the campus, work schedules, or other responsibilities. Thus the Open University offers courses designed for independent study away from the campus.

In presenting these courses, teaching is primarily done by the use of television, radio, and newspaper articles, augmented by self-study texts, telephone conferences, and group discussion meetings. Generally students are not required to come to the campus except for examinations. Help from the instructor or from tutors is available by telephone or on campus.

1) Learner activity may vary the same as in the traditional lecture.

[4] A very exhaustive discussion of the Open University concept and particularly its impact in the U.S. can be found in [13].

TABLE I
RELATIONSHIP BETWEEN TEACHING METHODS AND THEIR BUILDING BLOCKS

Building Blocks	Formal Lect.	Lect. Disc.	Disc. Group	Lab	Self Paced	Blooms Method	Progr Inst.	CAI	Open Univ.	Audio Tutor.
Active Learning	W	W+	P	P	P	P-	P	P		P
Feedback	W	W+	P	P-	P	P	P	P	W	P
Positive Reinforc.	W	W+	P	W+	P	W+	P	P	W	P
Distrib. Learning	W 1)	W	P-	W+	P		P	P	W	
Small Steps				W	P	P	P	P		P
Obj.			W	P-	P	P	2)	2)		P
Pacing*	I	I	I	I	S	I	S	S	I	S

Legend:

P - Prominent component

W - Weak component

Blank - variable, dependent on instructor or other factors

*Pacing - S: self-paced

I: Instructor paced

1) Because studentshave a tendency to study only in preparation for the usually infrequent tests.

2) Although generation of programs for these methods is impossible without a well defined set of objectives, these objectives are not always communicated to the student.

2) Feedback tends to be weak and delayed because of the lack of immediate personal interaction between student and staff.

3) The same holds for positive reinforcement.

4) Because examinations are even more infrequent than in the traditional lecture, distributed learning is apt to give way to periodic cramming.

5) Step size is not mandated by the method but is controlled by the individual course design.

6) This also holds for explicit instructional objectives.

7) Since presentations are scheduled, the method is instructor paced.

Table I summarizes the information presented in this section.

K. The Utility Matrix

In the foregoing study we have discussed the anatomy of several teaching methods, we have delineated the components which comprise these methods, and we have discussed the applicability of the various methods to some extent.

There are, however, other factors which help determine the usefulness of a particular method under any given set of circumstances. For example, we may wish to know how well a student may be expected to remember what he has learned and for how long; how well students like a given new method; whether a given method helps wean the student and gets him started to learn on his own; the faculty effort required in developing a course of study and in maintaining it once it is developed; the need for special equipment and the cost.

These factors are discussed in the literature cited and are summarized in Table II on the following page.

L. Discussion

A study of Table II leads to startling and disturbing conclusions. It seems that the tradiational formal lecture and lecture discussion methods score rather poorly in comparison to most other methods: the former are at the low end of the

TABLE II
THE UTILITY MATRIX

Utility Factor	Formal Lect.	Lect. Disc.	Disc. Group	Lab	Self Paced	Bloom's Method	Progr. Inst.	Audio Tutor	CAI	Open Univ.
Short Term Retention	M	M	M+		H	H	H	H	H	M
Long Term Retention	M	M	M+		H	H	M+	M+	M+	M
Student Acceptance	M	M+	M+		H	M+	3)	H	H	H 4)
Devel. of abil. to learn	L	L+	M	1)	H	M	L-	L	L	H
Req'd Faculty effort: (a) Development	M	M+	M-H	H+	H	M	H	H+	H+	H+
(b) SS use	L-M	M+	M-H	H	M	M	L	M	L	M
Need for special a/v facilities	L	L	L	H	M	L	L	M+	H	H+
Cost	L-M	L-M	L-M	H	2)	L-M	L	H	H+	H+

Legend:

H - high

M - medium

L - low

1) Depends very much on student motivation, self-confidence, and on whether students perform singly or in groups.

2) Depends on whether proctors are paid, whether lab experiments are part of the course, and whether optional facilities such as proctoring carrels and seperate test rooms are supplied.

3) Usually low for superior students, high for slower students.

4) We are here dealing with a special type of student whose reactions are not similar in all respects to those of the traditional students. Moreover, the professionally produced OU courses carry little resemblance to the so-called television courses, which were tried in the 1960's and which were met by low student acceptance.

scale for both short-term and long-term retention. They fare no better with student acceptance and are counterproductive in any attempt to instill in students an ability to learn on their own-and surely the students won't be able to count on a lecture presentation once they leave college!

If the upper part of the table presents some puzzling questions, the lower part supplies at least some hints to an answer: in the lecture and lecture discussion the required faculty effort is comparatively low, special equipment is not required, and the methods are inexpensive. Thus economics and the inertia of the educational establishment prevent the more widespread use of the new teaching methods, in spite of their proven superiority.

APPENDIX A
BLOOM'S TAXONOMY

Bloom lists six levels of learning:

1) knowledge
2) comprehension
3) application
4) analysis
5) synthesis
6) evaluation.

The first level, *knowledge*, implies the learning of facts, the acquisition of information. Nothing other than memory is involved.

The meaning of the term "comprehension" as used by Bloom goes beyond that in the commonly used "reading comprehension." Here we speak of the students' ability to rephrase a statement, to reorder ideas contained in a communication, to draw some inferences from a set of "comprehended" data, and to apply a given abstraction such as Ohm's law or Kirchhoff's laws.

"Application" in Bloom's sense requires more than the activity implied in the last statement. In the taxonomy, "application" implies the students' ability to choose the correct abstraction to use in the solution of a problem, without being given any clues. For example, in a network problem the power supplied by a source may be asked for, with no indication about a possible solution method.

The reader will have recognized that a large proportion of undergraduate engineering education ends at this third level of Bloom's Taxonomy. What often is called a course in analysis actually fits the "application" level perfectly. "Analysis" goes a step beyond this. The student must now be able to look at information, and draw valid conclusions based on it. One place where this level of learning may occasionally be reached is the undergraduate laboratory.

"Synthesis" in the context of the taxonomy refers to creative behavior such as design. This could be the design of a report (implying that a standard form is not specified), the design of an experiment, or the design of an engineering device, system, or task. Not much of this is found in the typical undergraduate curriculum.

The final level in the Taxonomy, "evaluation" implies a judgement based on a fairly broad cognitive background. It is a rare path in the undergraduate curriculum that leads to this cognitive summit.

REFERENCES

[1] J. H. Block, Ed., *Mastery Learning: Theory and Practice.* New York: Holt, Rinehart and Winston, 1971.
[2] B. S. Bloom, "Learning for mastery," in *University of California Evaluation Comment*, vol. 1, no. 2, 1968. Adaptation of article appears as Chapter 4 in [1].
[3] ——, *Taxonomy of Educational Objectives, Handbook 1: The Cognitive Domain.* New York: McKay, 1956.
[4] D. G. Born, *Proctor Manual.* Salt Lake City, UT: Center to Improve Learning and Instruction, University of Utah, 1971.
[5] B. R. Bugelski, *The Psychology of Learning Applied to Teaching* (2nd ed.). Indianapolis, IN: Bobbs-Merrill., 1971, pp. 162–164.
[6] *Ibid*, p. 164.
[7] J. B. Carroll, "A model of school learning," *Teachers College Rec.*, vol. 64, no. 8, pp. 723–733, May 1963.
[8] P. K. Cross, *Accent on Learning.* San Francisco, CA: Jossey-Bass, 1976.
[9] J. R. Davis, *Teaching Strategies for the College Classroom.* Boulder, CO: Westview Press, 1976, pp. 81–84.
[10] R. Dubin and T. C. Taveggia, *The Teaching-Learning Paradox: A Comparative Analysis of College Teaching Methods.* Eugene, OR: Center for the Advanced Study of Educational Administration, 1968.
[11] O. M. Fuller, R. Clift, R. W. K. Allen, D. Bhaja, R. Farber, D. T. F. Fung, R. J. Munz, and A. S. Oberoi, "A course in experimentation," *McGill J. Ed.*, vol. IX, no. 1, Spring 1974.
[12] L. P. Grayson, "A guide to uses of computers in engineering education," *Eng. Ed.*, vol. 60, no. 7, pp. 755–757, Mar. 1970.
[13] R. T. Hartnett *et al.*, *The British Open University in the United States.* Princeton NJ: Educational Testing Service, June 1974.
[14] G. P. Kearsley, "Some facts about CAI: Trends 1970–1976," *J. Educational Data Processing*, vol. 13, no. 3, 1976.
[15] F. S. Keller, "Goodbye, teacher. . . ," *J. Appl. Behavior Analysis*, vol. 1, no. 1, pp. 79–89, 1968.
[16] F. S. Keller and J. G. Sherman, *The Keller Plan Handbook.* Menlo Park, CA: W. A. Benjamin, 1974.
[17] J. A. Kulik, "PSI: A formative evaluation," address at the National Conference on Personalized Instruction in Higher Education, Los Angeles, CA, Mar. 1975.
[18] J. A. Kulik, C. L. C. Kulik, and B. B. Smith, "Research on the personalized system of instruction," *Programmed Learning and Educational Techol.*, vol. 13, pp. 23–30, Feb. 1976.
[19] J. A. Kulik, and C. L. C. Kulik, "Effectiveness of the personalized system of instruction," *Eng. Ed.*, vol. 66, no. 3, pp. 228–231, Dec. 1976.
[20] J. C. Lindenlaub, "Audio-tutorial instruction—What, why, and how," *Eng. Ed.*, vol. 60, no. 7, Mar. 1970.

[21] R. F. Mager, and K. M. Beach, Jr., *Developing Vocational Instruction.* Palo Alto CA: Fearon Publishers, 1967.
[22] A. H. Maslow, *Motivation and Personality* (2nd ed.). New York: Harper and Row, 1970, pp. 35-58.
[23] W. J. McKeachie, "Instructional psychology," *Annu. Rev. Psychol.*, vol. 25, pp. 161-193, 1974.
[24] W. J. McKeachie, *Teaching Tips: A Guidebook for the Beginning College Teacher* (6th ed.). Lexinton, MA: D. C. Heath and Co., 1969, pp. 61-62.
[25] R. A. Mischke, "A new look for the laboratory: Constrained freedom," *ERM Magazine*, vol. 9, no. 3, pp. 56-59, Spring 1977.
[26] G. Pask and B. C. E. Scott, "Learning strategies and individual competance," *Int. J. Man-Machine Studies*, vol. 4, no. 3, pp. 217-253, July 1972.
[27] G. F. Paskusz, "Procrastination and Motivation in SPI," *ERM*, vol. 6, no. 1, Oct. 1973.
[28] G. G. Stern, "Measuring noncognitive variables in research on teaching," in *Handbook of Research on Teaching*, N. L. Gage, Ed., Chicago, IL: Rand McNally, pp. 398-447, 1963.
[29] J. E. Stice, "Expansion of Keller plan instruction in engineering and selected other disciplines: A final report," Univ. Texas, Austin, Center for Teaching Effectiveness, pp. 44-48, 1975.
[30] T. C. Taveggia and R. A. Hedley, "Teaching really matters, or does it?" *Eng. Ed.*, vol. 62, no. 6, pp. 546-549, Mar. 1972.
[31] U. S. Civil Service Commission, Bureau of Training, "Programmed Instruction: A brief of its development and current status," May 1970.
[32] M. D. Weir, *Tutor's Manual for the Personalized System of Instruction.* Monterey, CA: Office of Continuing Education, Naval Postgraduate School.

The Art of Teaching

ANTHONY B. GIORDANO, FELLOW, IEEE

***Abstract*—Effective teaching for effective learning is of basic concern to everyone involved in the process of teaching. Yet, a reminder is often necessary to focus attention upon objectives and techniques to achieve such effectiveness.**

Points of views are presented to stimulate concern with the teaching function. These views are attempts to summarize a vast body of literature generated by individual efforts and group efforts. In this regard, special mention should be made of the initiatives being exerted by the American Society for Engineering Education through its regional Workshops on Effective Teaching to promote awareness, an initiative which began in 1960 as a Summer Institute.

Introduction

CICERO, about twenty centuries ago, stated—"Not only is there an art in knowing a thing, but also a certain art in teaching it"—and we are still uncertain about that "certain art of teaching."

Curiously, little time has been spent in academic circles discussing the techniques and refinements of that art or even defining it satisfactorily—although it is well recognized that the prime responsibility of every college professor should be to motivate effective learning in the classroom and in the laboratory.

Ideally, the college professor should seek to establish a positive contact between each student and himself. He should apply a variety of techniques all aimed toward stimulating understanding and creative attitudes. He should be mindful at all times of the importance of motivation, reinforcement, evaluation, and feedback in the process of learning.

In former years college degrees were supposed, in some miraculous manner, to qualify one for teaching. Now it is evident that knowledge of subject matter alone does not provide complete qualifications. Very often a good scholar finds himself unable to win and hold his students, to inspire and challenge them. Indeed, the need for free discussion to motivate teaching effectiveness has been long neglected.

College teachers seem to be involved in a very strange profession. It might be called the "hidden profession" because it is practiced as a secret rite behind closed doors and never alluded to in polite academic society. Why is it that in any faculty gathering you will frequently hear a group of experts arguing about their subject, but almost never about the best way of teaching it?

It is really strange that there are professors who never bother to study teaching as a skilled process and very often oppose programs for achieving good teaching, the assumption being that effective teaching cannot be taught. Such an insidious attitude is difficult to understand. The quality of teaching as a whole would certainly improve if members of the academic teaching profession took a more direct attitude toward their teaching, and if they discussed the art and its techniques freely, and if they invited others to criticize their own work.

As we attempt to answer the question—"How Effective am I as a Teacher?"—we soon learn that effectiveness varies from year to year and from class to class. This fact gives urgent rise to the need for reinterpretation, reevaluation, reexamination, and reconstruction of content and procedure. It is this urge for intense absorption toward better performance that keeps the good teacher alert and alive.

Each one of us has experienced the stimulation that comes with challenging instruction. But to analyze the situation and extract those qualities that made it memorable is indeed difficult. In essence, there is no one way to teach and no one way to learn. The proper method depends on the character of the teacher, the class, and the subject, and each teacher must find out with reference to each class, and each subject, and indeed each part of the subject, what is for him his best method.

The Teacher Speaks

Basic to teaching is speaking. Your ideas are transmitted as words with your voice. Through your voice, you also indicate whether you are bored, tired, irritated, impatient, and indifferent. This is particularly true when you are under stress. At that moment, you may unwittingly convey a spirit of anger, dislike, or irritation with your voice, far from the impression you intended. However, students do not ask for reasons or motives. They interpret the communication as it takes place at the moment, reflecting the negative attitude toward the subject or toward them. On the other hand, a warm, confident, friendly, pleasant voice arouses a positive response.

If you talk too slowly, the student, who can listen much faster than you utter your words, becomes bored or distracted and his mind wanders. On the other hand, if you talk too rapidly, you are likely to slur your words so that the student is unable to understand them. Ordinarily, one should speak approximately 140 to 160 words a minute.

If you are talking to classes of 40 or more students, you should use a microphone. It is easier for everyone to hear you, takes less effort on your part, and allows you to be more conversational and get greater variety in your speech. A lavaliere microphone is best for this purpose.

Manuscript received June 17, 1970.

The author is with Polytechnic Institute of Brooklyn, Brooklyn, N.Y.

Reprinted from *IEEE Trans. Educ.*, vol. E-13, no. 4, pp. 196–199, Nov. 1970.

In speaking, you can move as much as you wish as long as you do not move the same way all the time. You can move from side to side of the room. You can lean against a table or lectern. You can sit on the table, as long as you do not sit there all the time. A general rule, however, is to move if there is a change in content.

Use your hands to show size, shape, direction, or capacity. There is no pattern, plan, or rule for gesturing. Generally, you cannot gesture too much. But use your hands in a spontaneous manner, not jerky, nervous movements that mean nothing and distract from your effectiveness. This should be almost an unconscious gesture, not taught, not planned, not learned but a reflection of what you are saying.

Nobody likes a sphinx. Even though the material you are talking about is objective and not of an emotional nature, you are still talking to human beings who have feelings and likes or dislikes. So, smile! Show enthusiasm and friendliness. Smile, frown, or even sneer at the material if the content lends itself to it. Laugh once in a while. Good teachers are much more inclined to smile appreciatively than poor teachers.

To summarize, effective teachers speak in a logical and well-organized manner rather than a rambling or disorganized one. They clearly indicate the central points and adequately support them with illustrations and examples. They use language the student can understand. The effective teachers' deliveries are direct, confident, lively, friendly, and conversational and are accompanied by appropriate actions of the body, arms, and head and yet free from distracting movements.

The Lecture as a Method of Teaching

College teaching and lecturing have been associated with each other so long that when one pictures a professor in a classroom almost inevitably the picture is one of a lecturer. Small wonder, though, because so few college teachers receive any orientation or training in teaching methods. Consequently, they very likely resort to copying what they have observed most of their lives as students, namely the lecture. This method might be termed a comfortable one for most instructors because it is familiar to them and also because it provides for careful planning in advance, for control over almost every class session, and for a minimum chance of being put on the spot.

The ability to lecture well is a skill which many can develop if they are convinced of its validity. As already mentioned, it involves such matters as skill in voice control, modulation, enunciation, projection posture, and gesturing, all of which come under the heading of skill in delivery. In addition, it involves techniques of gaining and holding attention; of stimulating and inspiring; of informing, convincing, and prodding to action; of commanding participation in the process of stimulating thought by challenging creative applications and by making the class session and the course and meaningful and rewarding experience for a student.

In preparing a lecture session, the following prescription is recommended:

1) Before assembling for class: a) be aware of the nature of human needs and motive, b) determine the purpose of the lecture, c) be prepared with a lesson plan.

2) In class: a) indicate what you expect to accomplish, b) make your presentation, c) hold attention, d) maintain flexibility and spontaneity, e) ask questions, f) summarize your conclusions.

In lecturing, the teacher should wear the cloak of humility if he hopes to establish rapport and to communicate with his students. Any superiority that he may actually have should be manifested in the content of his lecture, not in his personal attitudes.

In asking questions, he will compliment wherever and whenever he can. He will temper necessary criticism with patience and understanding. He will never ridicule. He will respect even the less well-endowed student as an individual, knowing that his lack of intellectual capacity is no fault of his own. The wise instructor will recognize that it is often the need for a sense of personal worth that makes the student make audacious statements, draw premature conclusions with an absoluteness that is sometimes admirable, and be more intent on winning an argument than on discovering a fact.

Use questions to stimulate interest, encourage student participation, clarify understanding, spot-check the effectiveness of your lecture, and maintain attention. Your purpose is to stimulate the student's thinking rather than to test his knowledge. Force yourself to concentrate on every student in your classroom. Think about their reactions, watch for sleepiness, lack of attention, boredom or bewilderment.

Students are critical of instructors who read long tracts or entire lectures, or who refer to their notes too long. Some authorities maintain that one should never refer to notes. They say that the use of notes creates a barrier in communication because the lecturer's attention is focused on the material in front of him rather than on the students; that it encourages inadequate preparation, because notes take the place of practice; and that it restricts him in the use of visual aids because he must either stay glued to his notes or wave them before the class with one hand while trying to handle chalk, signs, or a projector with his other hand. Necessary notes underlining key ideas should be on inconspicuous cards.

One self-evaluating tool for the instructor is a simple checklist to ascertain if the lecture plan is adequate for the purposes desired. One could easily argue that teaching is not simply a mechanical process and overemphasis on planning might remove the vitality of teaching and the personal rapport between student and teacher. However, a checklist can, at least, give the instructor guidelines toward individual lecture preparation.

The following list is representative of the kind of checks the teacher might use in lecture planning. It is neither exhaustive nor applicable to all types of lectures. These questions can be answered only by the instructor planning the lecture and will be of use to him only if he takes a personal interest in lecture planning.

1) Does this lecture fit in with the general objectives of the course?
2) Can you state the objective of the lecture to the student and show at the conclusion of the lecture how the material presented points toward this objective?
3) What do you want the student to learn from the lecture?
4) Does this lecture have an introduction which ties it into previous lectures?
5) Can you motivate the students by giving them a preview of this lecture during the introduction and show why it is worthwhile to learn the material?
6) Is the alloted time sufficient to get your material across?
7) Is the development of the lecture adequate for the state of training of the students?
8) Have you built slack into the lecture in case the students have difficulty in following you?
9) Have you allowed time for questions from the students, time to sum up, and time to make assignments?
10) Are all the visual aids and other teaching aid equipment ready?
11) Does the lecture have a logical progression—from the known to the unknown, from the simple to the complex?
12) Is it possible to vary the teaching method and the kind of learning required of the student (assimilation, generalization, critical evaluation, etc.) for the material presented?
13) Do you have an "attention getting" interest arouser —analogy, example, or demonstration to pick up the class during a lull which might arise?
14) Do you have an idea of what blackboard sketches and illustrations you will use (if appropriate)?
15) Do you have all the facts you need for the lecture —reference material, etc.?
16) Do you have notes on specific facts, mathematical procedures, etc. in case your memory should suddenly fail you?
17) Have you arranged to hand out assignments, distribute class material, and handle other mechanical aspects of the lecture in such a way that it takes a minimum amount of valuable classroom time?
18) Can you review the material covered by a "new view" toward it instead of simply repeating the highlights?
19) Have you made arrangements to measure how well your material has gotten across—have questions about lecture content?
20) Have you established the bridge to the next lecture or at least given the students a sneak preview?
21) Can you show how this lecture ties into the reading or problem assignments?

These considerations form the basis for evaluating the effectiveness of instruction.

In most courses, there is a portion of the content which lends itself to presentation through lecture; whereas other parts might better be taught in discussion, demonstration, laboratory, or seminar sessions. A careful analysis, coupled with experience, should provide guidelines in determining how often lectures should be scheduled in a given course.

Learning by Discussion

Procedures of teaching by discussion have been widely acclaimed in educational circles as most effective from the standpoint of motivating active learning processes. While the lecture method is very efficient in developing information about a subject, discussion techniques are seemingly superior when the object is to develop understanding in the implications and applications of facts and principles.

In discussion techniques, since opportunity is provided for a good deal of student participation, essential feedback results. Such techniques tend to develop a keenness for extracting essential facts, formulating broad hypotheses, and evaluating conclusions. Although the process is rather slow, results are gratifying because of the inherent stress on critical thinking.

There are many varieties of discussion techniques. Some primarily involve group problem solving; others are gripe sessions; others may be pep meetings. Each of these is useful in its place, and the instructor's role will obviously depend on which of these types of sessions he is leading. In classrooms, however, the most commonly used method is the developmental discussion.

Like other methods, the developmental discussion implies active participation of group members. However, in developmental discussion, participation is directed towards a definite goal, such as finding the solution to a problem, and the instructor takes an active role in helping the group progress towards the goal. However, this does not imply a type of discussion leadership in which the leader manipulates the group to follow his own thinking.

Teaching through the discussion method as compared to the lecture method is less predictable and definite, offers less opportunity for control and, therefore, often feels less comfortable to the instructor. It requires not only considerable restraint but also great skill and much experience on the part of the instructor. He must realize that a certain amount of floundering by the participants in group discussions, for example, can be far more meaningful from an educational standpoint than if he were to inject a ready-made, sure-fire answer.

Further, the instructor must understand that a permissive attitude and atmosphere will enable much more creative thought-provoking discussion to take place. He must appreciate the fact that agendas are not sacred and that spontaneous discussion on a topic of interest may provide them with an opportunity for greater insight and incentive for further study than would rigid adherence to some carefully planned and structured program. At the same time, he dare not fall into a trap of insufficient preparation; for although the discussion method places great responsibility on the student, without a set of lecture notes or an outline which will limit what is considered in a given session, the instructor must be both broadly and thoroughly prepared and should attempt to anticipate a variety of possible classroom developments. He must seek to overcome the feeling that nothing useful or beneficial is happening unless he is talking.

The outline which follows is intended to provide some "how to do it" suggestions to aid the instructor in applying in the classroom some of the principles relating to the discussion method.

1) An informal group atmosphere should prevail; if possible, arrange group in a circle or hollow square.
2) All group members should have the opportunity to participate actively.
3) All group members should share responsibility for the success of the discussion.
4) All group members should share responsibility for decisions or conclusions reached.

The instructor should do the following:

1) Study and think in advance, then come prepared to "rethink" with other members of the group.
2) Set comfortable atmosphere of frankness and friendliness which will help every person to get into the discussion and enjoy it.
3) Serve as impartial guide rather than as active participant.
4) Anticipate a variety of possible classroom developments by having readily available reference books, pamphlets, etc.
5) Speak primarily to reflect feeling; to provide internal summaries; or to raise points missed by the group.
6) Comprehend the fact that it may take many sessions for students to understand the method and become adjusted to it.

Good discussions-teaching ultimately depends upon the character and values of the instructor. To be successful, he must sincerely respect and accept students as fellow human beings engaged with him in the common pursuit of education.

Conclusion

It must be understood that there are no mechanical devices which will insure successful teaching. It appears as a flower whose source of nourishment lies in unseen roots—the attitudes, the outlook, the personal philosophy, and the frame of reference of the teacher. Education is a process of personal growth and development. You cannot force any living thing to grow. The best you can do is to provide a climate and atmosphere most conducive to the full actualization of the potentials resident within the living thing, whether it is a mustard seed or a student.

HOW TO IMPROVE CLASSROOM LECTURES

By LOREN D. REID
University of Missouri

The offer to improve classroom lectures is, I am aware, an audacious undertaking. So that we may get on common ground I wish to make two preliminary observations.

The first is that I am not here concerned with the content of classroom lectures. The way to improve *what is said* in classroom lectures is to read widely, to conduct research, to exchange views with colleagues at staff seminars and professional meetings, to reflect, and to write. The content of lectures should improve as the teacher's knowledge becomes broader and deeper. If a teacher becomes an original thinker about his subject matter, the content of his lectures will not only improve, but may become brilliant. Yet because there is truth in the campus comment frequently heard, "He knows but he cannot teach," it is profitable for all teachers to consider ways of improving the *presentation* of subject matter.

This brings me to the second observation, namely, that the teacher should always ask himself this question, "Is the lecture the *best* way of presenting the subject matter to students?" Would it be better to plan a field trip, set up a demonstration, use slides or motion pictures, conduct a discussion, have four or five bright students present a panel, or even write out the materials and distribute them in mimeographed form? Even in courses where the lecture is traditional, the lecturer may use a different procedure for the sake of variety. The lecture is not the only way of transmitting information; in many instances it is not even the best way. This paper, accordingly, is further limited to those situations in which the lecture has a fair chance of success.

Any group of teachers could sit down and list the many ways in which classroom lecturing can be improved. The list would include such arts of language as vocabulary, imagery, syntax, parallelism, repetition; such matters of organization as preview, subordination, transition, climax, summary; such principles of delivery as voice quality, rate of utterance, general physical energy, and animation. As the list grew, the possibility of improving lectures would seem more and more likely. This paper will discuss four categories of improvement chosen in part from personal observation and in part from informal interviews with seventy-five students who have received college instruction on twenty different campuses.

The Lecturer's Personality

In his *Rhetoric*, Aristotle states that the speaker's character is one of his most effective agents of persuasion. Listeners believe men of good sense, good moral character, and good will more readily than they do men of opposite traits. When I asked students, "What are the characteristics of the best classroom lectures that you have heard?" or "What are the reasons explaining why certain lectures are ineffective?" the answers often reflected opinions about the character and personality of the lecturer.

The good lecturer, these students pointed out, shows that he has the interest of his listeners at heart. At the beginning of the lecture, for example, good teachers use various methods of arousing the interest of their students. Instead of plunging coldly into the topic, the lecturer might open by commenting upon a chapel talk that all had heard. He might refer to some campus or national incident. He might mention a pertinent clipping that he had run across, or a new book he had received. He might begin with a summary or forecast. He might tell a story. All of these methods start the students to thinking, in as painless a way as possible, about the subject before the group. Some of my informants had observed that experienced teachers were more likely to do these things than were younger faculty members. The younger teachers, they reported, are often more serious, solemn, formal, dignified; they are more likely to open up all their big guns promptly at the ringing of the bell.

The personality of the lecturer is further shown by the way he answers questions. A good teacher welcomes questions from the floor and answers them with completeness, often bringing in rare

Reprinted with permission from *American Association of University Professors Bulletin,* vol. 34, no. 3, pp. 576–584, Autumn 1948.

details that otherwise might not have come into the discussion at all. The student naturally feels pleased to have his question treated with so much respect. A few teachers seem unhappy when a question is asked, blurting out such brief and inadequate answers that students hate to offend by further inquiries. Some teachers say, in chilling tones, "I discussed that last hour." Others use the familiar dodge, "I'll take that up later on." In some instances, "later on" may actually be the logical time to consider the question; but experienced teachers know that important items can be successfully repeated two or three times anyway, and the question provides a good motivation for one of the repetitions. Furthermore, the lecturer receives more credit from his students for being able to handle a question on the spot than to take it up later after he has looked up the answer.

In many lesser ways a teacher can show good will towards his listeners. It may help if he says, "Now this is a complex principle; I'm going to try to make it clear, but I want you to feel free to ask questions about any point that you do not understand." It shows good spirit for him to say, "We've had to spend a long time on this classification, but another half hour will see us over the worst of it." Or his personality may express itself in entirely different ways; instead of using gentleness and patience, he may use humor, challenge, praise, mock seriousness, or some other approach.

Sometimes students are embarrassed when the teacher begins his lecture by apologizing for his shortcomings. The chairman of the, let us say, Sanskrit department, who has grown white-haired in the pursuit of knowledge and who has achieved renown for his scholarship, may in all truth open a class by saying, "I do not know anything about Sanskrit." Such a declaration would express the humility that comes to a scholar who has long pursued a difficult topic. It may even mark him as a man of wisdom and distinction. If, however, a beginning instructor makes such a statement, students will take it at face value and wonder why they are so unfortunate as to have to study under an ignoramus. A teacher need not reveal the full scope of his ignorance on the first day of the course; he may at least assure his students that he is interested in the subject, that he intends to give them personal attention, that

he invites them freely to express their questions and difficulties.

If teachers will treat a student exactly as they would a colleague, they will have the proper mental attitude for good lecturing. If one thinks of his listeners as fellow scholars he is less likely to scold, nag, heckle, bait, or patronize them.

The Use of Examples

Illustrations, anecdotes, specific instances, and practical applications all add to the effectiveness of a lecture. One student mentioned a professor of philosophy who had a large fund of examples to illustrate faults of reasoning and types of propaganda. Another mentioned a professor of history who frequently exemplified his points by parallel incidents from other centuries or countries. Another mentioned a professor of language who had at the tip of his tongue instances of all sorts of grammatical constructions. Another described a freshman English instructor with a ready supply of unusual ways of beginning themes, developing paragraphs, and ending themes. Another told of a scientist and his stock of interesting intellectual curiosities. Another related how a professor of sociology chose illustrations from many different trades and industries.

Academic circles give their widest applause to the professor who can discover great generalizations: new laws, principles, concepts, interpretations, theories. I recall a professor of Anglo-Saxon who with some feeling told a graduate seminar that he would consider his life on earth well spent if he could discover a linguistic principle as significant as Grimm's law. Although students appreciate the generalizations, they are particularly intrigued by the specific examples. They are beginners, not practitioners. The margin of knowledge between them and their teachers is very great. Largely through the examples do they learn to appreciate the generalizations.

Humorous examples have a special appeal for the student. The opportunity to laugh gives him a chance to relax and tackle anew the serious instruction to follow. Yet the use of humor can be overdone. Students may laugh from 10:00 to 10:50, then at 10:55 complain that the lecturer is just an entertainer who doesn't really teach anything. A teacher may get such a reputation for humor that no one will take him seriously. The best type of humor is that which grows naturally out of the subject—a turn of phrase or a flash of wit that illuminates a subject without distracting from it.

Improving Delivery

The students I interviewed did not seem especially sensitive to matters of bodily action. Posture and gesture did not impress them, though they noted the difference between an animated, dynamic lecturer and a lethargic one. They were, however, aware of the lecturer's voice, especially when it was not loud enough. Inexperienced teachers holding forth in large lecture rooms sometimes have difficulty in making themselves heard. The student wearies of the constant strain of hearing, and soon loses interest altogether. One teacher answered complaints by this statement: "I am glad that you have to exert yourselves in order to hear me. That extra exertion will make you give special attention to what I am saying." A lecturer with the interest of his students at heart, however, will try to speak distinctly and with sufficient volume to be heard.

To improve audibility is not a simple problem. The teacher may need clinical advice about his voice. The institution may need to study the acoustic qualities of its physical plant. If colleges and universities are to have permanently large enrollments, with the resulting necessity for large classes, they must give acoustic treatment to lecture rooms and in some instances they will need to install sound-amplifying systems.

Clear enunciation, the distinctness with which words are uttered, is another prime requisite of good delivery. "Be sure to tell the teachers to watch their *pronounciation* and *enounciation*," said one student. My interviewees did not appear to be distressed by regional dialects or foreign accents except when comprehension was difficult. What especially worried them was carelessness, slovenliness, and indistinctness. They praised highly the speech of some lecturers, but registered no strong complaint about others so long as they met respectable standards of agreeability and distinctness. Anything below the minimum standard reduces

effectiveness at an alarming rate—may, in fact, bring it almost to the zero point.

Forms of Presentation

Lectures may be delivered impromptu, from notes or outlines, from manuscript, from memory, or from various combinations of the above. Impromptu and memorized presentations will not be considered. The former are too hazardous; as the lawyers say, those who go into court empty-handed will come out empty-handed. The latter are rare; few teachers go so far as to write out their lectures and commit them to memory.

My interviewees had little objection to the use of notes or outlines. They realized that instructors have to present a great deal of factual material, complicated organizations and classifications, and intricate tables and formulas, and that accuracy is of first importance. They agreed, however, that an instructor can be unduly chained to his notes. They did not appear to be especially distressed when the lecturer spends considerable time dictating materials, though if verbatim dictation is carried on too long they began to wonder why he did not mimeograph his ideas. They liked to have the teacher sufficiently free from his notes so that he could answer questions without keeping his finger on his place.

Of the various methods of presentation, the students I questioned had least sympathy with the practice of reading from manuscript. Although they had heard lecturers on many campuses, they did not recall a single instance of a teacher who read lectures effectively. Yet teachers do not have to ask their students for proof that the reading of lectures is usually ineffective. Every one has attended conventions or convocations where what might have been an enjoyable occasion was ruined when the speaker pulled a manuscript out of his pocket. Monotonous vocal pattern, fixed facial expression, and general lack of energy and animation nearly always seem to accompany the reading of a paper.

Theoretically there is little reason why good lectures cannot be read interestingly. A few ministers, like Fosdick, read from manuscript with uncommon skill. A few political speakers, like Churchill, have the ability to bring typed words to life. But the art of reading well is more difficult than the art of speaking well.

The instructor who begins his teaching career by reading his lectures is less likely to develop a successful speaking style than one who begins by using notes or outlines, gradually training his memory and developing his fluency so that he can communicate more and more directly to his students.

Two prerequisites to good reading often escape the teacher. One is that the vocabulary, the sentence structure, and the organization of the lecture should be adapted to oral presentation. Sentences should be simple, language vivid and striking, and organization clear. The general tone should be more informal than that of the scholarly essay. A good way to prepare such a lecture is to follow the practice of the late President Roosevelt and dictate it to a secretary. Such a procedure will tend to assure that the language will be the language of speaking rather than the language of writing. The next step is to revise and re-revise the stenographer's transcript—manuscripts of Churchill and Roosevelt have gone through six to twenty revisions, each revision trying to make the wording more clear, colorful, and meaningful. There is some truth in the statement of Charles James Fox that a good speech does not read well. The speaker aims at something that listens well, not at something for the academic journals.

The second prerequisite is that the reader must so present his ideas as to show that he is actually recreating the thought as he goes along. An incident from the long speaking career of the late President Roosevelt illustrates this principle. On October 29, 1940, he explained to the country over the radio how the Selective Service Act was to be put into operation. The occasion was the drawing of blue capsules from the large glass bowl in the House of Representatives to determine the order in which the young men of the nation would be called to service. In the course of his address, Roosevelt read these words:

> And of the more than 16,000,000 names which will come out of the bowl more than half of them will soon know that the government does not require their service.

Then he paused; something in the sentence did not make sense to him; and in a moment he continued:

> I made a mistake there—I'm afraid it's the fault of the copy—of the more than 1,600,000 instead of 16,000,000 . . .

There had been some talk of "16,000,000" earlier in the speech; but just now the correct figure was "1,600,000." One who read mechanically would not have noticed that a mistake had been made. By contrast, one of the announcers on the same program was assigned the responsibility of reading the numbers over the microphone as fast as they were drawn from the bowl. The nineteenth number drawn—105—was his own draft number, but he did not realize he had read his own draft number until afterwards when a colleague commented upon it.

The good reader is keenly aware of the significance and meaning of what he is reading. The poor reader follows his manuscript word for word, giving the impression that if a student interrupted him and said, "Professor, what does that last sentence mean?" the lecturer would have to go back and reread the paragraph—this time with awareness of content—before he could answer the question.

Invite Student Comments

Although the real test of a lecturer's effectiveness is measured by the lasting quality of his instruction—the impressions, recollections, and habits of thinking that persist years after graduation—the opinions of students at the time they take the course are valuable. Some teachers hand out questionnaires at the time of the final examination, inviting frank comments. One way is to list the titles of typical lectures, and to ask the students whether each one was poor, average, or good; or whether it should be expanded, deleted, or left unchanged.

Last summer, at a military university, I sat across a discussion table from an army instructor who followed this procedure religiously. His first set of questionnaires, he said, contained many brutal criticisms. "This lecture stinks," said one student-officer, "this one stinks too; in fact, they all stink." "Where did they find you?" wrote a second. About fifty such comments led the instructor to feel that his lectures were not very satisfactory. He found a few helpful clues in the avalanche of ridicule, conferred with some of his more successful colleagues, and did a little private soul-searching. He showed me the returns from his last set of questionnaires; many of them were quite commendatory. He planned to study that set with intellectual detachment, trying to discover still other avenues of improvement.

Attending a good classroom lecture is a thrilling experience. It is stimulating to sit in the back of an auditorium and observe a good lecturer who by force of his personality, the vigor and originality of his ideas, and the clarity and animation of his presentation arouses the interest and intellectual curiosity of two or three hundred students. The favorable comments of students heard in the hallways following such a lecture are understandably gratifying. The results of effective lecturing are great as regards both educational welfare and personal satisfaction. The improvement of classroom lectures is a worth-while objective for any of us.

Increasing Engineering Teaching Effectiveness

WILS L. COOLEY, MEMBER, IEEE

Abstract—Graduate teaching assistants and young university faculty because of their lack of teaching experience, often have difficulty in providing a good classroom experience for their students. This paper discusses several characteristics of the engineering learning environment, and attempts to clarify their relationships to effective teaching, which has been defined as scoring well on student evaluations. The development is cursory, but indicates several more detailed discussions for the interested reader.

INCREASING ENGINEERING TEACHING EFFECTIVENESS

A SIGNIFICANT portion of the undergraduate teaching load at many of our engineering schools is carried by graduate students. These students have a number of general characteristics: a high intellectual ability, technical competence, and a desire to teach. Unfortunately, they usually have other characteristics also: no teaching experience, and little knowledge of the learning experience.

The first semester of teaching for a graduate student is often a painful experience for both himself and his students. Very few first-time teachers are regarded as highly by the students as the more experienced faculty who have learned how to teach.

On many campuses, student attitudes toward teachers are reflected in the results of formal teacher evaluations which are circulated to teachers and administrators, and sometimes to the entire campus community. While some may argue that there need be little correlation between teaching ability and evaluation scores, administrators often rely heavily on student opinion of teaching ability. And while questions are often poorly worded, they touch many of the important aspects of a good learning experience. It is, therefore, useful to consider a number of these questions in detail, in order to become familiar with the basic characteristics of a good learning situation. We believe that such study by inexperienced teachers can greatly diminish the difficulties involved in learning how to teach effectively.

The following paragraphs consider several questions found on a university teacher evaluation survey. The discussion was developed by a small group of faculty to aid a beginning teacher in the development of his instructional abilities.

A. ARE THE OBJECTIVES OF THE COURSE CLEAR?

It is difficult to see how a course can be successful in terms of accomplishing certain objectives unless the objectives are clear to both student and teacher. Robert F. Mager's little paperback, *Preparing Instructional Objectives*[1] is quite helpful.

Instructional objectives fill the same need in the design of a course as do specifications for the design of equipment. An engineer who is asked to design a piece of equipment considers it essential that he be given a list of the specifications he is expected to meet. The list covers all the minimal required characteristics. If the product meets or exceeds all specs it is acceptable. At the end of the project the goods are delivered to be tested by the consumer. It is considered unfair if the product is expected to perform in some way other than according to the specifications.

So it is with course objectives. If you are to intelligently design a course, then you should identify at the outset certain things that you expect the student to *do* after your course that he couldn't *do* before. These objectives cannot be too detailed, since one cannot anticipate everything that will happen. They *must be able to be tested*, however, or they are meaningless. Once the designer knows the standards he is to meet, then he can proceed to the task of meeting them—designing a course. If the objectives are stated clearly, then the teacher can allow someone else familiar with the course to make up the examination based on the objectives, and the score achieved by his students will reflect how well he achieved his objectives.

Course objectives can also reduce the ill feelings of some instructors that the teacher of a prerequisite course did not do his job well. If both instructors can agree as to the objectives of the first course, then the first instructor will know precisely what is expected, and the final exam will show whether or not he was successful.

The statement of objectives should also be sufficiently clear and complete to be useful to a student who may be trying to master the course on his own and receive credit by examination. (If not now possible, this procedure is likely to be so in the future.)

The student should be given a copy of the course objectives. If no stated goals exist for your course, it will be necessary to write them. Writing objectives seems to be harder for such things as "developing a problem-solving ability" than it is for "learning the resistor color code." They can be written at all levels, however, and categorized depending upon the type of intellectual activity desired. The following levels for categorizing instructional objectives are based on Bloom's Taxonomy of Educational Objectives[2].

Knowledge

Knowledge of specific facts, of methodology, of theories and structures. For example, "The student is to be able

Manuscript received August 14, 1974.
The author is with the Department of Electrical Engineering, West Virginia University, Morgantown, W. Va. 26506.

Reprinted from *IEEE Trans. Educ.*, vol. E-18, no. 2, pp. 94–100, May 1975.

to write from memory the defining equation for bipolar transistor 'β' ".

Comprehension

The individual knows what is being communicated and can make use of the material. For example, "The student is to be able to correctly distinguish between a circuit schematic and an equivalent circuit model".

Application

Problem solving, construction of equipment. For example, "Given a one stage transistor amplifier with values given for all components except one bias resistor, the student should be able to determine within 10 minutes the correct value of resistor to place the transistor at its proper operating point".

Analysis

The ability to penetrate the structure of a phenomenon. For example, "Given a small network with an identified input and output, the student should be able to determine the function of the circuit by analysis and measurement".

Synthesis

The putting together of parts of elements to form a whole. For example, "Given a multistage direct-coupled transistor amplifier which does not operate because of the failure of one component, plus a copy of the circuit schematic diagram and specifications, the student must be able to locate the problem and repair the circuit within 2 hours".

Evaluation

Judgment about the value of material and methods for given purposes. For example, "Given the specifications of a number of audio amplifiers, the specification of associated equipment (preamplifier, speakers, etc.), and certain desired characteristics of a theater sound system, the student must be able to choose that amplifier which best fits with other components to form the overall system. He must be able to defend his choice".

B. DOES THE COURSE MOVE TOO QUICKLY OR TOO SLOWLY FOR THE BEST UNDERSTANDING?

The average rate of the course is determined by the total amount of material covered in the course. Even though individual classes may proceed with supreme clarity at ideal speed, new concepts introduced by the text, homework, labs, etc. may overwhelm the student by their sheer volume. This can occur even though each concept was so well presented that the student seemed to grasp easily the material as it appeared. On the other hand, if the text covers exactly the same material as the classwork, if the homework and labs contribute nothing new, then the total effect (even if each type of work is correctly paced) will be too slow, producing boredom and disinterest. The average rate is set when the course is outlined. The collective experience of people who have taught the course before, including textbook authors, provides a valuable guide. Homework, lectures, labs and readings must all be considered as to their synergistic effect on the apparent rate.

Once average course rate is set, one must choose the proper rate for the classroom presentation (or other components) to proceed. Consider that we all make decisions based on incomplete data, operating on theories and conceptions which we continually revise and refine as new data come in. Therefore, it is proper for the student's thoughts to be ahead of the actual presentation. He thinks he knows what the answer will be; he guesses the next step in the process. If he is substantially correct he feels rewarded, his theory is reinforced, and he proceeds to new learning. Clearly the student who correctly anticipates all the teacher does will be bored, however. He must be kept active in revising or adding concepts. On the other hand, if the student continually fails to perceive what the teacher intends to do, he will not comprehend what the teacher has done once it is accomplished. If the teacher stopped abruptly during a derivation, for instance, each student should be able to finish the equation, or write the next term, or at least the next letter in the term. The student who anticipates nothing is already behind, because it is the interaction of his preconception with the actual presentation which produces learning.

The conclusion, then, is that the course must proceed slowly enough so that the student has a chance to formulate preconceptions, but quickly enough that his preconceptions may interact with the presentation while they are still fresh in his consciousness.

C. DOES THE INSTRUCTOR EXCESSIVELY EMPHASIZE OR IGNORE FACTS AS COMPARED TO BROAD IDEAS AND CONCEPTS?

The wording of the question implies that facts and broad ideas should be given somewhat equal emphasis. It is easy to see why an instructor might be criticized for ignoring specific details. Presumably knowledge of these details is necessary if the student is to develop the ability to do something in the field. The instructor who glosses over too many details may give the students the mistaken impression that it is very easy to do certain things. When the student tries it himself, which he must do to learn properly, he discovers that he is unable to proceed. An instructor cannot hope to consider all the details, but he must try to provide enough pragmatic information to prevent the student from experiencing a complete failure in his attempts to use his new concepts.

Why are instructors criticized for excess dwelling on facts? All our ideas and concepts are developed from a multitude of "facts". It seems that the belief that the instructor dwells excessively on facts can come in four ways:

a. Poor organization and/or lack of clear course objectives. The brain is easily capable of assimilating all the "facts" an instructor can deliver, provided it knows where to put them and what relationship they bear to one another. If each "fact" seems to stand as an entity unrelated to all others, then indeed the student will have more than he can handle.

b. The student never gets to put the material presented into use. A specific detail is exciting if it is needed, but very dull if no use is ever made of it. This is a retrospective evaluation—if at the time of the evaluation the facts were unused, then they were by definition excessive. It is therefore necessary for the instructor to design situations and problems in which the student can use all the facts.

c. The student is bored by a slow moving course. Most of the material presented is already well known, and hence it seems to be a series of very small steps rather than any great concept.

d. The student does not see what use these facts will be to him beyond the scope of the present problem or subject being discussed.

There are other considerations also. Capable graduates are produced when the teacher makes a distinction between the acquisition and the examination of information. Teaching success is manifested in having the students accept a large role in acquisition, while class activity is devoted to joint examination. The teacher should always be wary of "telling" facts to his students. He should strive to have his class develop critical attitudes which examine the full meaning of what is presented—its veracity, economic impact, social impact, usefulness, projections for future development, etc. It is much better to fill the class time with case studies, problem solving, and discussions than to lecture on the material in the text. Examinations also should be geared to minimize rote memorization and maximize critical thinking.

Therefore, the student should know what facts are required of him and be provided with a means of acquiring them (text, programmed instruction, library reading list, etc.), while the class time should be spent using and examining the facts in some way which actively involves the students.

D. DOES THE PROFESSOR'S PRESENTATION GENERATE INTEREST AND ENTHUSIASM ABOUT THE SUBJECT?

Whether we like it or not, we who call ourselves teachers are in the entertainment business. We are not truly professional teachers unless we see the classroom situation as "putting on a show". This doesn't mean that we have to be comedians or acrobats, but it does mean that we should be more interesting than the spider walking down the wall in the corner. The characteristics of a professional teacher are different from those of a professional engineer, and sometimes in opposition.

The characteristics of a good learning "show" are fairly easy to enumerate.

1. It should be sufficiently entertaining to induce the student from his bed at 8:30 A.M. Poorly attended classes are generally a waste of everyone's time, including the instructor's.

2. It should produce lasting memories of the concepts covered. It is unimportant whether or not lasting memories of the actual presentation are created, except where they would interfere with memory of the material presented. The goal is not to provide entertainment for its own sake, but to provide an entertaining learning experience.

3. It should proceed at the proper rate.

E. HOW MUCH DOES THE INSTRUCTOR ENCOURAGE THE STUDENTS TO THINK AND REASON CRITICALLY OR ANALYTICALLY ABOUT THE GENERAL SUBJECT MATTER OF THE COURSE?

If the student is forced into a completely passive role in the learning process, he will have no chance to verbalize his concepts to the instructor. In many situations, if a student is not asked to communicate his concepts, then he need not struggle to formulate them properly, and they are useless. Certainly an instructor who never allows interaction with his students to influence his presentation or thoughts has conveyed very strongly the impression that he doesn't care whether they have thoughts or not, just as long as they remember what he says. Failure to answer questions satisfactorily gives a similar impression. The instructor should encourage constructive questioning, and provide the students with numerous opportunities to do problems or talk in ways which require them to organize their thoughts about the material which has been presented.

Excessive dwelling on facts may also give the impression of discouraging thinking. If the student cannot integrate the material presented into some sort of unified concept, then he will be unable to do other than regurgitate the facts (if he has a good memory for nonsense syllables). It is important to provide the student both in class and in homework with problems which not only make use of the facts he has grasped, but also require him to do constructive thinking in how to use these facts to solve the problem.

F. HOW WELL DOES THE INSTRUCTOR ANSWER QUESTIONS?

Answering questions can be a tricky business. Certainly not all questions deserve answers, if they waste valuable class time on unimportant points. These must not be fielded in such a way as to belittle the student, however. It is simply best to explain to him that the question does not bear directly on the material, but that you will answer it after class. It is very important to respect the integrity of the student and at the same time keep him from blocking the learning of other students.

Many questions are important, however. The most

troublesome ones are those which indicate the student has not understood the material covered. This can be due to a temporary lapse of attention or to a genuine failure in communication (it can also be due to complete inattention, but this rarely prompts a question). A glance around the room should indicate whether this is a single individual's problem or one which affects a large portion of the class. If you're not sure how many people have similar questions, ask them. If many students are lost, the question must be answered carefully and immediately. The most important thing to remember is that the question has uncovered a communication failure. The presentation has failed to communicate, and it is unlikely that a repetition of the same presentation will yield much success. You must reassess the students' mental state and attempt a different approach which will reach them. Whatever do you do, do not try to proceed without answering this type of question, even if answering the question sets back your class schedule. You cannot build new knowledge on top of a shaky or nonexistent foundation.

A second type of question is also critical. This is the question asked at precisely the right time—the one which sets the stage for the next part of your presentation. If one is truly adept he can arrange situations which induce the students to ask the right questions at the right time. The students perceive that the instructor values their questions and are encouraged to give critical thought to the material. They will also learn more in a class where they can actively contribute to the development of the concepts.

A final type of question is also important. It is the type which refers to some esoteric point that is extremely interesting, but outside of the scope of the course. In such a case you can make a strong point of the value and importance of the question, and then suggest that the asker do a short project on it which he can ditto and hand out to the class. Such questions arouse the interest of the students and show them that there are interesting things awaiting them beyond the confines of the classroom and the course content.

G. DOES THE TEACHER TAKE A GENUINE INTEREST IN HIS STUDENTS: IS HE SINCERELY MOTIVATED TO HELP THEM LEARN?

The question is "Is the course objective that the instructor *presents* the required material, or that the students *learn* it?"

The instructor must continually keep in mind that the students are paying their money to have him facilitate their learning. In the final analysis, most students don't care whether the instructor achieved any of his personal goals, but only that he paved the way for them to learn. The teacher should take pride in the product rather than the process.

It is absurd for a teacher to treat every class with the same approach and techniques, for each class has its own character. A teacher should establish a pace for the progress of the class through student feedback and personal observation.

Many an obstacle in the path of learning could be easily removed if teachers made themselves available during office hours or in the hallway to discuss matters frankly and openly with students. A concerned teacher should periodically call on students who are not performing well or could perform better to meet with him and discuss reasons and possible solutions.

A student tries harder to excel when he is recognized as an individual rather than an indistinguishable part of a group. Creating a personal link with each and every student helps to establish this recognition. This is partly implemented by remembering student names and following their personal development in class.

H. IS THE AMOUNT OF WORK REQUIRED APPROPRIATE FOR THE CREDIT RECEIVED?

One of the biggest educational mistakes is to use homework or classwork as a form of punishment. We probably all know someone who became a poor speller because he repeatedly had to write his spelling lessons fifty times for talking in class. Any student who receives more work than he perceives is beneficial to the course begins to see it as a hardship. None of us wants his course to be considered a punishment for being so foolish as to pursue a career in engineering. There are two ways to avoid this problem. The first way is to give less work. Sometimes this is possible, but usually the instructor feels that each assignment is absolutely necessary to the education of the student. In this case it behooves the instructor to develop interesting problems and assignments and to loosen the requirements for brighter students so that the work becomes a reward (self-accomplishment, satisfaction of intellectual curiosity, etc.) instead of just a required assignment. It does not help to tell the student that sometimes we have to do "dog work", so that he might as well learn it now—this is just an excuse for the instructor's inabilities. If the student is really interested in the outcome of some problem, he will do fairly large amounts of "dog work" to get there. Even in this case, if all the problems given in a course require a great deal of work, the student will soon feel put upon.

I. ARE THE OUTSIDE ASSIGNMENTS (LABS, TERM PAPERS, ETC.) WORTH THE TIME SPENT ON THEM? DO THEY ACT AS A VALUABLE LEARNING EXPERIENCE?

The success of a teacher could be measured by the enthusiasm and interest he generates in his students to further pursue the concepts presented in the classroom. Indeed, the quantity of knowledge imparted by a lecturer to his audience within the bounds of the classroom is rather minuscule compared to that available in the literature. Studies have shown in the simple acquisition of facts students do as well, if not better, without the personal intercession of teachers than they do with it; that teachres may actually distract students in this pursuit[3].

Outside assignments may be viewed, perhaps, as the guidelines that aid the student in his quest for knowledge. In this respect, these assignments become indispensable to the educational process of the average student.

Many of the characteristics of good problems are similar to those of effective classroom sessions. There are two reasons for giving work of this type—*practice* and *acquisition* of new knowledge or concepts. Practice is almost never exciting in itself, and must rely on a strong goal orientation on the part of the one who is practicing—he knows exactly what he wants and is willing to practice to achieve it. A crowded engineering curriculum is weak on practice. The faculty neglects it in deference to more "professional" problems, and the students have neither the time nor the motivation to practice on their own. Yet it is lack of practice which results in many of the perceived inabilities of our students. In giving practice problems, however, one must be certain that the students perceive their need for practice, so that they may be properly motivated.

The second purpose of outside assignments is to induce the student to develop new concepts and enlarge the sphere of his knowledge above and beyond the limits of the lecture. "Professional" problems fulfill this requirement to a large extent. But because of their greater demand on the creativity of the student, they could, if not well designed, lead to a strong sense of frustration.

Problem design should never be the intellectual equivalent of throwing a non-swimmer into deep water. He will either sink or he will make it to shore, but vow never to go near the water again. The possibility that the experience will be considered delightful and worth repeating is remote indeed.

The material content of outside assignments used to consolidate and expand the knowledge and experience of the student contributes a great deal toward their sucess. The assignments should give the student a genuine feeling of accomplishment. Any problem which is wholly artificial cannot do this unless it is written as a game. There seems little point (to the student) in working long hours on a problem that produces results which are either completely useless or already well known. The maturity and social consciousness of today's college students promote interest in non-artifical real-life problems. Relevance and irrelevance compete with other traditional student descriptions in labeling outside assignments. A genuine interest in a task, because of its social value, goes a long way in propelling the student to gladly and enthusiastically spend time and effort toward its accomplishment.

The ideal outside assignment, then, provides touchstones and goals for the student, challenges him without overwhelming him, and provides him with both a feeling of personal intellectual achievement and sense of having contributed to his field.

The laboratory sessions often represent a significant portion of the course work, and should be given the necessary amount of attention. Because of equipment limitations, it is usually necessary to have students work in groups. There are a number of ways of forming the groups, but allowing students to work with partners of their own choosing seems best.

Given the laboratory group situation, it is the responsibility of the instructor to prevent one member of the group from doing all the work while the others sign their names to the finished product. Requiring separate write-ups is not an answer, since these can be easily copied. Probably the most effective way to conduct a lab is to have constant interaction between the instructor and the students. Interrogation of each partner during each session uncovers those students who are letting someone else do everything. Also, the students will be better prepared if the instructor questions them during the lab period. Proper choice of questions can make the laboratory session a much more valuable experience than is the case when no instructor interaction is present, for the instructor can often create interest, relieve frustration, and give the student a feeling of individual worth by his attention.

J. DID THE INSTRUCTOR GIVE YOU ADEQUATE FEEDBACK (THROUGH EXAMINATIONS, COMMENTS ON PAPERS, GRADED PROBLEMS, ETC.) IN EVALUATING YOUR KNOWLEDGE AND UNDERSTANDING OF THE COURSE MATTER?

There is overwhelming evidence which shows that feedback from the source of instruction to the student, and two-way interaction between the student and the material are absolutely necessary. Nearly all instructors in engineering are aware of the importance of feedback in a physical system. This same principle applies in the learning environment. Even the most experienced and mature engineer begins to lose his desire to produce if he receives no feedback. The situation is analogous to that of receiving the directions "turn right at the third light and follow the street until you come to a big pink house on the left". If one makes the turn and proceeds for $\frac{1}{2}$ mile without seeing a house, doubt arises. After 2 miles there may be genuine concern about being lost. The further one goes without seeing the house, the smaller becomes his confidence, and the more he curses the source of direction. Psychological experimentation along this line has shown that it is possible to produce a complete breakdown of all ability by total withdrawal of feedback.

Proper feedback should tell the student a number of things. It should indicate where he stands at the present time in relation to the work which is expected of him. It should also indicate to the student the direction he should take to improve his situation. An extremely important part of feedback on an exam, for instance, is not the grade, but the instructor's comments explaining the cause of the students mistakes and praising his good steps.

There is also a more subtle psychological reason for feedback from the instructor to the student. Often it is the very act of the instructor's communicating to the student, rather than the information that he conveys, to which the student responds favorably[4]. Most human beings crave attention and recognition, and they will function well when they receive it. Studies conducted at mental hospitals, for instance, have shown that patients who were given tranquilizers responded to treatment better than those who weren't; however, patients who were given placebos but were told that they were tranquilizers responded nearly as well as the patients who actually got tranquilizers. It was the attention the patients were responding to, not the drugs.

One must remember that feedback is virtually useless unless applied directly and quickly, for delays lead to instabilities. A day or two after having worked an exam, students are still excited and stimulated by it, and if you can return a graded exam paper to them it will enhance learning. Students benefit very little from homework or exams returned to them the day of their final exam. It is very easy to put off grading papers—don't! A substantial portion of your efforts is preparing lectures and exams is negated by this practice.

Concerning interaction between the student and the material, there is a natural human tendency to believe that things are easier than they really are at the early stages of learning, primarily because of an inadequate appreciation of the subtleties of the problem. By definition, early learning first grasps the basic fundamentals, which often are deceptively simple. Only by *doing* does the student fully grasp the situation, through an interactive process with the instructor.

A very interesting experiment was recently performed to demonstrate the value of interaction for visual-motor learning[5]. Special eyeglasses which produced gross visual distortions were fitted to two volunteers. One of the volunteers sat in a wheelchair which the other pushed around the room. After a few days, the chair pusher became completely accustomed to his new visual environment, that is, he learned how to deal with the new visual images as effectively as he had dealt with his natural environment. Although the wheelchair rider saw everything that was happening, felt the bumps, and observed all the mistakes of his driver, he was able to learn nothing and could interact no better after several days than at the beginning. Similar things were done to cats with the same effect. This indicates that one can learn very little as an observer, and that interaction is basic to learning.

This means that if you want a student to learn to bias transistors, he actually has to *do* it, to make mistakes and be corrected, etc. This is particularly important to remember during your class lectures. One should not assume that simply because he has stated a fact or demonstrated a technique on the blackboard that the students have mastered it. They must rehearse techniques and put facts into context, and they must do this themselves, although one can help them by encouraging them, suggesting exercises, and giving them the time to do exercises, whether during class or at home.

K. HOW WOULD YOU RATE THE CONTRIBUTION OF THE TEXTBOOKS AND ASSIGNED READINGS TO THE COURSE?

Studies have shown that visually associated memories are very much stronger than any others except smell. The textbook, then, is important because it provides visual accounts of the material to be learned. The textbook and outside readings should therefore be chosen by the following criteria:

a. They contain the necessary basic principles and related information; the point of view is appropriate to the course objectives.

b. They provide a good visual impact. A full page of complicated equations is visually discouraging. So are very long paragraphs and pieces of text. Pictures, graphs and sketches are almost always well received by engineering students.

c. The text should be written to produce easy associations for memorizations. This does not mean long words or clever turns of phrases but rather text which is rich with imagery.

d. It reinforces, but does not present in exactly the same way, the class lecture or recitation.

e. It provides easy access for reference. This fault is the chief drawback of the programmed instruction books we have seen.

The concept of visually associated memory applies elsewhere also. Very little of what is heard ever finds its way into long term memory, unless associated with some other sensory input. The teacher who uses no visual aids, seldom waves his hands, and who cannot conjure up mental visual images in his listeners as he talks, is literally pouring words in one ear to run out of the other. We have all gone to a lecture which we enjoyed immensely at the time (the speaker had a way with words), but about which we can remember nothing.

SUMMARY

The above discussions do not represent ready answers to the problems that one may encounter as a teacher. They are meant simply to indicate the nature of several aspects of human learning, thereby helping to state the problem as much as to solve it. You can not be shown or told how to teach, you must *do* it.

ACKNOWLEDGMENT

The author is indebted to several faculty and students of electrical engineering at Carnegie-Mellon University for their contributions, especially R. Frank Quick. Other contributors were B. R. Teare, A. G. Jordan, D. L. Feucht, S. Charap, J. Grason, D. Tuma, A. Sanderson and K. Wellnitz.

REFERENCES

1) Mager, Robert F., *Preparing Instructional Objectives*, Fearon Publishers, Belmont, Calif., 1962.

2) Bloom, B., "Taxonomy of Educational Objectives:" Handbook 1, *The Cognitive Domain*, David McKay Co., Inc., New York, 1956.

3) Hatch, Winslow R., "What Standards Do We Raise," *New Dimensions in Higher Education*, No. 12, U. S. Government Printing Office, Washington, D. C., 1964.

4) McLuhan, Marshall, and Quentin Fiore, *The Medium is the Message*, Boston Books, New York, 1966.

5) Held, Richard, and Alan Hein, "Movement-Produced Stimulation in the Development of Visually Guided Behavior," *J. Comp. Phys, Psych., Vol. 56*(5): 872–876, 1963.

Part VI
Laboratory Instruction

Scope and Purpose

THERE are two reasons for devoting a part of this volume to laboratory instruction. The first stems from the fact that the aims, techniques, outcomes, and evaluation procedures of laboratory instruction are sufficiently different from those of other forms of instruction that they must be separately discussed. The second reason relates to the special demands that laboratory instruction makes on the instructor. Laboratory instruction has always been an indispensable part of engineering education; indeed, early engineering education was mostly based on the laboratory apprenticeship mode of learning. Today, instructional laboratories have a less important role, and anecdotal evidence seems to suggest that the average amount of time that engineering students and faculty spend in the laboratory has steadily declined over the last several decades. There are numerous reasons for this, including the large investment of time and resources that state-of-the-art work in the laboratory demands, the perception of many students and faculty that the laboratory work is inessential in some areas, and the absence of a rigid requirement for laboratory work in the accreditation guidelines for engineering curricula, except as part of design work. A laboratory teacher therefore carries the dual responsibility for making the best use of the limited facilities and limited laboratory time in a student's program of study, and of constantly examining, articulating, and justifying the purpose of a teaching laboratory.

The scope of this part is kept discipline-independent, in keeping with the rest of this volume, and should not be taken as representative of what assistance may be available in the literature for a laboratory instructor. There is much practical advice available in the literature of engineering education on the organization, content, operation, and equipping of instructional laboratories in specific areas, e.g., in control systems, optics, or electronic circuits. In particular, there are numerous papers containing suggestions for individual experiments, and novel or inexpensive equipment, suitable for instructional purposes.

The Issues

An instructor of laboratory courses may find some of the following issues relevant to his teaching:

- What are the stated goals of laboratory instruction?
- What instructional objectives are met, or can only be met, by laboratory instruction?
- What laboratory exercises, activities, and procedures for the evaluation of laboratory work, can be used in order to meet the stated goals?
- What laboratory exercises, activities, and evaluation procedures are actually used, and are therefore the operational goals of laboratory instruction?
- In what way can the lab exercises be modified to emphasize one or more of the objectives of laboratory instruction?
- What exactly does a student learn in the laboratory, and how?
- How should the sequence of laboratory activities or courses be designed so that the responsibility of planning, organizing, and carrying out the laboratory work is gradually shifted from the instructor to the student as the student develops more maturity, skills, resourcefulness, and judgment?
- What are the relative advantages and disadvantages of directed vs. open-ended laboratories, or pre-planned vs. project laboratories?
- How can the effectiveness of laboratories be improved?
- How can one improve the cost-effectiveness of laboratories (the cost being measured in terms of students' time, instructor's time, investment in the laboratory equipment, etc.)?
- What are the costs and rewards for a teacher involved in laboratory development?
- What can be done to make the laboratory development and teaching more attractive to the faculty?

The Reprinted Papers

A single paper is reprinted in this part:

1. ''Assessment of Undergraduate Electrical Engineering Laboratory Studies,'' G. Carter, D. G. Armour, L. S. Lee, and R. Sharples, *IEE Proc. (U.K.)*, vol. 127A, no. 7, pp. 460–474, Sept. 1970.

This paper by Carter, Armour, Lee, and Sharples, is an extensive survey of the literature on instructional laboratories. It presents, in a convenient, tabulated form, compilations of instructional objectives of laboratory courses, for three different types of laboratories: structured, unstructured, and project. It then summarizes the current practice in laboratory instruction, as judged by a survey of the literature. Thereafter, the paper presents, at great length, the methods of assessing a student's performance in each of the above three types of laboratories. This discussion includes a consideration of whether the assessment method is in support of the stated objectives of the laboratory. Finally the paper describes some of the operational problems, such as staffing and student aptitude.

The Reading Resources

A. Laboratory Objectives.

[1] P. Hammond, "The Case for the Teaching Laboratory," *Electronics and Power,* vol. 17, no. 2, pp. 77–79, Feb. 1971. States why students, faculty, and administrators dislike laboratories; describes three functions of instructional laboratories: teaching the existence of objective facts, teaching the need for scientific laws, and providing motivation and purpose to the student; finally states some of the inappropriate goals for which laboratories are used, such as teaching group work or report writing.

[2] K. R. Sturley, "Is Laboratory Work Really Necessary?," *International Jour. Electrical Engineering Education,* vol. 5, no. 1, pp. 63–66, Jan. 1967. Presents arguments to refute the claims that the purpose of laboratory work is to provide vocational training, or verify theory, or acquaint students with limitations of the theory, or gain familiarity with experimental techniques, develop critical and logical abilities, or promote communication skills; proposes some rules for successful laboratory work.

[3] L. Flansburg, "Teaching Objectives for a Liberal Arts Physics Laboratory," *American Jour. Physics,* vol. 40, no. 11, pp. 1607–1615, Nov. 1972. A detailed discussion of accomplishable behavioral objectives for laboratory work; many examples from phenomenological physics; engineering educators should not be discouraged by the words "Liberal Arts Physics" in the title.

[4] E. W. Ernst, "A New Role for the Undergraduate Engineering Laboratory," *IEEE Trans. Education,* vol. E-26, no. 2, pp. 49–51, May 1983. Adds a new item to the usual list of objectives of undergraduate laboratories: serving as a means for continuing professional development of the faculty; suggests the use of laboratory instruction only in new, developing fields.

B. Operational Details of Laboratories.

[5] M. W. P. Strandberg, "Physics Laboratory Administration and Operation," *American Jour. Physics,* vol. 27, no. 7, pp. 503–507, Oct. 1959. States the general goals of instructional laboratories and the details of the many administrative and logistic matters, for a traditional, advanced undergraduate laboratory course.

[6] M. C. Robinson, "Undergraduate Laboratories in Physics: Two Philosophies," *American Jour. Physics,* vol. 47, no. 10, pp. 859–862, Oct. 1979. Compares the operational details, assumptions, and limitations of the traditional structured laboratories, and the experimental unstructured ones; evaluates them with respect to three goals: the student should learn more, should learn to learn, and should learn to think for himself.

[7] G. Weiss, R. J. Juels, L. Bergstein, and M. L. Shooman, "A Project Directed Laboratory Curriculum," *Proc. IEEE,* vol. 59, no. 6, pp. 900–907, June 1971. Describes (a) difficulties that led to the development of a graduate level project laboratory, (b) technical content of the laboratory, (c) the economics and logistics of the laboratory, and (d) its educational value.

[8] H. Aharoni and A. Cohen, "Manpower for Engineering Laboratory Courses," *International Jour. Electrical Engineering Education,* vol. 14, no. 4, pp. 293–300, Oct. 1977. Suggests operational and attitudinal changes necessary to make the teaching of engineering laboratories attractive, and to attract qualified laboratory instructors.

C. Laboratory Teaching Methods.

[9] *IEEE Trans. Education,* vol. E-14, no. 3, Aug. 1971. Special Issue on Undergraduate Laboratories. Contains several reports of experience with individual laboratory courses having some innovative aspect or approach, or teaching a specific subject matter.

[10] *Journal of Engineering Education,* vol. 58, no. 3, Nov. 1967. Special Issue on Engineering Laboratory Teaching, and *Engineering Education,* vol. 60, no. 9, May 1970. Special issue on Laboratory Instruction. Includes articles on self-instruction in the laboratory, and on audio-tutorial technique for laboratory instruction. The second item contains the report of the Commission on Engineering Education, entitled "New Directions in Laboratory Instruction for Engineering Students."

[11] E. W. Ernst and J. O. Kopplin, "Modern Objectives and Methods for Laboratory Instruction," *IRE Trans. Education,* vol. E-5, no. 3 & 4, pp. 168–172, Sept. 1962. States two broad objectives: supplementing the teaching of subject matter, and the teaching of the theory and practice of experimentation; describes essential features of demonstration laboratories; and proposes a program of deliberate teaching of the principles of experimentation.

[12] B. Hazeltine, "Minilabs in Engineering Education," *Proc. IEEE,* vol. 59, no. 6, pp. 980–981, June 1971. Describes a mode of operation in which laboratory exercises are assigned like homework; walk-in laboratory is open for long hours; and reports are de-emphasized.

[13] E. W. Ernst, "Laboratory Oriented Studies," in *Britannica Review of Developments in Engineering Education, Vol. 1,* N. A. Hall, Ed. Encyclopaedia Britannica, Chicago, 1970. After a listing of six principal objectives of laboratory instruction in engineering, describes a number of individual methods of laboratory instruction, including demonstrations, self-instruction, and projects.

[14] B. A. Blesser, "From the Abstract to the Practical. Teaching a Project Laboratory," *IEEE Trans. Education,* vol. E-14, no. 3, pp. 101–107, Aug. 1971. Discusses the pedagogical basis and philosophy of an unstructured laboratory in which independent student projects provide the basis for topic discussion.

[15] J. C. Lindenlaub, "Educational Technology Applied to an Electrical Engineering Laboratory Program," *IEEE Trans. Education,* vol. E-18, no. 2, pp. 73–77, May 1975. Describes the development of a laboratory program, starting from the instructional objectives.

[16] J. Wiechel, G. Kinzel, and J. Charles, "Laboratory Slide Presentations: A Means of Saving Faculty Time," *Engineering Education,* vol. 71, no. 5, pp. 341–344, Feb. 1981. Describes the use of slides for tutoring laboratory instructors, and the benefits of this arrangement.

D. Design.

[17] W. H. Middendorf, "Methods for Improving Design Procedures," *IEEE Trans. Education,* vol. E-19, no. 4, pp. 148–153, Nov. 1976. Proposes the use of noniterative synthesis procedures as a vehicle for training students in design; gives two examples from induction motor design.

[18] N. R. Malik, "System Design Projects for Seniors," *IEEE Trans. Education,* vol. E-18, no. 4, pp. 207–209, Nov. 1975. An example of a design project which emphasizes that a variety of skills and considerations are needed in an overall system design.

[19] J. Law, "An Electrical Engineering Undergraduate Design Course," *IEEE Trans. Education,* vol. E-22, no. 3, pp. 138–142, Aug. 1979. Describes a pair of design courses; in the first, the student completes several individual projects, and in the second, participates in a single team project; gives a list of topics discussed in accompanying lectures, and examples of completed projects, mostly of the circuit design variety.

[20] *Engineering Education,* vol. 64, no. 5, Feb. 1974. Special Issue on Creative Design: The Sloan Experience. Report from eleven universities on their approach to teaching engineering design.

[21] A. R. Cook and C. J. Turkstra, "Practicing Engineers in Undergraduate Education," *Engineering Education,* vol. 68, no. 2, pp. 189–191, Nov. 1977. Describes an arrangement in which the senior designs are supervised by practicing engineers in industry.

[22] C. E. Wales, "Does How You Teach Make A Difference," *Engineering Education,* vol. 69, no. 5, pp. 394–398, Feb. 1979. Introduces the so-called "guided design" approach, further described by the following item.

[23] R. W. Eck and W. J. Wilhelm, "Guided Design: An Approach to Education for the Practice of Engineering," *Engineering Education,* vol. 70, no. 2, pp. 191–197, 219, Nov. 1979. Detailed description of an approach in which a design project forms the basis of lectures, homeworks, etc., and where the design work is carried out much as a laboratory course associated with a lecture course would be.

[24] Committee on Engineering Design (E. S. Taylor, Chairman), "Report on Engineering Design," *Jour. Engineering Education,* vol. 51, no. 8, pp. 645–660, Apr. 1961. The final report of a faculty committee at M.I.T.; after mentioning some problems of semantics, discusses the essential elements of engineering design; the proper engineering attitudes and the effect of single-answer problems on them; the role of experiments, internships, and design experience; and the recruiting of suitable engineering faculty.

Assessment of undergraduate electrical engineering laboratory studies

G. Carter, B.Sc., Ph.D., F.Inst.P., D.G. Armour, B.Eng., Ph.D., M.Inst.P., L.S. Lee, B.Sc.(Eng.), M. Litt., C.Eng., and R. Sharples, B.Sc., Ph.D., M.Inst.P.

Indexing terms: *Education and training*

Abstract: A description of several forms of laboratory study is given, and the major educational objectives likely to be attained via each mode analysed. Current practice of laboratory work in British undergraduate electrical engineering courses is then reviewed. The published methods of assessment of these study modes are considered, and the problems of value judgments of human qualities as well as intellectual skill development are highlighted. Particular attention is paid to the project method and the difficulties of this approach for both staff and students, particularly if this approach is to become an increasingly important component of study methods, are examined.

1 Introduction

Since engineering, by its nature, is an experiential discipline, there should be little need for further advocacy of the pre-eminent role which should be accorded to laboratory work in an undergraduate curriculum. Several authors have argued this philosophy cogently and convincingly,[1–3] but, on the whole, much less attention has been paid to public discussion of this aspect of undergraduate studies, than to other areas. A review of journals devoted to educational aspects of electrical engineering soon shows the small minority of space devoted to this area. Recently, however, and particularly in the recommendations of the Finniston Committee,[4] the needs for greater attention to professional studies in engineering at undergraduate level, i.e. an increased commitment to experiential learning, have been more clearly recognised. Much of this component will need to be undertaken in the laboratory environment, since it is of a pragmatic nature, and so the present review, which considers current conduct and objectives and methods of assessment, may be apposite.

When solving any engineering problem, the rational approach is to (*a*) specify the problem, (*b*) determine the known parameters and constraints, (*c*) postulate a variety of solutions, (*d*) proceed towards an optimised solution, and (*e*) evaluate the solution. The problem here is of preparing a useful manuscript, which it is hoped will be read, discussed and criticised, on the topic of the 'Assessment of undergraduate laboratory studies'. This appears to be a well posed specification but when we move into the next stages the greater complexity of the task becomes clarified, since 'assessment' presupposes that there is something to assess, i.e. laboratory studies. In order to discuss assessment meaningfully it is necessary to know the nature and intent of laboratory work, as it is currently practised in undergraduate courses, i.e. to determine the parameters. There are many ways of obtaining such information including interviewing all (or a sample) of practicitioners (staff and students), questionnaire administration, literature survey etc. The former strategies can undoubtedly provide a broader and more detailed conspectus of contemporary activities, and two of the present authors have undertaken a limited survey of this nature in the past,[5] and intend to repeat the study on a wider scale.

Paper 774A, first received 11th February and in revised form 10th April 1980

The authors are with the Department of Electrical Engineering, University of Salford, Salford M5 4WT, England

However, these investigations are time consuming and in order to meet our, partly self-imposed, deadline for preparation of this communication we have therefore relied heavily on a literature survey to provide the necessary background information. In the following two Sections we present the results of this investigation, and although, not providing the detail of a wider study, it is believed that such a strategy gives evidence of the range of current philosophies and activities. Just as assessment implies the need for an activity to be assessed, so must the activity itself be purposeful with definable aims and objectives. Without such specification, either implicit or preferably explicit and articulated, both the activity and its assessment can become meaningless and random exercises without direction or apparent motive. Therefore, in the next Section we examine the range of educational skills which a number of authors have suggested should be developed by organised laboratory studies, and in the third Section the strategies of laboratory studies employed to enhance such skill development are outlined. Having discussed the purpose and nature of laboratory studies we are then in a position to consider the methods employed to assess the development of desirable abilities as undertaken in the fourth Section (i.e. the solutions are posed). In the final Section the problems arising from employment of various assessment procedures are examined with specific reference to the demands upon staff and student time, resources and educational abilities.

2 Aims and objectives of laboratory studies

We may consider the *aims* of laboratory studies as being to produce an 'expected student behaviour' as a result of having experienced a prescribed set of specific education conditions: it is an attempt to condition the student to think and behave in a manner which, by general consent is considered to be appropriate by professional practitioners in that discipline. 'Aim' is used in a directional sense in which a student is directed toward a goal.[6]

The term 'objective' is used in the operational or behavioural sense to indicate ways in which a student can demonstrate the extent of his progress towards specified goals.[6] The objectives[7] of laboratory studies may therefore be regarded as an attempt to classify the detailed intellectual abilities, attitudes and experimental dexterity which it is intended should be developed by carefully planned and evaluated mode of study.

We have classified laboratory studies into three types,[7]

Reprinted with permission from *IEE Proc.*, vol. 2A, no. 7, pp. 460–474, September 1970.

namely controlled assignments, experimental investigations or experimentation and projects.

This classification has been determined from a study of the published literature on laboratory studies and a simple definition of each strategy is given below, but it should be noted that such classifications and definitions are always of a somewhat arbitrary and subjective nature and to some extent we have adopted them for convenience.

2.1 Controlled assignments

Controlled assignments are normally short laboratory exercises, extending over one or possibly two periods. They are devised and controlled by the staff who may often issue instruction sheets which direct and contain the student's activities, enforce conformity, and implicitly give assurance that compliance with the methods prescribed and adherence to the courses of action indicated will inevitably ensure that the answer required (often predictable) will be achieved.

Some of the objectives which are thought to be achieved by controlled assignments are given in Table 1.

Table 1: Objectives of controlled assignments

(a)	Reinforcement of theory learned from lectures, tutorials, directed and nondirected reading etc.
(b)	Development of manual dexterity in the intelligent use of equipment, apparatus and tools
(c)	Development of the ability to observe, check and systematically record data
(d)	To provide training in formal report writing
(e)	To provide the means for development of awareness of the difference between theory and practice

2.2 Experimental investigations

Experimental investigations are normally longer exercises designed by the staff, but which include elements of choice and opportunity for student deviation in the pursuit of the prescribed goals. Sometimes they take the form of more loosely controlled assignments with provocative questions posed by the staff to encourage extrapolation or translation of the information and results obtained to other associated situations or problems. Generally the purposes of experimental investigations are to provide experience in the discipline and methods of scientific investigation, as well as experience in design and minor development exercises in the multisolution situation, as preambles to future excursions into major projects.

Table 2 lists some of the objectives which may be achieved by experimental investigations.

Table 2: Objectives of Experimental investigations

(a)	To encourage the understanding of, and ability to define precisely, those questions which require to be answered
(b)	To foster the ability to determine what measurements will be required
(c)	To ensure that the students may gain confidence in the selection of the appropriate apparatus and strategies required to obtain such measurements
(d)	To promote initiative in the ability to select the correct measurements, and to arrange them in such a form as will give precise answers to the questions posed
(e)	To inculcate an intelligent appraisal on a proper assessment of the significance of the results obtained

2.3 Projects

A project, the subject of which is frequently chosen by the students themselves often in consultation with supervisors, is an attempt to solve a practical problem and to assess the effectiveness of a solution to it within the limitations of time, material resources and experience available.

Students may work individually, in pairs or as members of a project team. They may be either designing, constructing and testing a device or system or considering the designing, conducting and the evaluating of the results of an investigation or feasibility study.

The project method appears to be the logical consequence of the endeavour to involve the student emotionally, retain his interest and help him to identify the exercise with the engineering field as well as exposing him to the demands, stresses and realities of the industrial scene.

Table 3 indicates some of the objectives which it is thought may be achieved by projects, but to these should be added those objectives already tabulated in Tables 1 and 2.

Table 3: Objectives of project exercies

(a)	To enable the student better to appreciate the existence of a problem capable of solution and the extent and nature of the likely scale of time and material resources required for such a solution
(b)	To demonstrate the greater range of thought that is necessary in determining the possible solutions to the problem as well as the extent of the difficulties which intervene
(c)	To underline the crucial role that decision making assumes in such exercises
(d)	To develop and improve interpersonal relations and the social skills essential for successful coordinated corporate activity
(e)	To demonstrate the depth of constructive criticism required, of both the solution adopted and the methods used.

The listings of objectives given in Tables 1, 2 and 3 have been selected to indicate some changes or growths in the level of intellectual demands made as the style of the activity develops. A more extensive list of the overall objectives is given in Table 4 which represents a wider selection drawn from a number of sources, and referenced in Table 4. Some of these objectives are, within our definitions of each type of laboratory activity, clearly more relevant to one or more of these strategies.

3 Current practice

Although it is comparatively easy to formulate the desirable educational objectives of undergraduate teaching laboratories, the development of laboratory schemes capable of achieving these objectives is unavoidably influenced by such diverse factors as resource limitations, technological change, differing staff commitment and widely varying student intake capabilities and attitudes. The importance of all these factors is clearly evident in much of the published work on laboratory teaching practice in electrical engineering even though most of the innovations described are based primarily on educational considerations. The fact that the relative significance of these factors must vary from institution to institution, and even course to course, makes it difficult to obtain a balanced view of current laboratory practice since details of many of the schemes in use have never been published other than in internal

Table 4: A selection of objectives of laboratory studies

		References
1	To illustrate, supplement, emphasise and reinforce material taught in lectures	[2, 8, 1, 9]
2	To illustrate topics more readily understood if introduced via laboratory studies rather than by means of lectures	[8, 10]
3	To achieve familiarity with and facility in the use of appropriate materials, techniques and instruments	[8, 9, 11, 12]
4	Acquisition of an appreciation of the fact that practical results do not necessarily correlate with the idealised theory	[8, 1, 9]
5	To encourage emotional involvement in the subject so resulting in greater personal motivation	[8, 1, 12, 13, 14]
6	To provide the student with a valuable stimulant to independent thinking	[2, 4]
7	To show the use of 'practicals' as a process of discovery	[2, 6]
8	To provide training in scientific method of inquiry	[7, 1]
9	To provide closer contact between student and staff, and a means for direct transfer of knowledge from the tutor to the student	[9, 12]
10	To stimulate an interest in design	[7, 15]
11	To give an appreciation of degrees of accuracy expected and tolerances allowable	[1]
12	Training the student to maintain concise and accurate records, in adequate detail under laboratory, field or workshop conditions	[1, 7, 9]
13	The development of interpersonal relations and social skills essential for co-ordinated corporate activity	[1, 8, 16, 17, 19, 36]
14	To simulate conditions of graduate engineering employment	[13, 20]
15	To further the ability accurately to define and specify a problem	[6, 7, 36]
16	To encourage the student to search out appropriate literature and other sources of information relative to the problem: involving search and retrieval and systematic recording of existing information obtained from library, research reports, professional journals, specifications, records etc.	[6, 7, 9, 12, 36]
17	To inculcate an awareness of the social, economic and other nonscientific restraints encountered in engineering problem solving	[6, 20, 21, 22]
18	To encourage critical analysis of relevant and nonrelevant information	[7, 8, 9, 12, 6]
19	To promote synthesis of knowledge from disciplines associated with the subject area	[7, 9]
20	To nurture an appreciation of the existence of 'objectives facts'	[1]
21	To encourage investigation of several methods of solution prior to selection of optimum	[6, 7, 8, 23, 36]
22	To encourage decision making on optimised solution strategy	[6, 7, 9]
23	To encourage creative ability and individual initiative in the postulation and development of optimised solution	[6, 7, 15, 24]
24	To encourage a logical procedure towards an optimised solution	[6, 7, 8, 17]
25	To encourage evaluation of solution strategies and their implications, and to be prepared to reconsider when faced with unsolvable difficulties	[7, 9]
26	To generate in the student persistence in the face of difficulties	[11]
27	To encourage evaluation of final optimised solution in terms of the initial specification and an understanding of the implications of the solution	[6, 7, 9]
28	To train the student in the compilation and mode of presentation of a report of methods used, results obtained and conclusions drawn: with appraisal of extrapolations to other problem areas, in a manner which is structured and intelligible; in language clearly comprehensible, to practitioners and nonpractioners alike, indicating the degree to which it is considered the problem has been solved	[1, 6, 7, 8, 9, 16, 25, 26]

laboratory manuals. A direct consequence of this is that the traditional, highly formalised approach to laboratory teaching, which, in the authors' experience, is still widely used in the first year of many courses, is only discussed in the literature in the context of its replacement by 'improved' schemes. Although the general criticism of this approach in the literature is not necessarily a true reflection of the extent to which its value is considered to be questionable, it does indicate that there is some change in emphasis in laboratory work towards the development of higher level skills such as problem solving and experimental design, and that there is a greater appreciation of the importance of student motivation.

The growing belief that the educational potential of the laboratory has been under exploited, and that the higher order objectives specified in Tables 2, 3 and 4 can be achieved by a more heuristic approach to laboratory teaching, has undoubtedly provided the major motivation for the introduction of carefully structured laboratory schemes designed to make optimum use of the three principal classes of laboratory exercise, i.e. controlled assignments, experimental investigations and projects. Although the educational arguments for the development of the various schemes for which details are available are similar, there are marked differences in the ways in which they are organised in terms of the nature of the various exercises, the style of the instructions given to the students, the relative amounts of time spent on the different types of exercise and the stage in the overall course at which they are introduced.

Controlled assignments have, despite their educational limitations, remained an important part of the majority of teaching laboratory schemes. However, whereas in traditional schemes they occupied at least the whole of the first year of a three year course, they are now increasingly used in an introductory capacity as a means of reinforcing the students' theoretical knowledge and enabling them to acquire rapidly the basic technical skills required for the more demanding experimental investigations and projects which, in some cases, also form part of the first year laboratory course. In the context of these objectives it has been argued that it is important that at this introductory stage the laboratory work and lectures are well coordinated,[8, 9, 27, 28, 29] and that all the students have sufficient time to acquire the necessary practical capabilities. In the electronics field, compliance with these requirements has been greatly assisted by the rapid technological advances which have occurred in recent years. The availability of cheap devices has made it possible from a resources point of view to provide sufficient components or experimental boards for all students to carry out the same exercise at the same time.[8, 9] This is in marked contrast to earlier approaches in which a selection of different exercises, each designed to illustrate a specific aspect of the theory or demonstrate the capabilities of a particular piece of apparatus, are carried out concurrently by different groups of students. Using this system it is obviously impossible to integrate the laboratory and lecture courses and the inherent time constraints tend to force students to follow the generally very detailed instruction sheets without necessarily understanding them or working beyond their narrow confines. By removing this time constraint, it is possible to introduce schemes which are essentially self-paced and to give the students more responsibility for their own rate of progress, a feature which has been found to contribute to improved students motivation.[8, 9]

The problem of stimulating or retaining student interest figures prominently in the discussion of most of the published laboratory schemes, and the emphasis on changes in the content of first year courses clearly illustrates that it is at this stage that the major difficulties occur. In fact, the largely experimental investigation and project content of the second and third year laboratory courses currently operating has remained more or less unchanged, at least in principle, for a number of years. Some innovations, for example, the introduction of short duration, industrial projects into a second year course,[30] the use of more multi-disciplinary projects in the third year (see for example Reference 32) and the integration of second and third year laboratory work into an overall, design based electronics course[31] have, however, been reported.

In contrast to the situation in the later years, first year laboratory schemes have, as already indicated, changed appreciably in recent years and the growing practice of using both experimental investigation and project exercises at this early stage has, at least in the opinion of the operators of the schemes reported, led to significant improvements in student motivation. In one case[33] where staff resources made it difficult to man the more open ended type of exercise in a satisfactory way for large first year classes, interest was retained by introducing a series of 'fun to do', programmed experiments, an important feature of which was the clear specification of the aims and objectives of each stage of the exercise in addition to the provision of detailed descriptions of the problem and the experimental procedure to be followed. In this scheme the students work through the programmed experiments at their own pace, and questionnaires, containing multiple choice questions where possible, constitute the marking medium. At the end of the series of six experiments the students undertake a twenty hour project. This scheme, in common with a number of others, places much less emphasis on report writing, in marked contrast to the more traditional situation in which at least as much time can be spent writing reports as is spent in the laboratory. The importance of technical communication and the maintenance of proper records is not, however, underestimated in laboratory schemes currently operating and, as will be discussed later, reports and log books play a major role in many assessment procedures.

Further efforts to teach basic practical skills without stifling interest have involved the use of circuit-construction exercises as part of the controlled-assignment section of the first year course.[29] Although these exercises only require the ability to assemble a kit of parts according to a detailed set of instructions, rather like the amateur constructor 'projects' commonly seen in electronics hobby magazines, the higher level of personal and technical involvement associated with the use of the self-built circuit in subsequent exercises does appear to generate considerable interest.

These changes in the general conduct of first-year laboratories inevitably affect the attitude of students towards subsequent laboratory work, and there has been a steady increase in the scope of the experimental investigations and projects offered to second- and third-year students. In one second-year laboratory, for example, following a reorganisation of the first-year course, group projects were introduced to complement sophisticated, open-ended experimental investigations[8] while in Salford, students who had been involved in a newly developed first-year laboratory scheme in which open-ended, ongoing experiments and miniprojects were introduced, expressed considerable dissatisfaction with a second-year scheme in which the experimental investigations were comparatively rigidly controlled even though they involved the use of sophisticated circuits and advanced test equipment.

Although a survey of the literature does not reveal any direct analysis of the influence of changes in first-year laboratory teaching practice on final-year projects, there is a clear indication that the introduction of such changes has not been considered in isolation from the activities in later years and the educational advisability of transferring students directly from controlled assignment activities typically lasting about three hours, to open-ended experimental investigations or projects, which typically last for twenty weeks at up to six allocated hour per week, is being increasingly questioned. Specifically, the development of more broadly-based project activities in engineering departments, for example the increased incidence of semi-technical[32] and even nontechnical, projects[8] must place additional demands on the laboratory work carried out in earlier years.

Thus, whereas it is obviously not possible to obtain a complete survey of current practice in undergraduate engineering teaching laboratories on the basis of a limited literature survey, there is a clear indication that the educational objectives of laboratory teaching, particularly in the first year, are being more explicitly stated and substantial attempts to match objectives and strategy undertaken. The evidence of success of such matching is still scarce but it should be noted that increasing attention is being paid to the important question of student motivation and many authors refer specifically to this beneficial feature of their activities. However, there are indications that the cost of providing a highly motivational, heuristic setting in which practical problem solving exercises can be undertaken is extremely high in terms of staff commitment and there is little doubt that the apparent success of many of the schemes reported in the literature is due almost entirely to the enthusiasm and dedication of the supervisory staff.

4 Assessment of laboratory studies

Although, as indicated in the preceding Sections, there is a reasonable source of information on the range of philosophies and strategic conduct of laboratory studies, there is substantially less information on assessment procedures. In particular, there is virtually no literature discussion of the evaluation of the somewhat more formalised types of laboratory work at undergraduate level, the major attention being paid to the assessment of projects. It is probably true that laboratory supervisors have implicit evaluation schemes which may also be explicitly communicated to the students, but that since such schemes are believed to be relatively common and widespread there is no need to rehearse them publicly. It is doubtful, however, if such schemes are detailed or precise, and this must give rise to uncertainty in the validity and reliability of subsequent assessment. It is therefore deemed worthwhile to give some account of existing detailed assessment schemes for more formalised types of study which, although not specific to undergraduate courses, are relevant to the closest point to undergraduate commencement, i.e. at GCE advanced level. In addition, detailed schemes of assessment of projects have been formulated for both advanced level and for undergraduate courses and so, in the present Section, we will discuss these together since, as will be shown, there are relatively minor differences with respect to course level.

In order to preserve a structure to this discussion, it will be helpful to employ the first approximation categorisation of laboratory study strategies outlined in the previous Sections, i.e. controlled assignments, experimental investigations and projects.

4.1 Controlled assignments

It is in this area that information is least available, including schemes for A-level studies. Indeed, although defining the nature and general conduct of such activities, the A-level assessment procedures[7] give no guidance to the style of assessment other than that a candidate should keep a journal or log book of all of his laboratory activities which will be submitted to the examiners for assessment. The maintenance of log books is probably a widely employed practice at undergraduate level also, and it is equally probable that a component, at least, of the assessment is based upon this record as well as upon written reports on specified experiments. Wilkins[34] indicated that the major assessment mode for prescribed experiments in 1st-year studies in his Department was via inspection of the laboratory journal, but also revealed that each experimental instruction sheet contained questions which the student must answer to demonstrate understanding of the work. McPhun[33] indicated, again for 1st-year studies, that although a journal was kept by students, this was not marked by supervisors. In this case, however, a substantial number of questions were written in to the detailed instruction sheet and student assessment was made from their answers. Dearden[35] has given probably the most detailed account of a first-year laboratory-study assessment procedure based upon staff evaluation during the course of each experimental period. This procedure was based upon two components, the laboratory log book and on-the-spot appraisal of student performance in the laboratory, each component carrying equal weight. The detailed assessment schedule, which was completed by a supervisor for each student at each laboratory period, is given in Table 5. It is immediately clear from this schedule that much of the assessment is of a subjective nature and is concerned with student attitude in addition to skill development. However, supervisors were given somewhat more detailed guidance of the factors to be considered when allocating a mark for each heading. For example, under the heading of 'intelligent approach' the extremes were considered to be 'original, full of ideas, questioning each step' and 'not contributing at all, no ideas'. Of some significance was the requirement that the assessment procedure was to be completed in the presence of the student, the supervisor indicating the justification of his assessment and outlining the students' strengths and weaknesses. It should be noted however than Dearden[35] describes the student reaction to this immediacy of feedback as less than whole-hearted approval. It must also be noted that the assessment scheme was not felt fully to reflect the aims of each experimental study; for example, if rather complete prescription of the conduct of the experiment was given it was difficult to assess a student's originality. This point of Dearden's[35] cannot be stressed too strongly since there is no virtue in constructing a detailed assessment strategy based upon a perceived battery of skill attainments and attitude developments if neither the educational aims nor the method of conduct of the work allowed for such developments to occur. There must be a match between definition, expectation and assessment as will be further highlighted in later discussion.

Although not providing a detailed assessment scheme, Pramanik and Dring[9] have also described how log books are employed as a component of student assessment in a 1st-year laboratory scheme which included both formal and open-ended styles of work. In the week following completion of a particular study students were expected to complete a discussion of the result entered into the log book, which was marked by a supervisor in consultation with the student. Students were then given the opportunity to amend any reports receiving adverse comment and the completed log books were remarked at the end of the year. These revised marks, together with marks from two

Table 5: Assessment schedule for a first year undergraduate laboratory in electrical and electronic engineering

Student		Experiment Poor	Doubtful	Satisfactory	Good	Excellent
Performance	Total	0	1	2	3	4
Planning and preparation						
Understanding of expt. and background						
Intelligent approach						
Application and effort						
Rate of progress and progress made						
Lab. Log.	Total	0	1	2	3	4
Sectioning and layout (ease of access)						
Data presentation and handling (tables and graphs)						
Experimental notes and observations						
Treatment of uncertainties						
Presentation of results and conclusions						
Grand total	Date					

completely written-up reports formed the basis of the laboratory evaluation. If a mean mark of 50% was not achieved the student was required to undertake a practical examination, which had to be passed before progress to the next year of study was allowed. The authors demonstrate that this strategy results in an overall improvement of performance of 2–3% (which, over a seven-year period, yielded an average mark of 66%) and that, importantly, students did not totally forget each laboratory study after its initial completion. Undoubtedly, laboratory courses often constitute a hurdle to successful completion of a course (frequently being written into departmental course requirements) but it would be interesting to know how widely the 'second chance' strategy, involving a considerable degree of teacher-student feedback, reported by Pramanik and Dring[9] is practised.

4.2 *Experimental investigations*

In the preceding Sections it has been outlined how experimental investigations (or experimentation) differ from controlled assignments in that students are expected to begin to develop strategies for more open-ended problem solving activities. This might require students to make reasoned choices for the types of instruments to be employed, to investigate the behaviour of an electronic circuit, or indeed to suggest circuit modifications to change a specification. To a certain extent, therefore, this type of study represents the beginnings of the project approach, although the level of, and responsibility for, major discretionary decision making is still only at a rudimentary stage. As outlined earlier this type of activity is beginning to find increasing favour in first, and particularly, second years of study, and may sometimes be designated as mini-project in nature, occupying usually only 6–12 hours of student time. Although the educational objectives of this type of work have been reasonably well documented in the higher education sector, and, as will be seen later, major attention has been given to the development of assessment schemes for full projects, we have been unable to find any equivalent assessment procedures for experimental investigations or mini projects in the literature referring to higher education. It is probably true that implicit schemes exist, and that some of the criteria employed are similar to those indicated in Dearden's[35] scheme for controlled assignment evaluation (Table 5), which are clearly appropriate for more open ended types of studies. In the absence of schemes specific to higher education, therefore, appeal is made to A-level guidance in this area. The most detailed assessment procedure has been established for the NUJMB Engineering Science A-level examination, and this should be noted in the context of guidance which is given to candidates on the role and nature of the experimental investigation. Candidates are required to submit reports on two experimental investigations each requiring an expected maximum of 12 hours study. The reports are expected to contain the following information, which should be understandable by an intelligent layman:

(*a*) A statement of the precise problem which has been tackled
(*b*) The possible solutions considered
(*c*) Any decisions taken with supporting arguments
(*d*) Important features of any design produced
(*e*) The major observations made
(*f*) The evaluation of observations
(*g*) Conclusions reached and assessment of the significance of the study
(*h*) Suggestions for further work.

The assessment schedule is divided into two components *A* and *B*, as shown in Table 6. The first part, *A*, consists of five fundamental questions which admit of Yes/No answers to determine whether the candidate possesses any acceptable experimental skill. If insufficient affirmatives are scored in this component the candidate cannot score on part *B*, i.e. a level of competence or hurdle is imposed. In the second component six questions are asked with a four-point ranking for each response. These questions are closely related to both the stated objectives of the activity and to the guidance given on the conduct of the work as outlined above. It is quite clear, particularly under headings (*b*), (*c*) and (*d*) of the guidance that at least a start in engineering decision making is demanded. More detailed assessment of decision making and design skills is made upon project work, as will be described in the following Subsection, but it is suggested that the style of assessment described here would be eminently suitable for the investigation and mini-project type of studies now becoming more widely employed in the first two years of undergraduate studies.

4.3 *Projects*

In marked contrast to the paucity of information available on controlled assignments and experimental investigations there is considerably more information available to the reviewer of the literature on projects and their assessment. There can be no doubting the increasing attention paid to project methods, not only in final-year studies, but also in earlier years of study.[9, 28] Detailed examination of the literature, however, reveals that many authors are conveying exactly the same message with perhaps some differences in emphases. In the following, therefore, some of the major issues which are raised by project assessment will be examined, and one detailed scheme of assessment discussed, which to a considerable degree can be said to subsume all other reported schemes. The aims and objectives of project studies have been discussed at length in earlier Sections, and these and their implementational consequences lead inevitably to two apparent difficulties. Firstly, students are *expected* to develop and display skills and attitudes not traditionally assessed by formal examinations, e.g. initiative, persistence, motivation etc., and serious doubts may be expressed as to whether the attainment of such skills and attributes admit to quantitative measure. Secondly, the general spirit and conduct of projects leads to a much closer interpersonal relationship between student and supervisor, and it has been argued[35, 36, 37] that if the supervisor is perceived as acting as assessor as well as mentor, guide and, at its best in the true concept of university education, co-learner, then the relationship will be less fruitful. Since, it is further argued, the role of the project in developing qualities of mind and action is valuable in its own right, and these qualities are difficult to assess quantitatively or even objectively, why inhibit the staff-student relationship by attempting assessment? Although there is undoubted intellectual merit in this argument there are powerful counter arguments. Engineering courses, by their very philosophy and nature, are preparative for professional employment and action, and possess an element of training. Thus it is desired that graduates possess specific intellectual skills, attitudes to work and their colleagues, and manipu-

Table 6: An assessment schedule for experimental investigations

(*A*)	Questions on the experimental investigations	
(i)	Has the candidate made safe use of the apparatus involved?	
(ii)	Has the candidate succeeded in making accurate observations within the limits of the apparatus used?	
(iii)	Has the candidate presented the observations in a clear and workmanlike manner?	
(iv)	Are the findings of the investigations consistent with the observations made?	
(v)	Does the final report contain an account of the essential features of the work?	
(*B*)	Criteria with respect to the two experimental investigations	
		Grade
(i)	*Theoretical understanding:* In relation to his depth of understanding of the theoretical aspects of the problem the candidate has shown	
	sufficient understanding of the problem to enable him/her to plan the approach competently	3
	sufficient understanding of the problem to be seldom in need of help	2
	limited understanding of the problem	1
	little or no understanding of the problem	0
(ii)	*Planning the investigation:* In determining the activities to be undertaken the candidate	
	considered a range of appropriate possibilities with respect to the type and scope of measurements to be made and came to a reasoned conclusion	3
	considered a range of possibilities and came to a less well reasoned conclusion	2
	considered an inadequate range of possibilities	1
	exercised little judgment	0
(iii)	*Procedures and equipment:* In selecting the experimental procedures and equipment to be used the candidate	
	made a reasoned assessment of the alternatives available and came to a well argued conclusion	3
	lacked depth in determining the final choice	2
	made some attempt to consider alternative approaches	1
	unthinkingly adopted standard procedures or relied entirely on the teacher's advice	0
(iv)	*Errors:* The report includes	
	a statement of errors together with estimates of magnitudes and a discussion of their relative significance	3
	a statement of errors and estimates of the magnitude of each error	2
	a statement of errors (including the most important errors)	1
	no explicit statement of errors	0
(v)	*Critical review:* In considering the investigation in terms of the results obtained and the conclusions reached the candidate	
	made a thorough appraisal including a thorough estimate of the effects of the assumptions made and including suggestions for improving the approach and/or taking the work further	3
	as above but with one of the features missing	2
	made a significant appraisal of the work done	1
	made no significant appraisal of the work done	0
(vi)	*Personal contribution:* In planning and executing the investigation the candidate	
	exercised initiative and judgement throughout	3
	lacked initiative and judgement at times	2
	made little personal contribution	1
	relied entirely on external help	0

lative ability, and it is valuable not only for the student to know his overall capabilities, but also for prospective employers to recognise his attainment and potential in these areas. Secondly, for the great majority of students, whether or not they have a correct attitude to higher (or indeed any) education, assessment provides a powerful incentive to endeavour. Thirdly, since the project is regarded as such an important component of study, why deny the possibility of some assessment, however imperfect, as providing important further information on the totality of the capabilities of the students, for the benefit of students, staff and employers. These arguments seem to have persuaded most authors[18] to attempt some form of assessment procedure and, as already indicated, the major differences between these authors seem to lie in emphases or weightings attached to different components of the assessment scheme and to different strategies of obtaining acceptably objective assessment. In the following we will first discuss the parameters which, it is generally conceded, are to be assessed, and then describe the modes of assessment. Before doing this, however, we will re-emphasise the need to identify the desired aims and objectives of project work before an appropriate and meaningful assessment procedure can be developed. It must also be emphasised that such desirable outcomes of the activity should be made explicitly clear to both students and supervisors *ab initio.* Unless this is done then, idiosyncratic differences between students and supervisors may lead to gross injustices, even with group and/or external moderation of supervisors assessment. They may equally lead to changing perceptions of potential outcomes from initiation, during progress and at finalisation of the 'contract of initial expectation' described by Black,[38] but which may admit of debated and agreed modification during progress of the project. Harris and Dowdeswell[39] suggest that this form of definition is task oriented and is somehow different from a project based upon specification of learning objectives. This distinction seems artificial since implicitly, and sometimes explicitly, the skill and attitude development (the definable educational objectives) can be stated as components to be suitably achieved in arriving at task solution. What essentially seems to be in debate here is the difference between 'focal' and 'subsidiary' aims discussed at length by Adderley *et al.*,[18] i.e. solution of the task is the focus of the endeavour, whereas necessary and desirable skills and attitudes are developed subsidiarily in reaching the solution. It seems quite reasonable to believe that both specific statements about the intended desirable behaviour and of the perceived nature of the final outcome can be articulated at the commencement of project studies. Indeed in Black's[38] description of project studies in a final-year mechanical engineering undergraduate course a list is given of eight specific objectives for students, five for supervisors and three concerned with the interaction of the student-supervisor team with colleagues. The list of objectives for students is reproduced in Table 7, since it is against this set that assessment is performed, as can be seen by comparing Table 7 with the assessment schedule shown in Table 8.

Table 7: Student objectives

(i)	To meet the challenge of a problem which is new to you, but which will make you use the knowledge you have acquired in your studies
(ii)	To search the existing knowledge, to establish what new knowledge is required, or to extend the existing knowledge to this new application
(iii)	To confine the problem to limits which will allow useful work to be achieved within constraints of time and resources available
(iv)	To learn how to plan the various segments of an investigation, and to adapt to the implications of changes as work proceeds
(v)	To design, construct, assemble and commission experimental apparatus and instrumentation, which will give maximum information for minimum demand on effort and resources
(vi)	To analyse experimental results, and compare with theoretical analyses, either existing, or produced by you
(vii)	To compare your results with those achieved elsewhere in order to establish or question them
(viii)	To communicate all the information and knowledge you have accumulated, not merely to your supervisor or the School of Engineering, but to the outside engineering world, i.e. to *write* briefly, explicitly, and with authority, to *present* your work to an audience and *reply* to a discussion of your results.

In this scheme, assessment is based upon performance during the project as deduced by observation and from a continuing journal of activity maintained by each student, upon the final report and seminar presentations by the students. Very clear guidance is initially given to students about the philosophy, role and expectation of their project work, and supervisors are required to make initial specific statements about their perceptions of expected final outcomes. Students are expected to prepare an initial outline plan of their anticipated activities for each week of a 20 week period and to update and amend this, in consultation with the supervisor during progress of the project. The requirement for submission for *approval* by internal (teacher) assessors and external moderators of an initial project outline and relatively detailed plan of attack is also included in the NUJMB engineering science A-level project studies.[7] Allanson[40] has also indicated the requirement for an evaluation of both an initial project plan and a mid-session interim report on progress in a final-year electrical engineering course. The role of student presentations at seminars as part of the assessment process has also been discussed by Beynon and Bailey[8] and by Benson[41] and Allanson.[40] It is probable that project outlines are required in a significant number of departments although our private experiences (as well as the literature evidences) would indicate that brief personal seminar-style exposition of project work is demanded in the majority of courses.

The assessment schedule of Table 8 is the most detailed published scheme that we have encountered and it is notable that for each assessable item there is a six-point scale allocated, ranging from outstanding to inadequate. The definition of these six categories is worth recording as in Table 9. Black[38] is careful to point out that these do not represent numerical scores but act as a guide to the supervisor, an assessor allocated to monitor progress on each project, and to the final moderation panel of all supervisors as to what degree of achievement is necessary to *score* an allocated mark. To a certain extent it may be considered that the items under 1–3 of Table 8 could admit of a numerical score and this has been undertaken by Benson[41] and Bowron[42] and in NUJMB A-level engineering science[7] with differing scale lengths (e.g. 0–10 or 15 by Benson[41] for a coherent group of Black's items,[38] 1–6 by Bowron[42] and 0–3 by the NUJMB[7]). It is more difficult to justify a numerical rating for such qualities, appearing under 5 of Table 8, as initiative, innovative ability, enthusiasm, perseverance etc., and we would tend to agree with Black's[38] strategy here in requiring commentary rather than quantitative evaluation as proposed by Benson[41] and Bowron.[42] It could be argued that if such components represent only a small fraction of the total assessment, then imperfections of quantitative judgment will be relatively unimportant. On the other hand these may be considered to be precisely the attributes admired in professional engineers and to accord them minor importance is to devalue their predictive capacity. If, however, it is recognised that these are describable qualities and not determinable quantities then it would seem permissible to employ them as additional *qualifying* information on both the success of the student on his course and of potentially predictive value to prospective employers. It thus seems desirable to begin thinking in terms of profiles of student capabilities, rather than adhering completely to the present system of classification based upon an assumed competence in quantitative grading. This is a theme which will be re-examined in the next section.

The tactics of assessment must be considered to be equally important to the details of the evaluation. We have already noted that in the scheme described by Black,[38] both the supervisor and an independent assessor contribute towards a continuing monitoring of student progress and a supervisors' general meeting at the end of the course translates performance grades into a mark scheme. The project report is initially evaluated by the supervisor and returned with comment and criticism for final completion and grading evaluation by individual members of a panel of senior staff.

The seminar presentations are also included in the grading procedure, but only in providing additional evidence, rather than contributing a specific mark component. The weightings given to the components of the assessment schedule are then applied in the proportions; log book 5%, performance assessment 50%, draft report 30%, final report 15%. Benson,[41] Bowron,[42] Adderley *et al.*[18] and Allanson[40] all describe similar systems employing progress assessors (in addition to supervisors) but with somewhat less frequent contact with their allocated students, and also note the importance of some form of final moderation of the written reports. In these reported schemes different weightings (but not widely variant) are given to the several components outlined by Black.[38] All of those authors acknowledge the vital roles played by assessors and moderating panels in ensuring consistency of standard and demand of different supervisors.

Harris and Dowdeswell[39] have discussed alternative methods of the evaluation of the final project report in which either, both individual assessors and their combined grouping arrive at an agreed ranking within broad divisions (associated with conventional classifications of 1: 2.1: 2:2 etc.) based upon internally agreed criteria and

Table 8: A grading assessment profile for final year projects

PROJECT MEMO

STUDENT'S NAME: SUPERVISOR:

PROJECT TITLE:

NO.:

Supervisors should enter their grading of the student's performance under these headings: they should indicate where they are inapplicable, and add any special aspects in section 4. The items are not weighted, so there is no requirement to convert the grading into marks.

O = Outstanding, E = Excellent, G = Good, S = Satisfactory, P = Poor and I = Indequate

	O	E	G	S	P	I	COMMENTS
1. CARRYING-OUT							
1.1 Approach							
Exploration & enquiry							
Literature & background search							
Setting objectives							
Preparation of programme							
1.2 Implementation							
Decision on test facilities & instrumentation							
Design of rigs and apparatus							
Building & commissioning rigs							
Setting-up calibration							
1.3 Experiments							
Logical planning							
Accuracy & relevance of measurements							
Overcoming difficulties							
Modification during progress							
1.4 Computing							
Modelling							
Programming							
Analysis							
Presentation							
1.5 Evaluation							
Study of previous theories							
Prediction based on theory							
Analysis of experimental results							
Relation between theory & results							
1.6 Log book							
Maintained as instructed							
Standard of entries							
2. MANUSCRIPT DRAFT REPORT							
Structure of report							
Clarity of argument							
Balance of sections							
Validity of results							
Justification of conclusions							
Achievement of objectives							
3. FINAL REPORT							
Response to criticism of draft							
Lay-out of report							
Standard of execution							
Style							
4. SPECIAL ASPECTS							

5. SUPERVISOR'S COMMENTS ON STUDENT'S PERFORMANCE AND ATTITUDES

e.g. Response to advice and criticism, initiative, determination to suceed, dependence on instruction, powers of innovation, enthusiasm, perseverance

MARKS ALLOCATION

Carrying-out (50%)	%
Log-book (5%)	%
Draft report (30%)	%
Final report (15%)	%
	%

Table 9: Definition of the six performance categories employed in table 8

Outstanding	Exceptional ability and achievement beyond the supervisor's expectation, and not requiring his direction
Excellent	Highly competent work, with achievement up to that of supervisor's expectation, minimum supervision
Good	High standard, which though failing to match up to total expectations, nevertheless contained much actual achievement: moderate supervision
Satisfactory	Some weaknesses and limited in achievement, but carried out supervisor's directions adequately and worked carefully
Poor	Little actual achievement: maximum supervisor's direction: limited understanding
Inadequate	Little or no serious work. Failure to respond to direction or instruction. Lack of effort.

perceptions or individual assessors rank reports on their own personal and subjective contracts of expectation and no moderation is attempted. Although Harris and Dowdeswell[39] indicate some possible advantages of such relatively or totally unstructured systems they point out the clear disadvantages and injustices which may result. Our own experiences would indicate the absolute *necessity* for moderation to ensure comparability and fairness even if the detailed schemes of evaluation described earlier are not pursued. We may finally note an interesting proposal for project assessment made by Adderley *et al.*[18] which does not, as yet, command any other attention in the published literature. This is the concept that a component of the evaluation may be via self assessment and/or peer assessment by fellow students, and Adderley *et al.*[18] provide persuasive arguments against the immediate reactions to such a scheme of insufficient student experience and vested student interest.

5 Problems and prospects

All the available evidence indicates that there is a continuing movement away from formalised stereotyped laboratory studies towards more challenging and meaningful self-directed open-ended and project work. This movement carries with it some important problems, some of which are clearly recognised, others less so. At the same time this growing belief in the benificence of active learning raises important new prospects for future educational patterns of engineers. In this final Section we examine some of the problems created by the introduction of strategically designed course work from the points of view of academic staff and students and consider some future prospects. Up to this point we have attempted, as far as possible, to record the statements and views of other authors with relatively little personal value judgment. In this final section however, whilst maintaining an objective approach to the work of others we will admit to not only drawing our own conclusions from such studies but also offering opinion based upon personal experience and observation. Such subjective statements will be recorded explicitly and should not be accorded the same status of perceived wisdom derived from the consensus evaluations preceding. Nevertheless, we offer some such views in order to be provocative and encourage further debate.

5.1 Staff problems

Whatever the reasons for the introduction of the sausage-machine style of formalised laboratories with full, printed directions issued to students, and certain apparent advantages can be perceived in processing larger numbers of students through a set routine with minimum wear and tear on equipment, i.e. matching experiment to currently learned theory, familiarisation with equipment, etc., we would suggest that it has often led to a minimum level of supervisor involvement. The demands on staff intellect tend to be repetitive rather than changing, and in many such laboratories staff attitudes can often be characterised by frequent boredom and even absence. The laboratory is largely designed to run itself with minimum intervention and infrequent change. There is substantial evidence to suggest[37] that the situation is very different in project-style studies where constantly changing demands are made upon staff intellectual capabilities and their own innovative and creative talents and many staff experience much greater satisfaction and motivation in this environment. Our own observations indicate that not all staff respond to such stimulating challenge in this way; some feel threatened by possible exposure of their incomplete knowledge of a topic, some even refuse to participate in these activities whilst others participate unwillingly or half-heartedly. Under such circumstances, unfairness and injustice to students can, and does, result. More fully structured (but not stereotyped) laboratory and project studies and assessment procedures certainly demand increased work load, commitment and enthusiasm from staff, and if this style of activity is to become even more widespread, and applied throughout all years of undergraduate courses, then even greater staff involvement will be required. It may be that unwilling staff should be more positively persuaded and even directed into such responsibilities. There are indications that the number of students undertaking projects at undergraduate[18] or A-level[7] that can be adequately and responsibly supervised by a member of staff is between 8 and 15. This band includes the approximate (nominal) student/staff ratio in many engineering departments of about 10:1 and therefore, theoretically should be capable of implementation. This implementation must rely upon commitment and competence by all staff, however, if loads are not to fall unequally and unjustly upon the 'apparently' able and willing.

With respect to academic competence to supervise project work, we would argue that it is surely not too much to expect of a member of academic staff, even if he believes himself to be currently unable to fulfil a guidance role in an innovative situation, that he be prepared to learn and develop his abilities in this direction. If constant learning, skills and attitude development is expected of students it must *a fortiori* be expected of staff.[37] However, merely insisting on self-development (or even guided development) of staff, is probably doomed to failure unless perceivable and tangible reward can be offered. Although increased motivation and intellectual stimulation are clear rewards of the project strategy and challenge, they must be set against the potential and indeed often real disadvantage that increased commitment to undergraduate teaching detracts from research time and is thus antagonistic towards promotion prospects. It is thus almost mandatory that *all* staff in a department should assume comparable responsibility for project studies unless the privileged few (or many) who escape are enabled to advance their personal careers more advantageously. It is not without significance, therefore, that a substantial fraction of the published literature on project studies is authored by heads of department, professors and other senior staff who are clearly able to generate a philosophy and ensure a whole-hearted and equitable commitment to these activities in their departments. Personal commitment is thus supported and stimulated by an institutional commitment.[12]

An alternative to demanding total staff commitment is to recognise more clearly in the promotion system, or via financial incentives and differentials as advocated in the Finniston report,[4] the activities of staff who devote their major energies and interests in this direction.

5.2 Student problems

5.2.1 Priorities: Our own experiences would strongly suggest that one of the more important problems for any student undertaking any type of open-ended, self-motivating work is that of maintaining an adequate balance between the project work and the rest of his studies. Although there is documented evidence[18, 37, 38, 40, 41, 43] that projects probably give the student more satisfaction than any other aspect of his engineering course, it may be considered unwise for students to spend too much time in this activity, fascinating though it may turn out to be, at the expense of other required and examinable aspects of the course. This question of maintaining the correct distribution of effort is impossible if the student does not know to what extent the laboratory work contributes to his final degree classification. External pressures also tend to increase the importance of the project particularly in final year, since we have obtained substantial feedback from our students that potential employers seem to be more interested in this type of activity than any other. The idea that it is easier to sell oneself with a good account of one's final year project seems to becoming entrenched in undergraduate folklore. A consequence of these pressures is that, in the authors' university at least, the overall standard of the project work has improved considerably over the past few years, and called into question the distribution of marks between project and results in the final examination, a factor which will be examined again later. In addition, the drive to improve standards in projects is satisfying in that it has been generated largely within the student body, with the staff often acting as willing accomplices. This situation can leave the staff in the difficult position of not wishing to dampen down the enthusiasm of the majority of students, but advocating a balance of study endeavours.

5.2.2 Aptitude: We have been able to identify two groups of students who give special cause for concern. The first group is of students who seem to have no conception about even how to begin solving practical problems at any stage of their undergraduate career, although quite competent in passing examination. The second extreme group is of students who consistently produce sound, imaginative solutions to practical problems but who perform badly in examinations. We have often observed that the second group attract more sympathetic consideration by the teaching staff during final assessment than do the former group. We would acknowledge that different priorities may well operate in different departments and institutions who may hold differing views of the desirable qualities of graduate electrical engineers.

For the former group of students who reveal limited aptitude for open-ended final-year project work, it might be argued that if such potential ineptitude were discovered earlier in their courses they would receive a more personally satisfying education by continuing standard laboratory experiments, or that remedial action may be undertaken at an earlier stage to improve preparation for the major project activity.

In the context of perceivable or developing competence to undertake open-ended laboratory studies it seems at least plausible to suggest that a student's personality and background may play a greater part in his performance in such studies than in any other aspect of his university education. This type of activity requires qualities of foresight and intuition, and an ability to synthesise ideas from many sources in order to achieve satisfactory results. In some cases the courage to stand ground and resist contrary opinions, particularly those of a supervisor with considerably more experience, and the tenacity and perseverance to arrive at a satisfactory outcome are also demanded. Many of these qualities are not 'taught', nor is the environment often provided to encourage their development in earlier years of many engineering courses. It is hardly surprising therefore that some students produce results widely disparate from their formal examination performances. We have no doubt that students in this group often experience depression at their lack of perceived progress and may demand inordinate assistance from supervisors which can condition their assessments. Thus since it is possible that the instruction often received in the first two years of an engineering degree course has little or no influence on the preparation of students for project work, the chief source of inadequacy may consequently lie in the pre-university education experience of the student. It may be that students who perform well receive a significant portion of this aspect of their education and training from their general cultural background in a way which we do not understand and do not recognise.

The situation with respect to students who perform well in open-ended tasks and poorly in examinations is probably less of a problem since there is less need to ponder on methods of improving the project performance but only of giving due weighting to the different types of skill shown in reaching degree status. This then shifts the problem from being student centred to being centred on the body of established engineering opinion and particularly to those of us engaged in educating future engineers. The question of an equitable distribution of credit for projects in association with other modes of assessment has been taken up by several authors,[18, 41, 44] but there seems to be no generally agreed pattern as to what fraction of credit should be assigned to projects. Thus it seems that, in electrical engineering at least, projects may carry from 10% to 50% of final year marks, or in other cases no formal credit at all, but merely an extra-qualitative factor to be considered when arriving at a classification. Our own and other[44] experiences suggest qualitatively that substantial changes in the quantitative weightings accorded to projects generally have little modifying influence on the best and worst students but will considerably change the aggregate scoring and rank ordering (and thus concomitantly the individual classifications) for students with mid-range marks (classes 2.2. and 2.1).

This region will become more sensitive in the future with the changed requirements for Institution membership.

An allied, and possibly more intractable concern, is the situation of the small fraction of students on an ordinary degree course who turn out to have a talent for the solution of practical problems which is far and away greater than their examination performance would indicate. In our experiences such students are few in number, but occur from time to time, and show a degree of understanding and ability in the problem solving area in excess of that of many honours degree students. Because of the structure of the degree courses at most universities such students have no opportunity to display this talent at the time when the composition of the ordinary degree class is determined, and it seems inappropriate that the aspects of a students performance which could account for up to a half of the degree assessment should not enter into the selection for the honours of ordinary degree streams.

The idea that the general cultural background provides an unknown and unspecified component of a student's ability in open ended laboratory work is reinforced by the frequently but privately voiced opinion of engineering lecturers that some groups of overseas students are poor at project work. As a statement this is almost meaningless, but from comparisons attempted in this university over several years we can say that whereas 75% of European students performed better in project work than in examinations in final year, using the same criteria of measurement, only 40% of non-Europeans (mostly Asians) performed better in project work. It should be noted that the 'Europeans' included groups of Norwegian and Greek students not educated in Great Britain but indistinguishable from the British students in their relative performance and the non-European group contained several students educated in Great Britain. In consequence any differences in performance could not be easily related to pre-university education nor, probably, to differences in intelligence. It would appear that the unspecified and untaught aspects of European culture contribute in a certain way to aptitude for problem solving, and it presents a considerable challenge to identify these aspects and possibly modify our education structures to take them into account.

This cultural bias is not just confined to final-year students, but has also been noted in laboratory work at first-year level. The work of Freeman *et al.*[45] also undertaken at this university, shows that the response to the type of laboratory work offered at first-year level is probably culturally linked. British students were shown to make faster and better progress with a more open-ended and less structured approach to laboratories whereas the opposite was true for overseas students.

Similar culturally related factors may also be influential among European students since our own culture is not static. Personal observations suggest that many British born and educated students do not seem to have the same background in scientific and technical awareness at entry to the university as did students ten or fifteen years ago. Our first-year laboratory demonstrators seem to spend a considerable amount of time and effort trying to convince new entrants that work in engineering is something to be enjoyed, and not just a mindless task to be completed with the minimum of effort. Many students also seem to spend much of their time outside their studies in a purely passive role as spectators rather than participators. It may well be that students' attitudes are merely reflections of those of society as a whole, and that the sense of wonder engendered by inquiry for its own sake, and satisfaction in active participation and accomplishment, is vitiated by confinement to a narrow path of classroom study or by changing patterns of childhood activities (i.e. more passive pastimes rather than making things work). Such activities or at least the attendant attitudes may be detrimental to the pursuit of an open-ended task.

5.3 *Prospects*

5.3.1. Justification for projects: All the evidence presented in earlier paragraphs illustrates the increasing attention paid to project activities or efforts devoted to assessment of such study modes. The motives for introducing projects can be fairly associated *inter alia* with the increased motivation of the majority of students and staff with more self-directed learning, the enhanced opportunities to develop higher-order educational skills and qualities and attempts to simulate the active and experiential learning of the employment situation. Some of these motives have sound bases in learning theory but others are founded on the personal beliefs and experiences of educators. There is, however, no incontrovertible proof that the project method is of lasting benificence to all, or even some, students. We know of no longitudinal study of the effects of specific electrical engineering undergraduate curriculum styles and strategies on the short- or long-term performance of graduates in career situations. It cannot be indisputably claimed that study modes based more upon active learning will give more self-satisfaction to graduates or benefit to their employers in two, five or twenty years from graduation. Until such investigations are carried out, and they are difficult, since experimental rigour requires simultaneous studies of control groups with, for example, highly formalised modes of instruction but *no* other differences in parameters such as syllabus, student intelligence, long term goals etc., the expansion of project activity must be regarded as an act of faith. Other disciplines such as medicine and dentistry, in which there is a similar element of professional training although not necessarily employing widespread project strategies, do demand experiential study before professional practice is condoned. It may be argued, therefore, that, in engineering, experimental study, i.e. the project or real-life problem solving activity, should occupy an important component of the curriculum, and that it can be fairly predicted that such activities will be beneficial.

5.3.2 Assessment: Accepting the premise that such activities are valuable in their own right, it must be clearly recognised, as has been illustrated already, that active, self-directed, experimental learning makes greatly increased demands upon students and staff alike. It has been argued that if the pressure for assessment of project studies were removed, projects would flourish as an admirable educational experience, would encourage improved staff-student co-learning, and the disquiet engendered by the imprecision innate in any current project evaluation scheme would be removed. The counter argument that the assessment of project work should be attempted in an endeavour to gain additional information about a student performance is, in our view, more persuasive, but the attempt should be approached with extreme caution.

Open-ended problem solving certainly allows more scope than many of the more traditional styles of laboratory work for the development of not only skills but also qualities and attitudes. All learning and judgments are idiosyncratic, although in formal examinations the levels of idiosyncracy may be minimised by well defined and previously accepted solution strategies, by the use of panels of examiners and external examiners, and by statistical treatment of marks. When human qualities such as initiative, intuition, creativity, perseverance etc. are under scruitiny, however, it becomes much more difficult to postulate any norms of behaviour, since each supervisor will be as different in his expectations and judgments as are the students he supervises, and may often examine. If any level of credible pseudo-quantification is to be attached to such abilities of students, value judgments must be made on both the written reports from students and on the 'on the job' performance of the students, and groups of supervisors must be able to agree among themselves and, if necessary, externally with other groups on their own judgment criteria.

Even then it must be doubtful if qualities can be 'scored' quantitatively, as are formal examinations, although some rank ordering within a yearly cohort of students should be possible. If the assessor panels can work for a number of years and build up a continuing and transferable 'wisdom' to successors, a continuing year-by-year comparable ranking should also be possible. In this way desirable qualities may become defined by consensus and custom.

It is still highly questionable, however, if such quality assessment should constitute a quantifiable and numerical component of a final degree classification since different university departments are every bit as idiosyncratic as the individual staff and students within them. It would be probably be far superior if a degree classification were awarded on the basis of traditionally established examination together with an assessment of agreeed semi-quantifiable skills (e.g. the ability to evaluate errors, to handle equipment safely and competently) derived from laboratory studies and a statement or profile of those other admirable engineering qualities and attitudes which a Department believes it can truthfully and fairly state as its consensus view about a student. It can be argued that this strategy may already be possible from the references provided by university staff to employers. The proposed consensus view, however, can potentially remove some injustice whilst providing a broader view of measured student performance and predicted student capabilities. In this way a positive benefit to potential employers would ensue.

As pointed out earlier, such commitment to more strategically planned laboratory studies and carefully designed and evaluated assessment methods will present significantly increased demands upon staff resources, and in order that justice be done (and seen to be done) departments and all individuals within departments must participate willingly, enthusiastically and equitably. To be provocative, if such commitments were already devoted to more traditional learning methods great improvements in student learning may result anyway, and the arguments of this paper could be at risk. However, we see good reason for putting faith in an increasing level of better designed laboratory studies, i.e. in a method which embodies and gives practice in the values we hold, and are supported in this view by the considerations of the Finniston Committee.[4] Two paragraphs in that report are particularly opposite. These are Para. 4.49 which states in part (referring to B. Eng. courses):

'The whole B. Eng. programme should be firmly set in a context of purpose, with the primary emphasis on the synthesis of basic subjects and on developing students design and problem solving capabilities. This should be taught through progressive project work incorporating such elements of engineering science and mathematics as are required to provide a foundation for instruction in practice and applications. Design practice should be a prominent and unifying theme of the course (as is achieved in continental engineering formation) and not an isolated expertise through the strong emphasis, particularly in the later years of their courses, on a specialised project which focuses the more general engineering theory and applications previously taught. .....'

and in Para. 4.55 which states in part (referring to M. Eng. courses):

'The M. Eng. degree course would have the following characteristics:

(a) Selection would usually follow a diagnostic first year on a common course with B. Eng. entrants, designed *inter alia* to identify those with the potential to undertake the M. Eng. programme and to succeed as engineers at the R. Eng. (Dip) level. There is evidence that ultimate attainment, at least in academic terms, is better predicted on the basis of first-year performance than it is on entry qualifications and for this reason selection upon entry to higher education would be premature; selection after the first year should take account of more than academic attainment, considering also candidates' personal aptitudes.'

In Para. 4.62 the committee also comments on the advisability of providing bridging points between the different types of engineering courses advocated, and in Para. 4.59 points out that it has not sought to prescribe outline curricula of syllabi for B. Eng. or M. Eng. courses.

These are fine sentiments and concur with the general ethos developed in the present paper. The problems of course arise in translating these philosophies into practice, an exercise from which the Finniston committee has carefully refrained. Thus, what are the modes of assessment which will give early and continuing information on the 'personal aptitudes' other than academic attainment of students and which will determine not only their educational patterns but also their career aspirations and patterns? These are the problems which have been, to some extent, addressed in this paper and it should be abundantly clear that although progress is being made towards better defined laboratory learning strategies and more meaningful assessment modes, we are far from generally acceptable solutions. Much still needs to be done in this area, not least of which is the consideration of the forms of acceptable predictive entrance requirements which will better measure student potential for problem solving, than do current A-levels and similar qualifications. Equally important is the recognition of those students who may after admission exhibit early signs of being unable to adequately cope with the problem-solving demands of the new course philosophies and structures. Will such students still qualify as some form of engineer, or will they be advised to seek alternative education, or will special endeavours be made to give remedial help? This brings into focus the real problem that all learning is idiosyncratic, and skill and attitude development are not time and experience independent. Should, therefore, degree award be fixed at some arbitrary period, 3, 4, *n* years after entry, and if not, what are more appropriate educational strategies and diagnostic and assessment procedures which will determine a student's arrival at an agreed definition of a graduate engineer. Some of the major problems for the future thus appear to be:

(i) What is the consensus agreement on the desirable skills, attitudes and experience of graduate engineers of different types?

(ii) How best can complex learning strategies be developed to give opportunities for all potential engineering graduates to excel in their own individual ways?

(iii) What are the optimal assessment modes for evaluating the historic, current and predictive attainments of students?

We are thus almost back where we started in attempting to define a well posed engineering problem. The optimal solution and strategies to be employed will require major thought and action but the prospects should be exciting for the staff member committed to his role as teacher, mentor and co-learner. The Finniston committee[4] advocate that closer education-industry contacts and a course-licensing authority may enhance the introduction of problem-solving activities and curriculum development generally, but our own[3] and other[4,6] observations of the desire for and effectiveness of such interactions on both sides, are far from optimistic. Perhaps the injection of money and resources as well as the growth of increasing awareness of our problems will prove persuasive.

6 Summary

The important role accorded to laboratory work in undergraduate engineering curricula reflects the general acceptance of the educational value of experiential learning, and the present paper contains an objective analysis of current understanding of the aims and objectives of laboratory work, the development of detailed laboratory schemes, and the assessment of the various types of study strategy.

The review of the aims and objectives for the different types of laboratory activity, classified into controlled assignments, experimental investigations and projects, clearly illustrates the wide range of educational skills involved, and there is, if the published literature gives an accurate reflection of the trends in current practice, a steady move towards a more heuristic approach to laboratory teaching. The growing use of experimental investigation and projects in the second and in some cases first year of undergraduate courses has presented increasing difficulties in assessment, and the validity and reliability of current schemes is discussed in detail.

Although the major part of this review is concerned with an objective analysis of current literature and practice, it would not be complete without some, unavoidably subjective, discussion of the problems and prospects associated with the future development of undergraduate engineering courses. The major conclusion, that there is a clear need for better-designed laboratory studies and more meaningful modes of assessment are particularly important in view of the recent recommendations made by the Finniston Committee concerning the conduct of undergraduate teaching strategies.

7 References

1 HAMMOND, P.: 'The case for the teaching laboratory', *Electron. and Power*, 1971, **17**, pp. 77–79

2 STURLEY, K.S.: 'Is laboratory work really necessary?', *Int. J. Electr. Eng. Educ.*, 1967, **5**, pp. 63–66

3 HALDER, A.K., JORDAN, T.A., and CARTER, G.: 1977. 'A survey of industrial attitudes to undergraduate electrical engineering courses', *ibid.*, 1977, **14**, pp. 11–16

4 FINNISTON SIR M.: 'Engineering our future'. Report of the Committee of inquiry into the engineering profession, Cmnd. 7794 (HMSO, London, 1960)

5 LEE, L.S., and CARTER, G.: 'A sample survey of departments of electrical engineering to determine recent significant changes in laboratory work pattern at first-year level', *Int. J. Electr. Eng. Educ.*, 1973, **10**, pp. 131–135

6 ADDERLEY, K.J.: 'The aims, objectives and criteria of undergraduate project work incorporating CA1', *ibid.*, 1975, **12**, pp. 165–176

7 Northern Universities Joint Matriculation Board. 'Engineering science (Advanced) syllabus and notes for the guidance of schools' ES/N2 1972, and Coursework Assessment ES/CWA/1 1975, and ES/CWA2 1977

8 BEYNON, J.D.E., and BAILEY, A.G.: 'An appraisal of undergraduate practical work with particular reference to the first year of an electronics course', *Int. J. Electr. Eng. Educ.*, 1971, **9**, pp. 49–53

9 PRAMANIK, A., and DRING, D.: 'An evolving teaching laboratory for first year students of electrical and electronic engineering', *ibid.*, 1977, **14**, pp. 17–25

10 ERNST, E.W., and KOPPLIN, J.O.: 'Modern methods and objectives of laboratory instruction', *IRE Trans.*, 1962, **E5**, (3), pp. 168–172

11 BEAKLEY, G.C., and PRICE, T.W.: 'Introduction to engineering', *IEEE. Trans.*, 1966, **E9**, (4), pp. 187–192

12 AHARONI, H., and COHEN, A.: 'Manpower for engineering laboratory courses', *Int. J. Electr. Eng. Educ.*, 1977, **14**, pp. 293–300

13 KENT, G., and CARD, W.H.: 'An experiment in laboratory education', *IRE Trans.*, 1961, **E4**, pp. 62–66

14 BRADSHAW, E.: 'Laboratory work in electrical engineering courses'. *Proc. IEE*, 1955, **102**, pp. 23–24

15 BAILEY, A.G., BEYNON, J.D.E., and SIMS, G.C.: 'Examinations and continuous assessment' *Int. J. Electr. Eng. Educ.*, 1975, **12**, pp. 13–24

16 BALDWIN, C.T.: 'Industry, universities and engineers'. *Proc. IEE*, 1970, **117**, (3), pp. 661–663

17 PALMER, C.J., and HADEN, C.R.: 'The changing role of the laboratory in the electrical engineering curriculum', *IEEE Trans.*, 1961, **E9**, (21), pp. 187–192

18 ADDERLEY, K., ASHWIN, C., BRADBURY, P., FREEMAN, J., GOODLAND, S., GREENE, J., JENKINS, RAL, J., and UREN, O.: 'Project methods in higher education'. Society for Research into Higher Education Ltd., London, 1975

19 SMART, G.C.: 'A students view of project based higher education'. Proceedings of the international seminar, 'Project-orientation in higher education in science', University of Bremen, 1976. University teaching methods unit, University of London, 1977, pp. 61–63

20 LEE, L.S.: 'Research report - Projects or controlled assignments in the laboratory?', *Further Education*, 1970, **1**, (3), pp. 117–118

21 MARSHALL, A. (Ed.): 'School technology in action', (English Universities Press, London 1974) pp. 180–186
22 SIMS, G.D.: 'Electronics and education – here and elsewhere', *Proc. IEE,* 1967, **114,** (2), pp. 205–206
23 FRANKSEN, O.I.: 'Closed and open-ended design projects in the education of Engineers', *IEEE Trans.,* 1965, **PAS-3,** pp. 228–231
24 PRINGLE, E.J.: 'The art of inventing', *Trans. Amer. Inst. Elect. Engrs.* 1906, **25,** pp. 519–547
25 HAMILTON, F.L.: 'Technical writing and the engineer', *Proc. IEE,* 1965, **112,** (2), p. 310
26 HEYWOOD, J., and KELLY, D.T.: 'A study of engineering science among schools in England and Wales'. Proceedings of the IEEE/ASEE Conference, 'The frontiers of education', Purdue University, 1973
27 THOMAS, A.F.: 'A pilot scheme in controlled laboratory assignments', *Int. J. Electr. Eng. Educ.,* 1976, **14,** pp. 101–103
28 BETHAND, L.A.: 'A first year electrical engineering laboratory to satisfy all requirements', *ibid.,* 1977, **14,** pp. 301–306
29 PUGH, A.: 'A new introductory electronics laboratory at Nottingham', *ibid.,* 1974, **11,** pp. 293–299
30 BROWN, R.: 'Second-year think tank industrial projects', Proceedings of the Conference on the teaching of electronic engineering in degree courses, University of Hull, 1973, pp. 13/1–13/8
31 CUTHBERT, L.G., and COWARD, P.R.: 'Undergraduate electronics courses with emphasis on design', Proceedings of the conference on, 'The teaching of electronic engineering in degree courses', University of Hull, 1973, pp. 26/1–26/9
32 GOODLAD, S.: 'Socio-technical projects in engineering education', The general education in engineering (GEE) project, University of Sterling, Stirling, 1977, see also Proceedings of seminar on 'Project orientation in higher education in science', University of Bremen, 1976, (Eds. M. Cornwall, F. Schmidthalz, and D. Jaques)
33 McPHUN, M.K.: 'A 'Programmed laboratory' course in electronics for first year students'. Proceedings of the conference on 'The teaching of electronic engineering in degree courses', University of Hull, 1973, pp. 27/1–27/10
34 WILKINS, B.R.: 'The design of a first-year laboratory course'. Proceedings of the conference on 'The teaching of electronic engineering in degree courses', University of Hull, 1973, pp. 10/1–10/11
35 DEARDEN, G.J.: 'Laboratory assessment' some generalised findings'. Proceedings of the conference on 'The teaching of electronic engineering in degree courses', Universtiy of Hull, 1976, pp. 12/1–12/9
36 HARDING, A.G.: 'The project: its place as a learning situation', *Br. J. Ed. Techn.,* 1973, **4,** pp. 216–229
37 HARDING, A.G.., and SAYER, S.: 'Prelude to 'project work', Proceedings of the seminar on 'Project orientation in higher education', University of Bremen, 1976, (Eds. Cornwall, M., and Schmidthalz, F.) 1977, pp. 116–122
38 BLACK, J.' 'Allocation and assessment of project work in the final year of the engineering degree course in the University of Bath', *Assessment in Higher Education,* 1975, **1,** pp. 35–53
39 HARRIS, N.D.C., and DOWDESWELL, W.H.: 'Assessment of projects in university science', *ibid.,* 1979, **4,** (2) pp. 94–118
40 ALLANSON, J.T.: 'Private communication
41 BENSON, F.A.: 'Assessment of student project', Proceedings of the conference on 'The teaching of electronic engineering in degree courses', University of Hull, 1976, pp. 10/1–10/7
42 BOWRON, P.: 'Final year projects: towards more objective assessment'. Proceedings of the conference on 'The teaching electronic engineering in degree courses', University of Hull, 1976, pp. 11/1–11/9
43 CARTER, G., and LEE, L.S.: 'A study of attitudes to first-year undergraduate electrical engineering laboratory work at the University of Salford', *Int. J. Electr. Eng. Educ.,* 1975, **12,** pp. 278–289
44 CLEMSON, D.: 'Project assessment – a sample analysis', *Assessment in Higher Education,* 1977, **2,** pp. 196–221
45 FREEMAN, J., CARTER, G., and JORDAN, T.A.: 'Cognitive styles, personality factors, problem solving skills and teaching approach in electrical engineering', *Assessment in Higher Education,* 1978, **3,** (2) pp. 86–121
46 GOSLING, W.: 'An end to half-hearted engineering education', Proceedings of the conference on 'The teaching of electronic engineering in degree courses', University of Hull, 1976, pp. 9/1–9/9

Part VII
Examinations and Grading

Scope and Purpose

EVERY classroom teacher must construct quizzes and examinations to measure and evaluate the learning of the course material by the learners, and usually must assign grades to the students based on that evaluation. These two tasks, examination and grading, are the focus of this part.

The distinctions between terms like educational evaluation, measurement, and testing are subtle for a novice, and some definitions will be helpful in following the literature on this subject. *Evaluation* refers to a broad appraisal of an individual's present level of proficiency in a particular area. Evaluation is called *summative evaluation* when it is carried out at the end of a course of study for the purpose of determining and recording the student's level of proficiency. It is called *formative evaluation* when it is carried out during the course of study for the purpose of providing feedback to the student or the teacher. Evaluation can take many forms, such as ratings, written or verbal comments, reference letters, and remarks written on examination papers, and it can be based on a number of factors, including the evaluator's opinion and observation, and the individual's performance on some tasks like a quiz or seminar participation. One of the factors (and for classroom teachers, typically the only factor) on which the evaluation is based is educational measurement. *Educational measurement* is a quantitative assessment of some ability of an individual, carried out with the help of an *instrument* such as an intelligence test, a classroom test, an oral examination, or an assigned laboratory task. Of most interest to teachers is one variety of these instruments called achievement tests. An *achievement test* is any device or method for measuring an individual's present competence in some particular field. For classroom teachers, the competence is usually a cognitive or intellectual skill acquired as a result of instruction in the subject of the course. *Grading* is a term which describes, in its most direct form, the process of scoring, i.e., assigning scores, marks, or some indicator of the quality of work, such as an alphabet, to the work on an achievement test, based on some scale or criteria. The scale used to relate the work performance to grades may be of two kinds: *norm-referenced*, which compares an individual's performance to that of other individuals in some specified group; and *criterion-referenced*, which compares the individual's performance to some absolute standard, stated in terms of the work alone. When these scores are the only form of appraisal reported to students and others, grading effectively becomes identical with evaluation.

The Issues

The following are some of the issues that have been raised in the professional literature on educational measurement, and that a beginning teacher may face in carrying out the evaluation and grading:

- What is the purpose of testing and evaluation?
- What constitutes a good test?
- What are the limitations of testing?
- Does the test evaluate the learning that was intended (is it "valid"); are the test scores non-random, meaningful measures of learning (are they "reliable")?
- What type of questions are suitable for measuring each type of learning?
- How can the test be scored with uniformity, fairness, and with minimum subjective bias?
- What is the appropriate level of difficulty of test questions and of tests?
- If and how should the preparation of the tests for in-class and take-home tests (or closed-book and open-book tests) be different?
- How should the test questions be stated to make them less vague/more precise?
- What information about the class or the test itself can be gathered from the test scores?
- Do grades and grading have any legitimate functions, and what are those functions?
- How should the course grades be assigned in a fair manner?
- Should the grades depend on the level of achievement, or on the amount of progress?
- Does the variability among instructors and courses, concerning the meaning attached to a grade, make the process of grading suspect?

The Reprinted Papers

The following five papers are reprinted in this part:

1. "Marks and Marking Systems," R. L. Ebel, *IEEE Trans. Education,* vol. E-17, no. 2, pp. 76–92, May 1974.
2. "The Test: Uses, Construction and Evaluation," M. D. Svinicki, *Engineering Education,* vol. 66, no. 5, pp. 408–411, Feb. 1976.
3. "A Test is Not Necessarily a Test," M. D. Svinicki, *Engineering Education,* vol. 68, no. 5, pp. 410–413, Feb. 1978.
4. "Suggestions on the Construction of Multiple Choice Tests," H. T. Hudson and C. K. Hudson, *American Journal of Physics,* vol. 49, no. 9, pp. 838–841, Sept. 1981.
5. "Design of Examinations and Interpretation of Grades," M. W. P. Strandberg, *American Journal of Physics,* vol. 26, no. 8, pp. 555–558, Nov. 1958.

The first of the reprinted papers is authored by Ebel, a noted educational psychologist who is also the author of a highly-regarded textbook [2] on this subject. This paper is valuable, not simply because it is authoritative, but because it is an unusually clear and rational statement of the purpose and meaning of grades, and the merits of the various systems of grading. It should help dispel many of the prevalent myths and misunderstandings, and clarify the thinking on this subject.

Given an understanding of the grading process, the next important skill for an instructor to learn is that of constructing good tests. The next several papers are devoted to this skill.

The pair of papers by Svinicki is an elementary, nontechnical introduction to the construction and characteristics of tests. The first paper describes (i) some of the uses to which a test could be put, (ii) a systematic procedure for ensuring that the test broadly samples the subject matter as well as the cognitive skills, and (iii) a very simple form of item analysis for determining the adequacy of individual test questions. The second paper defines two of the basic measures of the quality of a test: its validity and its reliability, and then describes simplified procedures for determining these two parameters for a given test that has already been administered. All of these subjects are discussed in much detail, and at a more sophisticated level, in many textbooks, such as [2] and [3]. The primary virtue of the papers reprinted here is their brevity and simplicity, which should increase the likelihood that these ideas are actually used.

The use of multiple-choice questions in engineering is much less prevalent than in other fields. The reason is in part that many engineering educators feel that such questions are incompatible with the goals of engineering education, and in part that most engineering educators find themselves unable to construct multiple-choice questions that can test anything but the recall of memorized facts. Nevertheless, such tests are indeed in use; the Graduate Record Examination in engineering is an example of such a test. The large engineering enrollments are also a possible factor that may promote wider use of this form of tests. The reprinted paper by Hudson and Hudson is a very brief account of how to construct meaningful multiple-choice questions.

Teachers, and especially beginning teachers, often discover (alternatively, the item analysis mentioned above may reveal) that the students find their test problems too difficult, even though the subject matter on which the test problems are based was discussed thoroughly in the class, every student seemed to understand it, and nobody asked any questions. And after the teacher solves the test problems in the classroom, the students agree that they knew all the steps and therefore should have been able to solve the problems. In the last paper reprinted in this part, Strandberg discusses this mystery, and presents two yardsticks by which the level of difficulty of a problem may be estimated by the instructor; he calls these the ''concept of logical steps,'' and the ''concept of avenues of approach.''

The Reading Resources

The literature on the subject of educational measurement, testing, and grading is very extensive, and by one estimate, includes several thousand papers, a couple of hundred books, and several professional research journals. The first item below is an annotated bibliography on achievement tests, which, although dated, and centered on physics teaching, can serve as a good starting point for further exploration of the field, and for locating some of the basic literature in the field.

[1] H. Kruglak, ''Resource Letter AT-1 on Achievement Testing,'' *American Journal of Physics,* vol. 33, no. 4, pp. 255–260, Apr. 1965. This annotated bibliography is a part of a booklet of reprinted papers, entitled ''Achievement Testing in Physics,'' edited by Kruglak, and published by the American Institute of Physics, New York.

[2] R. L. Ebel, *Essentials of Educational Measurement.* Prentice-Hall, Englewood Cliffs, NJ, 1972. A standard textbook in this area, particularly valuable from the viewpoint of classroom teachers, with much practical, useful information on the construction of tests.

[3] R. L. Thorndike, Ed., *Educational Measurement.* American Council on Education, Washington, DC, 1971. A standard reference work for the professionals in the field, with state-of-the-art discussions and technical fine points.

[4] A. J. Lien, *Measurement and Evaluation of Learning.* W. C. Brown Co., Dubuque, IA, 1967. A highly condensed, factual, practical guide to the field, suitable for beginners.

[5] *Engineering Education,* vol. 68, no. 5, Feb. 1978. Special Issue on Evaluation of Learning. Articles on criterion-referencing, evaluation procedures, contract grading, assessment in affective domain and in competency-based education.

[6] B. L. Coulter, ''Improving Tests in Basic Physics,'' *American Jour. Physics,* vol. 43, no. 6, pp. 499–501, June 1975. Practical, non-technical suggestions for (i) improving the reliability and validity of test questions, (ii) selecting the appropriate type of test and appropriate level of difficulty, and (iii) accumulating test items.

[7] B. J. Ley, ''Grades, Quizzes, Motivation, and Computers,'' *IEEE Trans. Education,* vol. E-13, no. 3, pp. 176–180, Sept. 1970. A computer program for predicting the course grade for a student at any time during a course is proposed as a means of providing periodical progress review, and for motivating the students.

[8] H. Lin, ''The Hidden Curriculum of the Introductory Physics Classroom,'' *Engineering Education*, vol. 70, no. 3, pp. 289–294, Dec. 1979. Proposes the hypothesis that, given the pressures of time and grades, the students learn what is explicitly graded; reports students' comments and strategies to support this hypothesis.

[9] B. S. Bloom, J. T. Hastings, and G. F. Madaus, *Handbook on Formative and Summative Evaluation of Student Learning.* McGraw-Hill, New York, 1971. A technical reference work for the professional, but also a mine of authoritative information on the subject that can be useful to the classroom teacher.

Marks and Marking Systems

ROBERT L. EBEL

Invited Paper

Abstract—**Marking systems are frequent subjects of educational controversy because the process is difficult, because different educational philosophies call for different marking systems, and because the task is sometimes disagreeable. Claims that marks diminish the effectiveness of the educational system do not seem to be generally valid, and properly assigned marks do measure the degree of attainment of the basic objectives of education. Measurements and reports of achievement are essential in education and no better means than marks seems likely to appear.**

In this paper the basic purposes of marks and their effects on student attitude and motivation are discussed. Two basic systems of marking, relative and absolute, are compared, and means of improving marking through institutional standardization are discussed. Pass-fail grading and other alternatives to conventional marking systems are evaluated. Weighting of various measures of achievement in assigning marks is discussed, and detailed computational procedures for assigning marks are developed.

THE PROBLEM OF MARKING

THE problem of marking student achievement has been persistently troublesome at all levels of education. Hardly a month goes by without the appearance in some popular magazine or professional journal of an article criticizing current practices or suggesting some new approach. Progressive schools and colleges are constantly experimenting with new systems of marking, or sometimes of not marking. And still the problem seems to remain.

One of the reasons why it remains is that marking is a complex and difficult problem. From some points of view it is even more complex and difficult than the problem of building a good test and using it properly. In an early classic of educational measurement, Thorndike explained some of the reasons why educational achievement often is difficult to measure.

> Measurements which involve human capacities and acts are subject to special difficulties due chiefly to:
> 1. The absence or imperfection of units in which to measure.
> 2. The lack of constancy in the facts to be measured, and
> 3. The extreme complexity of the measurements to be made [1].

Marks are, of course, measurements of educational achievement.

A second reason why problems of marking are difficult to solve permanently is because marking systems tend to become issues in educational controversies. The rise of progressive education in the third and fourth decades of this century, with its emphasis on the uniqueness of the individual, the wholeness of his mental life, freedom and democracy in the classroom, and the child's need for loving reassurance, led to criticisms of the academic narrowness, the competitive pressures, and the common standards of achievement for all pupils implicit in many marking systems. In the sixth and seventh decades renewed emphasis on what is called "basic education" and on the pursuit of academic excellence has been accompanied by pleas for more formal evaluations of achievement and more rigorous standards of attainment.

A third reason why marking systems present perennial problems is that they require teachers, whose natural instincts incline them to be helpful guides and counsellors,

Manuscript received January 7, 1974. This paper is adapted from Chapter 12 of the author's book, *Essentials of Educational Measurement* (c) 1972, and appears with the permission of Prentice-Hall, Inc., Englewood Cliffs, N. J.

The author is with the Department of Counseling, Personnel Services and Educational Psychology, Michigan State University, East Lansing, Mich. 48823.

Reprinted from *IEEE Trans. Educ.*, vol. E-17, no. 2, pp. 76–92, May 1974.

to stand in judgment over some of their fellow men. This is not the role of friendship and may carry somewhat antisocial overtones.

For all these reasons, no system of marking is likely to be found that will make the process of marking easy and painless and generally satisfactory. This is not to say that present marking practices are beyond improvement. It is to say that no new marking system, however cleverly devised and conscientiously followed, is likely to solve the basic problems of marking. The real need is not for some new system. Good systems already exist. Reviewing articles on marking problems and practices that were written a half century ago, one is struck by their pertinence to the present day. The same problems that were troublesome then are still troublesome. Some of the same remedies that were being proposed then are still being proposed.

Marking procedures, Hadley has pointed out, are about as good or as weak as the teachers who apply them [2]. Few teachers mark as well as they could, or should. The more confident a teacher is that he is doing a good job of marking, the less likely he is to be aware of the difficulties of marking, the fallibility of his judgments, and the personal biases he may be reflecting in his marks. Most teachers' marks, says Hadley, are partly fact and partly fancy. The beginning of wisdom in marking is to recognize these shortcomings. The cultivation of wisdom is to work to improve them. For measurements and reports of achievement are essential in education, and no better alternative to marks seems likely to appear.

THE NEED FOR MARKS

Most instructors, at all levels of education, seem to agree that marks are necessary. Occasionally a voice is raised to cry that marks are educationally vicious, that they are relics of the dark ages of education. However, as Madsen has pointed out, the claim that abolition of marks would lead to better achievement is, by its very nature, impossible to demonstrate [3]. If we forego measurements of relative achievement, what basis remains for demonstrating that one set of circumstances produces better educational results than another? Comparison of achievement between persons or between methods of teaching is inevitable. It is the misuse of marks, not their use, that is in need of censure.

The uses made of marks are numerous and crucial. They are used to report a student's educational status to him, to his parents, to his future teachers, and to his prospective employers. They provide a basis for important decisions concerning his educational plans and his occupational career. Education is expensive. To make the best possible use of educational facilities and student talent, it is essential that each student's educational progress be watched carefully and reported as accurately as possible. Reports of course marks serve somewhat the same function in education that financial statements serve in business. In either case, if the reports are inaccurate or unavailable, the venture may become inefficient.

Marks also provide an important means for stimulating, directing, and rewarding the educational efforts of students. This use of marks has been attacked on the ground that it provides extrinsic, artificial, and hence undesirable stimuli and rewards. Indeed, marks are extrinsic, but so are most other tangible rewards of effort and achievement. Most workers, including professional workers, are grateful for the intrinsic rewards that sometimes accompany their efforts. But most of them are even more grateful that these are not the only rewards. Few organized, efficient human enterprises can be conducted successfully on the basis of intrinsic rewards alone.

To serve effectively the purpose of stimulating, directing, and rewarding student efforts to learn marks must be valid. The highest marks must go to those students who have achieved to the highest degree the objectives of instruction in a course. Marks must be based on sufficient evidence. They must report the degree of achievement as precisely as possible under the circumstances. If marks are assigned on the basis of trivial, incidental, or irrelevant achievements or if they are assigned carelessly, their long-run effects on the educational efforts of students cannot be good.

Some students and instructors minimize the importance of marks, suggesting that *what* a person learns is more important than the *mark* he gets. This conception rests on the assumption that there generally is not a high relationship between the amount of useful learning a student achieves and the mark he receives. Others have made the same point by noting that marks should not be regarded as ends in themselves, and by questioning the use of examinations "merely" for the purpose of assigning marks.

It is true that the mark a student receives is not in itself an important educational outcome—by the same token, neither is the degree toward which the student is working, nor the academic rank or scholarly reputation of the professors who teach him. But all of these symbols can be and should be valid indications of important educational attainments. It is desirable, and not impossibly difficult, to make the goal of maximum educational achievement compatible with the goal of highest possible marks. If these two goals are not closely related, the fault would seem to rest with those who teach the courses and who assign the marks. From the point of view of students, parents, teachers, and employers there is nothing "mere" about the marking process and the marks it yields.

Marks are necessary. If they are inaccurate, invalid, or meaningless, the remedy lies less in deemphasizing marks than in assigning them more carefully so that they more truly report the extent of important achievements. Instead of seeking to minimize their importance or seeking to find some less painful substitute, perhaps instructors should devote more attention to improving the validity and precision of the marks they assign and to minimizing misinterpretations of marks by students, faculty, and others who use them.

SHORTCOMINGS OF MARKS

The major shortcomings of marks, as they are assigned by many institutions, are twofold: (1) the lack of clearly defined, generally accepted, scrupulously observed definitions of what the various marks should mean and (2) the lack of sufficient relevant, objective evidence to use as a basis for assigning marks. One consequence of the first shortcoming is that marking standards and the meanings of marks tend to vary from instructor to instructor, from course to course, from department to department, and from school to school. Another consequence is that instructor biases and idiosyncrasies tend to reduce the validity of the marks. A consequence of the second shortcoming is that the marks tend to be unreliable.

Variability in marking standards and practices has been reported by many investigators. Travers and Gronlund, for example, found wide differences of opinion among the members of a graduate faculty on what various marks should mean and the standards that should be followed in assigning them [4]. Odell reported,

> Where a typical five-letter system is used, the percents of the highest letter are likely to vary from 0 or near 0 to 40 or more; of the next to the highest from about 10 to 50 or more; and of the failure mark from 0 up to 25 or more [5].

Some schools and colleges publish, for internal guidance and therapy, periodic summaries of the marks assigned in various courses and departments. In one such unpublished study, instructors in one department were found to be awarding 63 percent A's or B's whereas those in another awarded only 26 percent A's or B's. Course X in one department granted 66 percent A's and B's, whereas Course Y in the same department granted only 28 percent. Each of these two courses, incidentally, enrolled more than 50 students.

Lack of stability in the assignment of grades from year to year is also apparent. Specifically, there appear to be long-run tendencies in many institutions to increase the proportion of high grades issued and to decrease the proportion of low grades. For example, during the years 1936 to 1939 one large institution issued 33 percent A and B grades. From 1951 to 1954 the corresponding proportion of A's and B's had increased to 44 percent. In the same two periods the proportion of D's and F's dropped from 30 to 18 percent.

On the other hand, as Hills and Gladney [6] have reported, the level of marks issued in a college tends to remain about the same even when, as a result of changes in admission policies, the level of ability of the student body has increased substantially. This is an inherent limitation of the kind of relative marking that letter marks were designed to provide. An A among highly able students does not reflect the same level of achievement as an A among less able students. This limitation needs to be recognized. It also needs to be tolerated since there is no good way to overcome it without losing the values of relative marking, which will be discussed later.

The lack of clearly defined, uniform bases for marking and standards for the meanings of various marks tends to allow biases to lower the validity of marks. Often a student's mark has been influenced by the pleasantness of his manner, his willingness to participate in class discussions, his skill in expressing ideas orally or in writing, or his success in building an image of himself as an eager, capable student. Some of these things should not ordinarily be allowed to influence the mark he receives. Further, as Palmer has noted, some instructors deliberately use high marks as rewards and low marks as punishments for behavior quite unrelated to the attainment of the objectives of instruction in a course [7].

The studies of Starch and Elliott on the unreliability of teachers' marks on examination papers are classic demonstrations of the instability of judgments based on presumably absolute standards [8]. Identical copies of an English examination paper were given to 142 English teachers, with instructions to score it on the basis of 100 percent for perfection. Since each teacher looked at only one paper, no relative basis for judgment was available. The scores assigned by the teachers to the same paper ranged all the way from 98 to 50 percent. Similar results were obtained with examination papers in geometry and in history.

INSTITUTIONAL MARKING SYSTEMS

An obvious method for dealing with one of the major shortcomings of marks—the lack of a clearly defined, scrupulously observed definition of what the various marks should mean—is for a school faculty to develop, adopt, and enforce an institutional marking system. As Odell observes,

> . . . when serious attention has been given to the matter and the general principles that should govern marking are agreed upon by a group of teachers, the marks the members of the group assign tend to form distributions much more similar than if this had not been done and their reliability is somewhat improved [5].

If an institution lacks a clearly defined marking system or if instructors do not assign marks in conformity with the policies that define the system, then the marks will tend to lose their meaning and the marking system will fail to perform its essential functions adequately. A marking system is basically a system of communication. It involves the use of a set of specialized symbols whose meanings ought to be clearly defined and uniformly understood by all concerned. Only to the degree that the marks do have the same meaning for all who use them is it possible for them to serve the purposes of communication meaningfully and precisely.

The meaning of a mark should depend as little as possible on the teacher who issued it or the course to which it pertains. This means that the marking practices of an instructor, of a department, or indeed of an entire educational institution are matters of legitimate concern to other instructors, other departments, and other institutions. It means that a general system of marking ought to be adopted by the faculty and the administration of a school or college. It requires that the meaning of each mark must be clearly defined. General adherence to this system and to these meanings ought to be expected of all staff members. Such a requirement would in no way infringe the right of each instructor to determine which mark to give to a particular student. But it would limit the right (which some instructors have claimed) to set his own standards or to invent his own meanings for each of the marks issued.

An educational institution that sets out to improve its educational effectiveness by improving its marking practices and to improve its marking practices by developing an institutional marking system is likely to encounter a number of questions such as these on which opinions will differ.

1. Should marks report absolute achievement, in terms of content mastery, or achievements relative to those of other comparable students?
2. Should marks be regarded as measurements or as evaluations?
3. Should marks simply indicate achievement in learning or should they be affected by the student's attitude, effort, character, and similar traits?
4. Should marks report status achieved or amount of growth in achievement?
5. Should a student receive a single composite mark or multiple marks on separate aspects of achievement?
6. Should the marking system report few or many different degrees of achievement?

Each of these questions will be discussed in turn in the sections that follow.

ABSOLUTE VERSUS RELATIVE MARKING

Two major types of marking systems have been in use in the United States since 1900. In the early years of the century almost all marking was in percents. A student who learned all that anyone could learn in a course, whose achievement could therefore be regarded as flawless, could expect a mark of 100 (percent). A student who learned nothing at all would, theoretically, be given a mark of zero. A definite percent of "perfection" usually between 60 and 75 percent was ordinarily regarded as the minimum passing score. Because a student's percent mark is presumably independent of any other student's mark, percent marking is sometimes characterized as "absolute marking."

The second major type of marking system is based on the use of a small number of letter marks, often five, to express various levels of achievement. In the five-letter A-B-C-D-F system, truly outstanding achievement is rewarded with a mark of A. A mark of B indicates above average achievement; C is the average mark; D indicates below average achievement; and F is used to report failure (that is, achievement insufficient to warrant credit for completing a course of study). Because a letter mark is intended to indicate a student's achievement relative to that of his peers, letter marking is sometimes characterized as "relative marking."

A popular term for one variety of relative marking is "grading on the curve." The "curve" referred to is the curve of normal distribution. One method for grading or marking on the curve, using five letter marks, is to determine from the ideal normal curve what proportion of the marks should fall at each of five levels and to follow these proportions as closely as possible in assigning marks. For example, the best 7 percent might get A's, the next best 23 percent get B's, and so on. Another process is to define the limits of the score intervals corresponding to various marks in terms of the mean and standard deviation of the distribution of achievement scores. Those whose scores are more than 1.5 standard deviations above the mean might get A's, those between .5 and 1.5 standard deviations above the mean get B's, and so on. This second process does not guarantee in advance that 7 percent of the students must get A's and 7 percent must fail. If the distribution of achievement is skewed, as it may be, or irregular, as it often is, these characteristics will be reflected in variations in the proportions of each mark assigned.

In the early decades of this century letter marking began to supersede percent marking. A clear majority of educational institutions now uses letter marks. But percent marking is by no means dead. Some schools still use percents instead of letters in their marking systems. Others indicate percent equivalents of the various letter marks they issue. Official examining bodies still prefer to define passing scores in terms of percent, though they often transform test scores, or control the scoring process, to avoid any significant change in the ratio of failures to passes. The controversy that began when percent marking was first seriously attacked, and when the movement to substitute letter marks gained support, continues today. The issue of absolute versus relative marking is still a live one, particularly when relative marking is identified as "grading on the curve."

PERCENT MARKING PROBLEMS

Ruch has critized percent marking on the ground that it implies a precision that is seldom actually attainable [9].

> ...a large body of experimental evidence points to the fact that from but five to seven levels of ability are ordinarily recognizable by teachers in marking pupils....The difference between an "85" and an "86" is a difference at least five times as fine as the human judgment can ordinarily distinguish.

There is substantial basis for this criticism, but it should not be taken too literally. How many levels of ability can be recognized depends on how much good information the teacher has available. No doubt an attempt to decide whether to give a pupil an 85 or an 86 has seemed ridiculous to many teachers. Seldom is a teacher in a position to say, "I am 100 percent certain that this pupil's true mark is 85, not 86 or 84." But given more information as a basis for judgment, the teacher's task might seem less ridiculous. And if a teacher recognizes that he will always have to settle for less than 100 percent confidence in the accuracy of his marks, he may be willing to trade a little of that confidence for a little more precise indication of his best guess at a student's level of achievement. As pointed out elsewhere in this paper, the use of broad categories in marking tends to reduce the number of wrong marks, but it always sacrifices so much precision in the process that the overall reliability of the marks declines.

The widespread initial use of percent marking and its persistence in modern times suggest that there must be things to be said in its favor. Two advantages are mentioned most often. The first is that percent marking clearly relates achievement to degree of mastery of what was set out to be learned. It does not give high marks to incompetent students simply because they happen to be the best of a bad lot. It does not require that some students receive low marks when they and all their classmates have done well in learning what they were supposed to learn. To know what proportion a student has mastered of a defined set of learning objectives is to know something more, and something less, than to know that he was outstanding, or average, or poor, compared with some other students, in mastering the same set of learning objectives. Achievement cannot be reported fully without reference to what (that is, command of knowledge and utilizable skill) has been achieved.

A second reputed advantage of percent marking is that it provides fixed, standard measures of achievement. This is in contrast to relative grading, where the achievements of the group set standards for what is excellent, average, or very poor in individual achievement. A student whose achievements result in his designation as valedictorian in one school might be regarded as only averaged in another school. Relative standards are likely to be variable standards.

Unfortunately, percent marking often fails to live up to its promise of providing truly meaningful and stable measures of achievement. To calculate a meaningful percent, two quantities must be measured in common units with reasonable accuracy: (1) the amount available to be learned in a course, and (2) the part of that amount that a particular student did, in fact, learn. These things are so difficult to measure that they are almost never measured. In what units can the material available to be learned be measured? On what basis can an instructor estimate how much might be learned in a particular course?

A mark of 90 used to be thought to mean that the student who received it had learned 90 percent of what he was expected to learn in a given course. But it takes very little analysis to discover how little meaning there is in such a statement. In what units does one measure amount of knowledge? On what basis can an instructor say how much *should* be learned in a given course? By what operations can one measure the total to be learned and the part the student did learn?

Sometimes those who defend so-called absolute standards argue that the experienced teacher knows what these standards are and that the inexperienced teacher can quickly learn them if he will only try. But this is obviously just a roundabout way of saying that the standards are derived from the observed performances of the students themselves, which means that they are relative and are not absolute measures of subject matter achievement at all. In spite of the arguments in favor of absolute, teacher-determined marking standards, no one has ever described in concrete, operational terms exactly how such standards could be defined and applied.

Few percent marks are ever calculated by division of a measurement of the part learned by a measurement of the total available to be learned. Instead, percent marks almost always reflect a highly subjective judgment, against a somewhat vague and individualistic standard, of the *relative* quality of a student's performance. If he is very good, relative to other students the instructor has, has had, or can imagine having, his percent mark is likely to be high. If not, it is likely to be lower.

RELATIVE MARKING PROBLEMS

One of the major problems of relative marking arises from the fact that not all class groups to which marks must be assigned are typical. Some are above and others are below average in general ability and achievement simply because of sampling fluctuations. There may be almost as much chance variation in potential learning ability from one class of 26 students to another class of 26 students as there is in trick-taking potential from one hand of 13 bridge cards to another. From one point of view it is absurd to give an A for the best achievement in a low-ability group if identical achievement would have received a mark of C in a group of higher ability. But from another point of view it is equally absurd to deny the possibility of recognizing outstanding achievement among low-ability students with a mark of A, or of recognizing, with marks of C, D, or even F, that students of high ability can do what is for them mediocre, poor, or even failing work.

Granting the generalizations that most classes in schools and colleges are atypical and that seriously atypical classes can put serious strains on systems of relative marking, one can still make a case for the advantages in reliability and meaningfulness of a marking system that places more emphasis on relative than on absolute achievement. Absolute standards seem much too difficult to define clearly and with general acceptability to provide in themselves an adequate basis for reliable, meaningful marks.

Variations from class to class in ability and achievement do not preclude reasonable marking on a relative basis. Allowances can be made for the sampling fluctuations in small classes by not insisting on rigorous adherence to a designated ideal distribution in each small, atypical class, but only in the composite distribution of marks for many such classes. One can make provision for a somewhat different "curve" in classes of high ability than in classes of lower ability. But such differences should be public knowledge and should be sanctioned by the whole faculty. Ordinarily they should not be extreme differences. It is unrealistic and psychologically unwise to foster the belief that capable students are entitled to higher marks simply because they are capable. It is equally unrealistic and unwise to create conditions that predestine some students to do below average work and to receive below-average marks in every course they undertake to study. Human beings, fortunately, exhibit a diversity of talents. A good school or college makes provision for developing diverse special talents and for recognizing and rewarding them.

CRITICISMS OF RELATIVE MARKING

One common criticism of relative marking is that it permits the students rather than the teachers to set the standards. This is true. There is the further implication that student-set standards are likely to be low—lower, at least, than those most teachers would set. But this appears not to be true. For when the teachers depart from the proportions of marks recommended in a system of relative marking, they usually seem to do it by giving too many high marks rather than too many low marks.

Another way of stating essentially the same criticism is to claim that relative marking encourages a general slow-down of student effort. The argument is that under a relative system the members of a class can earn just as good marks on the average by taking it easy as by putting forth maximum effort. There is nothing wrong with the abstract logic of this argument, but there is something wrong with its practical psychology. How can any student who agrees to this slow-down be sure that some other student may not work just a bit harder and wind up receiving a higher mark?

Yet another criticism of relative marking is that it requires an instructor to give some low marks and some average marks, even if most of the students in the class learn practically all that he was trying to teach them. Do they not all deserve high marks in such a case? How frequently this hypothetical situation of nearly universal, nearly maximum achievement can be expected to occur is not specified. Some good instructors claim they have never encountered such a case. But even if purely hypothetical, the issue it raises is important and deserves an answer. Actually, two answers can be offered.

The first is that if a high mark implies outstanding, and hence unusual, achievement and if an average mark implies normal, and hence typical, achievement, it is logically inconsistent to give most of the class members a high mark. They should all get average marks since their achievements are about average for the group. None of them have shown outstanding achievement.

The second is that if the students in a class differ appreciably in abilities, preparation, and interests, as they almost always do, the only way that nearly all of them can achieve all that is asked of them is to ask much less than the best could achieve. To equalize achievement by minimizing it deprives some students of the education they could get and may deprive society of the benefits of their fully developed talents. Most good teachers prefer to challenge and to help each pupil to learn as much as he can.

Some instructors have expressed fear that relative marking would lead to such a slow-down. A few students have claimed to know of classes where it happened. But there seems to be little evidence that it ever actually did happen. If it did, the blame might lie more with uninspiring course content or with poor student motivation than with relative marking. Ordinarily the personal competition implicit in relative marking will stimulate as much or more effort to achieve than the impersonal stimulus of an absolute standard of achievement.

In most areas of human activity awards go to individuals who are outstanding in relative, not absolute, terms. There are no absolute standards for speed in running the mile or for distance in throwing the javelin. The winner in any race is determined on a relative basis. Runners on a starting line seldom agree to loaf along simply because there is no absolute standard of speed they have to meet. From the point of view of the individual runner in the 100-yard dash, as well as from that of the individual student majoring in history or chemistry, the best way to achieve oustanding success is to put forth outstanding effort.

A marking system cannot be all things to all men. A single symbol cannot represent low achievement from one point of view (that is, actual degree of content mastery) and high achievement from another (that is, progress in relation to reasonable expectation). What it can and should have is one clearly defined and jealously guarded kind of meaning. School and college faculties have the opportunity and the obligation to establish and maintain clearly defined meanings for the symbols used in their marking systems.

MEASUREMENT OR EVALUATION

As the term is used here, a measurement in education is a quantitative description of how much a student has achieved. A measurement is objective and impersonal, and it can be quite precisely defined in operational terms. An evaluation, on the other hand, is a qualitative judgment of *how good or how satisfactory* the student's performance has been. Evaluations are often based in part on measurements of achievement, but they are also based on many other kinds of evidence. Measurement can describe how much of this ability or that characteristic an individual possesses. But to tell how well educated he is or how well prepared for a particular job, an evaluation is required.

There are several advantages in treating a marking system as a means of reporting measurements of achievement rather than as a means of reporting evaluations.

In the first place, evaluations are complex, involving many variables and many considerations that are unique to a particular student. This makes it difficult to report evaluations adequately in a standardized marking system. Judgments of how good a student's educational achievement has been depend not only on how much he has achieved, but also on his opportunity for achievement, the effort he has put forth, and the need that this achievement is likely to serve in his educational and vocational future. It is not easy to make marks valid as *measures* of achievement, but it is next to impossible to make them valid as *evaluations*.

In the second place, because of the many poorly defined and highly individual factors which must be considered in making an evaluation, few teachers have a sufficient basis for making fair evaluations. Some teachers may be well informed about the home background and the personal problems of many of their students and hence can judge quite accurately each student's opportunity to learn and the real effort he has made to learn. But many teachers lack some of this background essential to sound evaluation. Such teachers can help a student to make an honest evaluation of his own achievements, but the teacher can seldom do the whole job alone. And perhaps the teacher should not, even if he could. Imposed evaluations may be not only less accurate than self-evaluations, but less effective also. If students and teachers regard marks as objective measurements of achievement rather than as subjective evaluations, there is greater likelihood that the teachers will assign fair and accurate marks, and there is less likelihood that students will react emotionally so that the relations between student and teacher are damaged.

Consider this analogy. If the scale shows that a man weighs 210 pounds, it is reporting a measurement. The fact may be unpleasant, but there is nothing personal about it and no good cause for anger at the scale. But if a tactless acquaintance suggests that he is getting too fat, the acquaintance is making an evaluation. It can be taken as a personal affront, and a natural reaction is to resent it. The fact that the heavy man may have been thinking the same thing himself does not soothe his feelings very much. In somewhat the same way, the interpretation of marks as evaluations rather than as measurements may have been responsible for some of the tension and unpleasantness associated with their use.

Finally, there may be less need or justification for making a formal report and a permanent record of an evaluation than there is for recording a measurement. Usually it is less important for a future teacher or employer to know that Henry did as well as could be expected of him or that he failed to live up to expectations than to know that he was outstanding in his ability to handle language or mediocre in his mathematical ability. Evaluations are instrumental to specific decisions. Once the decision has been made, there is little to be gained by basking in a favorable evaluation or in agonizing over an unfavorable one. To be successful and to maintain emotional stability, a person must not cherish too long the evaluations, favorable or unfavorable, that others have made of him or that he has made of himself. Measured achievement, on the other hand, often represents a more permanent foundation on which future education and success depend. The more accurately it is reported and the more completely it is recorded, the more soundly a student and his advisors can judge which choices he should make.

A teacher may thus be well advised to regard the marks he assigns in a course as objective, impersonal measures of achievement rather than as subjective, personal evaluations. He may even be able, happily, to persuade his students to accept them on this basis.

ACHIEVEMENT OR ATTITUDE AND EFFORT

Studies such as those by Hadley, Travers and Gronlund, and others indicate that teachers often base the marks they issue on factors other than degree of achievement of the objectives of instruction. No doubt they will continue to do so, since marks can be useful instruments of social control in the classroom and since some degree of such control is essential to effective teaching. But the use of marks for these purposes must be limited, for it can easily be abused and tends to distort the intended meaning of the mark.

One of the important requirements of a good marking system is that the marks indicate as accurately as possible the extent to which the student has achieved the objectives of instruction in the particular course of study. If improving the student's attitude toward something or improving his willingness to put forth effort for educational achievement is one of the specific objectives of the course and if the instructor has planned specific educational procedures in the course to attain this goal, then it is quite appropriate to consider these things in assigning marks. But often this is not the case. When it is not, attitude and effort probably should be excluded from consideration in determining the mark to be assigned.

Involving judgments of character and citizenship in marking is even more hazardous. Such judgments tend to be impressionistic evaluations rather than objective descriptions. If we like the behavior we call it straightforward, or perhaps thoughtful. If not, we are more likely to call it tactless, or perhaps indecisive. Seldom are the traits of good character or good citizenship defined objectively, without the use of value-loaded and question-begging modifiers like "good," "desirable," "effective," or "appropriate."

The result of these difficulties is that valid assessments of character and citizenship are not easy to secure and not often secured.

STATUS OR GROWTH

Some instructors, seeking to improve the fairness of the marks they issue, attempt to base them on the amount of improvement the student has made rather than on the level of achievement he has reached. Scores on a pretest, and other preliminary observations, are used to provide a basis for estimates of initial status. The difference between these and subsequent test scores and other indications of achievement permits estimates of the amount of change or growth.

Unfortunately, these growth measures usually are quite unreliable. Each test score or observation includes its own error of measurement. When these are subtracted from other measurements, the errors tend to accumulate instead of to cancel out. Consequently, the difference scores sometimes consist mainly of errors of measurement. If his tests are appropriate and reliable, an instructor may safely use the difference between mean pretest and posttest scores as one measure of the effectiveness of his instruction. But few educational tests are good enough to reliably measure short-run gains in educational achievement for individual students.

From some points of view it may seem fairer to use growth rather than final status as a measure of achievement. But, apart from the characteristic unreliability of growth scores just mentioned, there are other problems. One is that for many educational purposes, knowledge that a student is good, average, or poor when compared with his peers is more important than knowledge that he changed more or less rapidly than they did in a certain period of time. Another is that students who get low scores on the pretest have a considerably greater likelihood of showing subsequent large gains in achievement than their classmates who earned higher initial scores. Students are not slow to grasp this fact when their achievement is judged on the basis of gains. The course of wisdom is for them to make sure that their pretest performance is not so good as to constitute a handicap later on.

One rather strong incentive for marking students on the basis of growth rather than status is to give all students a more nearly equal chance to earn good marks. A student who makes good marks on the basis of status in one course is likely to make good marks in other courses. A student whose marks are high one semester is likely to get high marks the next semester. The other, darker side of this picture is that status marking condemns some students to low marks in most subjects, semester after semester. Low marks discourage effort. Lack of effort increases the probability of more low marks. So the vicious cycle continues, bringing dislike of learning and early withdrawal from school.

The debilitating effect of low marks on educational interest and effort is probably sufficient to constitute a major educational problem. But whether marking on the basis of growth rather than status provides an effective solution to the problem may be open to question. For one thing, growth measures are usually of rather low reliability, as already indicated. Few students, even poor students, would really favor the substitution of more or less randomly distributed (and hence rather meaningless) praise or blame for consistently dependable measures of status, however discouraging that status might seem to be. Students are not likely to forget, nor should they, that in the long run it is competence achieved that will count and that rate of growth is important only as it contributes to status.

What, then, is the answer? Success is important to all of us. None of us should expect it all of the time, but we should not expect it to be denied all of the time either. If students are taught to dislike school by constant reminders of their low achievement, the remedy probably is not to try to persuade them that rate of growth toward achievement is more important than status achieved, for that is a transparent falsehood. The remedy probably is to provide varied opportunities to excel in various kinds of worthwhile achievement. Certainly this can be done within a comprehensive school. It may even be done within a single classroom by an alert, versatile, dedicated teacher. When it is done, marking on the basis of status achieved will no longer mean that some students always win and others always lose. Each can enjoy, as he should, some of the rewards of excellence in his own specialty.

SINGLE OR MULTIPLE MARKS

Achievement in most subjects of study in schools and colleges is complex. There is knowledge to be imparted, understanding to be cultivated, abilities and skills to be developed, attitudes to be fostered, interests to be encouraged, and ideals to be exemplified. Correspondingly, the bases used for determining marks include many aspects or indications of achievement: homework, class participation, test scores, apparent attitude, interest and effort, and even regularity of attendance and helpfulness to the teacher. How can a single symbol do justice to these various aspects of achievement?

The answer of some observers is that it cannot. A mark, they say, is a hodegepodge of uncertain and variable composition. They suggest that the essential step in improving marks is to make them more analytical and descriptive. Multiple marks or written reports have been proposed as improvements over the traditional single letter or number.

There is considerable merit in these suggestions, and under favorable conditions they can improve marking considerably. But they do involve problems. For one thing, they multiply considerably the already irksome chores of marking. For another, they create additional problems of defining precisely what is to be marked and of distinguishing clearly among the different aspects of achievement. An even more serious problem is that of obtaining sufficient evidence, specific to each aspect of achievement, on which to base a reliable mark. Finally, and largely as a result of the preceding difficulties, the multiple marks exhibit considerable "halo effect." That is, they seem to be determined more by the instructor's overall impression of the student than by his successful analysis and independent measurement of various components of achievement. Multiple marking is not the only road to improvement in marking and probably not the best road currently available. Much can be done to make single marks more meaningful and more reliable. Perhaps those possibilities should be exploited before the more complex problems of multiple marking are tackled.

HOW MANY STEPS ON THE GRADE SCALE?

A major difference between two systems of marking is that letter marks are usually few in number (five, most commonly) whereas percent marks provide up to 100 different values, of which about 30 are commonly used. Those who advocated letter marks when they were first introduced suggested that the bases on which marks are usually determined are not reliable enough to justify the apparent precision of percent marking. They claimed that the best that most instructors can do is to distinguish about five different levels of achievement. Many instructors seemed to agree with this view.

Some proposals for improving marks have gone even further than the five-letter system in reducing marking categories. The use of only two marks such as "S" for "satisfactory" and "U" for "unsatisfactory," "P" for "pass" and "F" for "fail," or "credit"–"no credit" has been suggested and adopted by some institutions [10], [11]. Pass–fail grading enjoyed considerable popularity during the late 1960's, particularly on the more liberally inclined campuses. At the same time there was increased interest in refining the grade scale by adding plus and minus signs to the basic letters, or decimal fractions to the basic numbers.

The notion that marking problems can be simplified and marking errors reduced by using fewer marking categories is an attractive one. Its weakness is exposed by carrying it to the limit. If only one category is used, if everyone is given the same mark, all marking problems vanish, but so does the value of marking. A major shortcoming of two-category marking, and to some degree of five category marking as well, is this same kind of loss of information. To trade more precisely meaningful marks for marks easier to assign may be a bad bargain for education.

The use of fewer, broader categories in marking does indeed reduce the frequency of errors in marking. That is, with a few broad categories more of the students receive the marks they deserve because fewer wrong marks are available to give them. But each error becomes more crucial. The apparent difference between satisfactory and unsatisfactory, or between a B and a C, is greater than the difference between 87 percent and 88 percent. If a fallible instructor (and all of them, being human, are fallible) gives a student a mark of 86 percent when omniscient widom would have assigned a mark of 89 percent, the error has less consequence than if the instructor assigns a C when a B should have been given, or an "unsatisfactory" mark when the mark should have been "satisfactory." Hence the use of fewer categories is no royal road to more reliable marking. And, as noted previously, reducing the number of categories reduces the information conveyed by the mark.

The more reliable the information on which marks are based, the greater the value of a large number of marking categories. But no matter how unreliable that information may be, it is *never* true that few categories report the information more accurately than many categories. This is illustrated in Table 1, which may be read as follows: "If marks are based on information having a reliability of 0.95, the use of 2 categories in marking would reduce the reliability to 0.63, of 5 categories to 0.85, and of 10 categories to 0.92. In the case of 15 categories the reduction is very slight, only from 0.95 to 0.94." Other rows in the table may be read similarly. The data in Table 1 were prepared from formulas derived and explained by Peters and Van Voorhis [12]. Values in the table assume equally spaced categories and normal distributions of scores and marks.

Table 1 illustrates a general proposition of considerable importance.

> *Regardless of the inaccuracy of the basis for grading, the finer the scale used for reporting the grades, that is the more different grade levels it provides, the more accurate the grade reports will be.*

Whenever scores are grouped for purposes of grading, errors are introduced. In some cases the grouping errors will offset, or correct for, measurement errors in the data on which grades are based. But on the whole, the grouping adds more errors than it cancels [13]. The use of very few categories in grading aggravates the problem of unreliability. If maximum reliability of information is the goal, a 5-letter system is better than a 2-letter system, and 10 categories in marking is better than 5. The main arguments for fewer categories in marking must be on grounds of convenience and simplicity, not on grounds of the unreliability of the basis for marking.

PASS–FAIL GRADING

During the late 1960's successful efforts were made on a number of college campuses and in some public school systems to supplement or supplant conventional letter grades with a pass–fail grading system. Several pressures initiated and sustained these efforts: the pressure from faculty members to get rid of grading problems, the pressure from students to get rid of the threat of low grades, and the pressure in the academic community to try something new.

A number of arguments in favor of pass–fail grading were advanced.

1. By the time a student gets to graduate school, or into college, or even into high school, he has proved his ability. He should not be called upon to prove it again and again in every course he takes.
2. Grades are unimportant outside of school. The typical employer doesn't care what kind of grades a person got in school. Grades are poor predictors of later achievement.
3. Many instructors do such a poor job of grading that

TABLE 1
LOSS OF RELIABILITY FROM USE OF BROAD CATEGORIES IN MARKING

Reliability of Marking Basis	*Reliability of Marks* Number of Categories 2	5	10	15
0.95	0.63	0.85	0.92	0.94
0.90	0.60	0.80	0.87	0.89
0.80	0.53	0.71	0.78	0.79
0.70	0.47	0.62	0.68	0.69
0.50	0.33	0.45	0.48	0.49

the grades they issue are almost without meaning.

4. The need to protect his grade point average deters a student from taking courses he want to take and ought to take outside his major field.
5. The pressure to make high grades forces students into bad study practices such as rote learning and all-night cramming, and drives some to cheat on examinations.
6. The threat of low grades destroys the love of learning that schools ought to foster.

There is some merit in these arguments, but there are also a number of flaws.

1. No man, however successful, ever becomes immune to failure. Those who live good lives never stop trying to do their best, never limit themselves to enterprises whose success is assured in advance. Past success promises future success, but never can guarantee it.
2. Many personal qualities other than academic success contribute to success on the job. We should not expect the first to predict the second infallibly. Nor should we expect success or failure in appreciation of poetry to have much to do with success or failure in managing a grocery. But if there is no relation at all between the competence a student shows in learning how to be a good teacher and his subsequent competence in the classroom, something is seriously wrong with the teacher training program.
3. The better remedy for incompetent grading is to get rid of the incompetence, not to get rid of the grades.
4. The student who is deterred from taking courses outside his major may be acting wisely, if the course he wishes to take is in fact likely to be quite difficult for him to master. The fear of a low grade may be well justified by the student's lack of adequate preparation. Pass–fail grading may encourage him to take the course, but it will not help him at all to master it. On the other hand, if the course is one he really needs and feels able to handle satisfactorily, and if his grade point average is not already dangerously low for other reasons, he is acting quite unwisely if he lets the possibility of a B or C deter him from taking it.
5. If grades are properly given they will not reward rote learning or cramming. Even if badly given they do not justify cheating.
6. Low achievement does more to destroy the love of learning than do the low grades that report that low achievement. Most people come to love doing the things they know they can do well. The grades they get help them to know this.

The arguments against pass–fail grading are less numerous, but may be more substantial.

1. Pass–fail grading removes much of the immediate motivation and reward for efforts to excel.
2. Pass–fail grading leaves the student with an incomplete or an inaccurate record of his achievements.

The practical force of the first of these arguments is such that most systems of exclusive pass–fail grading tend to be short-lived. The force of the second is such that when a pass–fail option is offered to students, few of them decide to make use of it. They know that a transcript loaded with A's and B's will look better to a graduate school admission committee or to a prospective employer than one loaded with pass marks.

Almost every school or college offers some courses in which the aim is to provide certain experiences rather than to develop certain competencies. For such courses neither grades nor pass–fail decisions seem appropriate. Instead, the student who attends enough to get a large proportion of the desired experience should simply be given credit for his attendance. Courses in the appreciation of art, music, or literature, in recreational pursuits, in social problems, or in great issues may belong in this category of ungraded courses. But most other courses do not. Most courses do aim to develop competencies. Such courses call for assessments of achievement, and for them pass–fail grading is a poor substitute for more detailed and precise reporting of achievements.

Most of us want to be valued as persons. Most of us don't particularly want to be evaluated. But we can't enjoy the first without enduring the second. The weakness of pass–fail grading is that by doing a poor job of evaluating it keeps us from doing a good one of valuing.

QUALITY CONTROL IN A MARKING SYSTEM

What a mark means is determined not only by how it was defined when the marking system was adopted, but also, and perhaps more importantly, by the way it is actually used. If an instructor assigns some A's, many B's, some C's, and very few lower marks, then B has become his average mark, not C, as the marking system may have specified. Thus institutional control of marking requires surveillance of the results of the marking process and may require corrective action.

The temptations for instructors to depart from institutional policy in marking are many, and the rationalizations for doing so are not hard to find. Some instructors regard marking as the personal prerogative of the instructor. They may not distinguish between their very considerable freedom to determine which mark a particular student shall receive and their very considerable responsibility to make the meaning of their marks consistent with those of other instructors. To rationalize deviations from overall institutional policy in distribution of marks, they may claim unusual ability or disability in their students, special interest or aptitude in the subject of study, or (usually only by implication) exceptionally fine teaching.

Some instructors yield to the subtle pressures to give more high and fewer low marks. Perhaps they feel inclined to temper justice with mercy. Perhaps they wish to avoid controversy. An instructor seldom has to explain or justify a high mark or to calm the anger of the student who received it. Some instructors may feel that the favorable reputations of their courses among students depend on their generosity with high marks. Many good instructors like their students so much as persons that they find it difficult to disappoint any of them with a low mark, particularly if the student seems to have been trying to learn. These temptations to depart from standard marking practices are understandable as temptations, but most of them do not carry much weight as reasonable justifications. There are indeed some situations that do warrant departure from general institutional policies in marking. But the determination of which situations those are probably cannot be left to the individual instructor concerned if uniformly meaningful marks are desired.

There are several things an educational institution can do to maintain the meaningfulness of the marks issued by its instructors. One is to publish each semester summary distributions of the marks issued in each course by each instructor. This is done systematically by some colleges and has been found quite effective. Another is to record alongside each mark reported to a student or his parents a set of numbers showing the distribution of marks to the student's classmates. The purpose of this is to make the relative meaning of the mark immediately apparent to all concerned.

A somewhat simpler variant of the procedure just described is to accompany each marking symbol by a fraction, the numerator of which shows what percent of the class received higher marks, while the denominator shows what percent received lower marks.

These fractional interpretations may be required only when the instructor has exceeded or fallen short of specified limits for the proportion of marks above or below each category. Such a requirement tend to encourage observance of institutional regulations without preventing necessary exceptions.

Finally, an institution can return to the instructor a set of marks whose distribution among the marking categories is unsatisfactory and ask him to resubmit a revised set of marks.

SYSTEMATIC MARKING PROCEDURES

This and the next section are concerned with a particular set of systematic procedures for converting test scores, or composite numerical measures of achievement, into marks. The method is built around the five-unit scale of letter marks that most schools and colleges use currently.

One of the purposes of any systematic method of assigning marks is to establish greater uniformity among instructors in their marking practices, and hence in the meaning of the marks they issue. A school or college faculty that adopts such a system and requires all faculty members to conform to it in issuing marks will almost certainly improve the uniformity of marking practices and hence make the marks issued much more consistently meaningful. Another purpose is to make the systematic conversion of numerical measures into course marks simple enough to compete successfully with the unsystematic, hit-or-miss procedures that some instructors actually do use. To this end the procedures make use of statistics that are easy to determine or can be estimated with sufficient accuracy by short-cut methods.

One basic assumption of the method is that the five-letter marks should represent equal intervals on the score scale. This is an alternative to strict *grading on the curve*, which disregards numerical score values, considers only the rank order of the scores, and gives the top 7 percent A's, the next 23 percent B's, and so on. It is also an alternative to the use of unequal numerical intervals, which usually result when the end points of these intervals are located at gaps or natural breaks in the distribution of scores. Most such gaps are chance affairs, attributable largely to chance errors of measurement. Hence they seldom reflect natural points of division between discrete levels of ability.

Acceptance of this assumption means that there can be no *a priori* certainty that expected percentages of A's, B's, or any other mark will be assigned. Indeed, in a particular class group there might be no A's or no F's at all. However, in large class groups (and, in the long run, in small class groups) the distribution of letter marks will ordinarily approximate a normal distribution.

The method makes use of the median score as the basic reference point or origin of the letter mark scale. Since there is seldom a meaningful *absolute zero* on any scale of academic achievement, it is almost always necessary to use some other reference point in setting the scale. The median or middle score provides a reference point that is reasonably easy to determine and reasonably stable from one sample to another. If the distribution of scores is skewed, the median is a more typical or representative measure than the mean.

When a five-unit scale is used for measuring achievement, the standard deviation of the test scores provides a unit of convenient size. The usual range of scores in a distribution of 20 to 40 scores equals about four or five standard deviation units. Although the standard deviation is tedious to calculate without machine assistance, it can be estimated quite simply, with reasonable accuracy, for purposes of assigning marks.

Finally, the method of mark assignment here described makes a provision for different distributions of marks in classes having different levels of average academic ability.

The method does not require such differences, but it does allow for them if the faculty decides in favor of them. Mention was made earlier in the paper of the differences of opinion that exist in school and college faculties on this question. Probably most faculties would favor giving more high marks in classes of high ability. But when they vote for this policy they also vote, whether explicitly or not, for giving lower than average marks in some other classes. There are both advantages and disadvantages in differentiating levels of marking to correspond with ability levels in various classes.

The means by which marks are adjusted to reflect class ability levels is illustrated in Figure 1. Essentially it involves moving the scale of measures on which the marks are to be based up or down in relation to the marking scale. The marking base measures are usually obtained by adding test scores and the numerical equivalents of marks on papers, projects, recitations, and so on. Figure 1 shows how marks might be adjusted for classes of two different ability levels, average and exceptional. The marking scale is shown in the center. Distributions of marking base measures are shown as normal distributions. For the class of exceptional ability the distribution of marking base measures is higher on the marking scale than it is for the class of average ability. Some details of Figure 1 are worth noting.

For a class of average ability the lower limit of the A's is located 1.5 standard deviations above the mean of the measures on which the marks are to be based. With mark intervals of one standard deviation, the lower limit of the B's would be 0.5 standard deviations above the mean. The lower limit of the C's would be 0.5 standard deviations *below* the mean, and the lower limit of the D's standard deviations below the mean. From a table of areas under the normal curve one can determine that with these limits, 7 percent of the marks would be A's, 24 percent B's, 38 percent C's, 24 percent D's, and 7 percent F's. Using numerical values of 4 for A, 3 for B, 2 for C, 1 for D, and 0 for F, the grade point average for this average level grade distribution is 2.00. Since that grade point average falls exactly in the middle of the distribution of numerical values, its percentile equivalent is 50. The foregoing explains what each of the figures in the fifth row (average ability level) of Table 2 means, and how all were derived from the first (lower limit of the A's). But perhaps a second illustration would still be helpful.

For a class of exceptional ability the lower limit of the A's is located only 0.7 standard deviations above the mean of the measures on which the marks are to be based. With mark intervals of one standard deviation, the lower limit of the B's would be 0.3 standard deviations *below* the mean. The lower limit of the C's would be 1.3 standard deviations below the mean, and the lower limit of the D's 2.3 standard deviations below it. From a table of areas under the normal curve one can determine that with these limits 24 percent of the marks would be A's, 38 percent B's, 29 percent C's, 8 percent D's, and 1

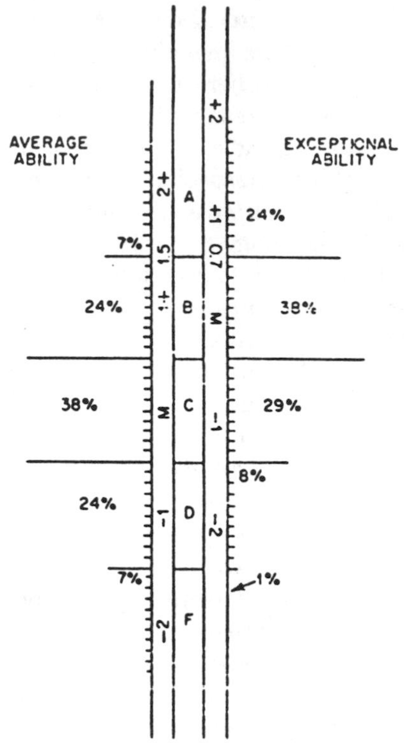

Fig. 1. Adjusting grades for different ability levels.

TABLE 2
LETTER MARK DISTRIBUTION STATISTICS FOR CLASSES AT SEVEN LEVELS OF ABILITY

Ability Level	*Lower Limit of A's*	*Percent of Marks* A	B	C	D	F	*Ability Measures* GPA	Percentile
Exceptional	0.7	24	38	29	8	1	2.80	79
Superior	0.9	18	36	32	12	2	2.60	73
Good	1.1	14	32	36	15	3	2.40	66
Fair	1.3	10	29	37	30	4	2.20	58
Average	1.5	7	24	38	24	7	2.00	50
Weak	1.7	4	20	37	29	10	1.80	42
Poor	1.9	3	15	36	32	14	1.60	34

percent F's. Using the same numerical values as before (A = 4, B = 3, C = 2, D = 1, F = 0) the grade point average for this exceptional distribution of grades is 2.80. This average is 0.8 standard deviations above the mean (2.00) of a distribution of average (not exceptional) grades. Thus it corresponds to a percentile of 79, for in a normal distribution 79 percent of the measures lie below a point which is 0.8 standard deviations above the mean. Again these statements explain the source and meaning of the figures in the first row (exceptional ability) of Table 2.

Mark adjustment data for classes at these two and at five other levels of ability, ranging from somewhat below to well above average, are presented in Table 2. The difference between successive levels in the lower limit of the A's is 0.2 standard deviation units. As a necessary but not easily demonstrable consequence of these lower limit differences there are corresponding differences of 0.2 between successive levels in the grade point averages. These differences seem sufficiently small, and the range of levels sufficiently wide, to accommodate most situations in which mark adjustments are needed. Of course the differences could be made smaller or larger, or the range of levels extended if either should seem necessary. The determining value in each row of the table is the first, which tells where the lower limit of the A's is located in standard deviation units above the mean. Given that value and tables of areas under the normal curve, all other figures in the row can be calculated.

Differentiating levels of marking requires uniform ability measures for the pupils in various classes. This could be provided either by scores on some test of academic aptitude or by grade-point averages in previous courses. Of course the mark a particular student receives should not be directly affected by his aptitude test score or his previous grade-point average. Few instructors would argue that one student should get a higher mark than another simply because he is thought to have more ability. His achievement should determine the mark he gets. But since there is a substantial correlation between prior measures of ability and subsequent measures of achievement, it seems reasonable that the average mark in a class of more able students should be higher than the average mark in a class of less able students.

ASSIGNING LETTER MARKS

Four steps are involved in this process of assigning marks.

1. Select from Table 2 a distribution of marks appropriate to the level of ability of the class being graded.
2. Calculate the median and the standard deviation of the scores on which the marks are to be based.
3. Determine the lower score limits of the A, B, C, and D mark intervals, using the median, the standard deviation, and the appropriate lower limit factor from Table 2.
4. Assign the designated marks to the students whose scores fall in intervals determined for each mark.

Table 2 presents mark distribution statistics for classes at seven different levels of academic ability. The first column lists descriptive labels of the ability levels. The last two columns, headed "ability measures," provide means for deciding which level is appropriate for a particular class. If the grade-point averages (GPA) of the class members in their previous course work is known, the mean of these GPA's indicates which ability level is appropriate for the class. If, for example, the class mean of those GPA's was found to be 2.24, the teacher could conclude that this class is slightly above average in ability, so that the level designated "fair" would be appropriate.

If grade point averages are unavailable, inconvenient to use, or undesirable for some other reason, average aptitude test scores can be used in place of the grade point averages. For this purpose, all students in the school or college must have taken the same test or battery of tests, which yields a measure of the academic ability of each student. If the scores of those students are available in the form of local school percentile ranks, then the average of those percentile ranks could also be used to select the appropriate ability level. If, for the hypothetical class we have in mind, that average turned out to be 45, the instructor could conclude that it is below average in ability and that the mark distribution for a *weak* class would be appropriate.

Another possibility, particularly appropriate at the college level, would be to use different distributions of marks for different class levels, from freshman to senior. It is generally assumed that senior classes are more able than freshman classes, simply because the less able students have been weeded out by the inevitable process of attrition. Most teachers take this into account at least informally, tending to assign higher proportions of A's and B's to higher-level classes. Schools wishing to obtain better control over grading, as discussed earlier, might specify different grade distributions from Table 2 to the various class levels. Before making such a decision, however, it would be desirable to make studies of the variations in average GPA or aptitude test scores from one class to another.

The five columns in the center of Table 2 indicate, for each ability level, what percent of the marks would be A's, B's, and so on if the distribution of numerical measures being converted to grades were perfectly normal. Since few distributions are likely to be perfectly normal, the percentage of each mark assigned in any actual case will usually differ somewhat from the percentage indicated in Table 2. One could, of course, arrange the numerical measures in rank order and convert them to letter marks on the basis of the "ideal" percentages for a class of the specified level of ability. But, as suggested earlier, this process is open to more of the criticism of grading on the curve than is the process being described here.

The second step in the process requires calculation of the median and the standard deviation. To calculate the median, follow the steps outlined below.

1. Arrange the scores in order from high to low.
2. If the number of scores is odd, the middle score is the median. For example, in a set of 25 scores the median is the thirteenth score.
3. If the number of scores is even, the median is the average of the two scores closest to the middle of the distribution. For example, in a set of 26 scores the median is the average of the thirteenth and fourteenth scores.

To estimate the standard deviation this short-cut approximation is recommended.

> Divide the difference between the sums of scores in the upper and lower one-sixths of the distribution of scores by one-half of the number of scores in the distribution.

Suppose we have twenty-six scores. The sum of the top four scores (upper one-sixth of the distribution) might be 137. The sum of the bottom four scores (lower one-sixth) might be 69. Then the estimated standard deviation would be

$$\frac{137 - 69}{13} = \frac{68}{13} = 5.23.$$

The third step in the process, determining the lower score limits, makes use of the second column of Table 2 headed "Lower Limit of A's." The values in this column show how far the lower limit of the score interval for A marks lies above the median, in standard deviation units. Since the score interval that corresponds to each mark is one standard deviation in extent, once the lower limit of the A interval is determined, the lower limits of the B, C, and D intervals can be found by successive subtractions of the standard deviation from the lower limit of the A's. The fourth step, assigning the marks, is facilitated by the recording of the score intervals for each letter grade in whole score units.

Table 3 illustrates the application of this method of mark assignment to a sample problem. The previous grade point averages of the students in this class, as well as their aptitude test percentiles, indicate that the grade distribution for a class slightly better than average, designated in Table 2 as "fair," would be appropriate. Since there are 38 students in the class, the median is the average of the nineteenth and twentieth scores. The top and bottom sixths in a class of 38 include six scores each. Hence the six highest scores, from 100 to 112, are added to obtain the sum of 636. The sum of the six lowest scores, from 44 to 59, is 318. The difference between 636 and 318, divided by half the number of scores, gives the estimated standard deviation.

Using the lower limit factor of 1.3 (obtained from Table 2 for a "fair" class) in conjunction with a standard deviation of 16.7 and a median of 80.5, it is determined that the lower limit of the A mark interval is 102.2. Successive subtractions of the standard deviation give the lower limits of the other mark intervals. From these lower limits the whole number score intervals are easily determined and the appropriate letter mark can be assigned to each numerical score. Note that the actual percentage of scores to which each mark was assigned differs somewhat from the ideal values of Table 2 reflecting the fact that he distribution of scores given in Table 3 was not perfectly normal.

TABLE 3
SAMPLE PROBLEM IN LETTER MARK ASSIGNMENT

A. Data for the problem
1. Class ability level measures
a. Mean GPA on previous years' courses 2.17
b. Mean percentile on aptitude test 56.3
c. Appropriate grade distribution (Table 2) *Fair*
2. Achievement scores (number of students = 38)

112	100	93	84	78	72	66	51
109	97	91	83	75	71	62	47
106	97	90	82	75	70	59	44
105	95	89	81	75	69	59	
104	95	84	80	74	68	58	

B. Calculations from the data

1. Median $\frac{81 + 80}{2} = 80.5$

2. Standard deviation $\frac{636 - 318}{19} = 16.7$

Marks	*Lower Limits*	*Intervals*	*Number*	*Percent*
A	80.5 + 1.3 × 16.7 = 102.2	103–112	5	13
B	102.2 − 16.7 = 85.5	86–102	9	24
C	85.5 − 16.7 = 68.8	69–85	15	39
D	68.8 − 16.7 = 52.1	53–68	6	16
F		44–52	3	8
			38	100

THE BASIS FOR MARKS

When the instructor determines a course mark, as he usually does, by combining marks on daily recitations, homework, term papers, and scores on quizzes and tests, each of the components carries more or less weight in determining the final mark. To obtain marks of maximum validity, the instructor must give each component the proper weight, neither too much nor too little. How can he determine what those weights ought to be? How can he determine what they actually are? If what they are does not correspond to what they should be, what can he do?

It is not easy to give a firm, precise answer to the question of how much influence each component ought to have in determining the final mark. But several guiding principles can be suggested.

In general, the use of several different kinds of indicators of competence is better than use of only one, provided that each of the indicators is relevant to the objectives of the course and provided also that it can be observed or measured with reasonable reliability.

Exclusive reliance on tests, for example, may give an unfair advantage to students who have special test-taking skills and may unfairly handicap students who give the best account of their achievements in discussions, on projects, or in other situations. But irrelevant accomplishments, such as mere glibness, personal charm, or self-assurance, should not be mistaken for solid command of knowledge. Nor should much weight be placed on vague intangibles or subjective impressions that cannot be quantified reliably.

If measures of each component aspect of achievement are highly correlated, the problem of weighting them properly is far less critical than if they are quite unrelated. For most courses the various measurable aspects of achievement are related closely enough so that proper weighting is not a critical problem. The natural "unweighted" weighting will give marks almost as valid as those resulting from more sophisticated statistical procedures.

The actual weight that a component of the final mark does carry depends on the variability of its measures and the correlations of those measures with measures of the other components. This makes the precise influence of a component quite difficult to determine. As a first approximation to the weight of a component, the standard deviation of the measures of that component serves quite well. If one set of scores is twice as variable as another, the first set is likely to carry about twice the weight of the second.

Table 4 shows that the influence (weight) of one component (for example, scores on one test) on a composite (the sum of scores on three tests, in this example) depends on the variability of the test scores. The top section of the table displays the scores of three students, Tom, Dick, and Harry, on three tests, X, Y, and Z, along with their total scores on the three tests. Dick has the highest total and Tom the lowest. The next section shows how the students ranked on the three tests. Each of them made the highest score on one test, middle score on a second, and lowest score on the third. Thus the totals of their ranks, and their average ranks, are exactly the same. But note, for future reference, that the ranks of their total scores on the three tests are the same as their ranks on Test Z.

The third section of the table gives the maximum possible scores (total points), the mean scores, and the standard deviations of the scores on the three tests. Test X has the highest number of total points. Test Y has the highest mean score. Test Z has scores with the greatest variability.

On which test was it most important to do well? On which was the payoff for ranking first the highest, and the penalty for ranking last the heaviest? Clearly on Test Z, the test with the greatest variability of scores. Which test ranked the students in the same order as their final ranking, based on total scores? Again the answer is Test Z. Thus the influence of one component on a composite depends not on total points or mean score but on score variability.

Now if the three tests should have carried equal weight, they can be made to do so by weighting their scores to make the standard deviations equal. This is illustrated in the last section of the table. Scores on Test X are multiplied by 4, to change their standard deviation from 2.5 to 10, the same as on Test Z. Scores on Test Y are multiplied by 2, to change their standard deviation to 10 also. With equal standard deviations the tests carry equal weight, and give students having the same average rank on the tests the same total scores.

A more sophisticated method of assuring equal weighting is to convert raw scores into standard scores. If X represents a student's raw score on a given test, M_x the mean score on that test, and σ_x the standard deviation of the scores, then the student's standard z-score is found from the formula:

$$z = \frac{X - M_x}{\sigma_x}.$$

This formula transforms a set of raw scores into a set of scores with a mean of 0.0 and a standard deviation of 1.0. While such scores are perfectly valid and significant, they tend to look unfamiliar, as we are used to scores with a range of 0 to 100. To correct this deficiency, we make a second transformation to produce the T-score, by the formula:

$$T = 10z + 50.$$

This transformation produces a set of scores with a mean of 50 and standard deviation of 10.

TABLE 4
WEIGHTED TEST SCORES

Tests	*X*	*Y*	*Z*	*Total*
Student scores				
Tom	53	65	18	136
Dick	50	59	42	151
Harry	47	71	30	148
Student ranks				
Tom	1	2	3	3
Dick	2	3	1	1
Harry	3	1	2	2
Test characteristics				
Total points	100.0	75	50	225.0
Mean score	50.0	65	30	145.0
Standard Deviation	2.5	5	10	6.5
Weighted scores	×4	×2	×1	
Tom	212	130	18	360
Dick	200	118	42	360
Harry	188	142	30	360

The standard scores corresponding to the example of Table 4 are shown in Table 5. Here we see, as shown in the last section of Table 4, that with equal weighting the students have equal scores. We also note a fact not obvious before, that with equal weighting all three students have an average equal to the class average. This is another advantage of standard scores, that a student's performance relative to the class performance is evident by inspection. However, a considerable amount of computation is required to convert raw scores to standard scores so their use may not be considered practical, unless the teacher has access to a computer, in which case the process may be easily mechanized.

When the whole possible range of scores is used, score variability is closely related to the extent of the available score scale. This means that scores on a 40-item objective test are likely to carry about four times the weight of scores on a 10-point essay test question, provided that scores extend across the whole range in both cases. But if only a small part of the possible scale of scores is actually used, the length of that scale can be a very misleading guide to the variability of the scores.

In view of the difficulty of determining precisely how much weight each component ought to carry, the difficulty of determining precisely how much weight each component does, in fact, carry seems less serious as an obstacle to valid marks. Further, as we have noted, if the components are quite highly related, the difference between optimum and accidental weighting may be hard to detect, as it affects the validity of the marks. But if the instructor finds a serious discrepancy between what he

TABLE 5
STANDARD TEST SCORES

Tests	*X*	*Y*	*Z*	*Average*
z-scores				
Tom	1.2	0	−1.2	0
Dick	0	−1.2	1.2	0
Harry	−1.2	1.2	0	0
T-scores				
Tom	62	50	38	50
Dick	50	38	62	50
Harry	38	62	50	50

thinks the component weights ought to be and what they in fact are, two courses are open to him.

One is to adjust the scores, by means of compensating weighting factors or conversion to standard scores. The other is to increase the number of observations of the underweighted component, or the precision with which it is measured, and hence also to increase the weight it carries. Of the two methods, the first is likely to be more convenient. The second is likely to yield the more reliable, and in this case the more valid, marks.

If an instructor has promised his class, for example, that the final mark will be based on five components, weighted as follows:

Contributions in class	15%
Daily assignments	20%
Term paper or project	15%.
Midterm test	20%
Final test	30%

Then he should plan to obtain enough independent scores on "contributions in class" so that the variability of the total of those scores is about half the variability of the scores on the final test. By the same token, the final test should be half again as long (or include half again as many items) as the midterm test.

A wise instructor will warn his students that the actual weight of each component may differ somewhat from the intended weight. But he can assure them, and rest assured himself, that if he has planned carefully to make the weights what he intended, the inevitable deviations of the actual from the ideal weights will not affect the validity of his marks appreciably.

One final admonition. It is a mistake to convert test scores to letter marks, record these in the grade book, and then reconvert the letter marks to numbers for purposes of calculating the final average. A better procedure is to record the test scores and other numerical measures directly. These can be added, with whatever weighting seems appropriate, to obtain a composite score which can then be converted into the final mark.

Not only does the recording of scores rather than letters usually save time in the long run, it also contributes to accuracy. Whenever a range of scores, some higher, others lower, is converted to the same letter mark, information is lost. Usually this information is not retrieved when the letter marks are changed back to numbers so they can be added or averaged. Each B, whether a high B or a low B in terms of the score on which it was based, is given the same value in the reconversion. Hence to avoid the loss of score information it is usually desirable to record the raw scores, not the scores after conversion to letter marks.

ONE SOLUTION TO TWO PRACTICAL PROBLEMS

Instructors sometimes feel inclined to include marks on daily work in the final course work, not so much to improve the validity of the final mark as to influence students to do the daily work. The daily assignments, after all, are intended to serve primarily as learning exercises, not as measures of achievement. Including these daily work marks also adds considerable labor, often without improving final mark validity appreciably. There is a way of maintaining the incentive to do the daily work while avoiding the labor of adding a multitude of numbers for each student. This way helps to solve another problem too: the problem of the anguished student who just missed a higher mark by a point or two. Here is the way.

Announce to students that their final marks will be determined primarily by their scores on major tests (midterm and final perhaps) and one or two major papers or projects. After points on these few major measures of achievement have been added, lower limits for each of the marks will be determined. Then for any students whose total puts him within three (or perhaps five) points of the next higher mark, the record of daily work will be reviewed. If the record shows that the student has done the assignments conscientiously and reasonably well, he will be given the next higher mark.

Under this system most students will feel sufficiently motivated to do the daily assignments. They are less likely to feel that only an inconsequential deficiency in achievement kept them from getting a higher mark. Respect for the fairness of the instructor's marking procedures is likely to be enhanced.

REFERENCES

[1] E. L. Thorndike, *Mental and Social Measurements*, 2d ed (New York: Teachers College, Columbia University, 1912), chap. 2.

[2] S. Trevor Hadley, "A School Mark—Fact or Fancy?" *Educational Administration and Supervision*, vol. 40 (1954), 305–12.

[3] I. N. Madsen, "To Mark or Not to Mark," *Elementary School Journal*, vol. 31 (June 1931), 747–55.

[4] Robert M. W. Travers and Norman E. Gronlund, "Meaning of Marks," *Journal of Higher Education*, vol. 21 (1950), 369–74.

[5] C. W. Odell, "Marks and Marking Systems," in *Encyclopedia of Educational Research*, ed. Walter S. Monroe (New York: The Macmillan Company, 1950), pp. 711–17.

[6] John R. Hills and Marilyn B. Gladney, "Factors Influencing College Grading Standards," *Journal of Educational Measurement* vol. 5 (1968), 31–39.

[7] Orville Palmer, "Seven Classic Ways of Grading Dishonestly," *The English Journal* (October 1962), pp. 464–67.

[8] Daniel Starch and E. C. Elliott, "Reliability of Grading Work in History," *School Review*, vol. 21 (1913), 676–81; "Reliability of Grading Work in Mathematics," *School Review*, vol. 21 (1913), 254–59; "Reliability of the Grading of High School Work in English," *School Review*, vol. 20 (1912), 442–57.

[9] G. M. Ruch, *The Objective or New-type Examination* (Chicago: Scott, Forestman & Company, 1929), pp. 369–402.

[10] W. M. Stallings and others, "Pass-Fail Grading Option," *School and Society*, vol. 96 (1968), 179–80.

[11] M. R. Sgan, "First Year of Pass-Fail at Brandeis University: A Report," *Journal of Higher Education*, vol. 40 (1969), 135–44.

[12] Charles C. Peters and Walter R. Van Voorhis, *Statistical Procedures and Their Mathematical Bases* (New York: McGraw-Hill Book Company, 1940), pp. 393–99.

[13] Robert L. Ebel, "The Relation of Scale Fineness to Grade Accuracy," *Journal of Educational Measurement*, vol. 6 (1969), 217–21.

Marilla D. Svinicki
The University of Texas at Austin

The Test: Uses, Construction And Evaluation

"For many instructors, concern with a test ends when the last paper is returned. This is most unfortunate, because further analysis can provide valuable information about the questions and instruction."

One of the important events in the learning-teaching cycle is the test. It is important to the learner because it offers the opportunity to demonstrate what skills have been developed and receive feedback on those skills. It communicates more than anything else what the instructor's goals are. The grades received on tests contribute significantly to the course grade and thus play a part in determining the learner's future. The test is important for the teacher because it provides a basis for the evaluation of student progress and feedback on the adequacy of instruction.

Given that the test is of such importance, it seems logical that it should receive a great deal of attention. Unfortunately, too many of us, teachers and learners alike, wait until the night before to prepare for a test. Planning the instructional role of tests more carefully is one thing a teacher can do immediately to improve teaching.

Uses of a Test

The most common and most obvious use of the test is the evaluation of student progress in order to assign a grade. This use, however, does little to make the test an active part of the learning process, since it occurs after learning has taken place. Although we hope the learner revises his conceptions of the material according to feedback from the test, there is little motivation for continuing to study that material further once the test is complete.

With a little thought, tests can be made a more integral part of the learning process by being used for several types of diagnosis as well as evaluation. In the first diagnostic case, the instructor uses the test to improve efficiency of instruction. For example, a *pretest* at the outset of the semester can give the instructor valuable information about a class's or an individual's grasp of prerequisite and planned course material. To assume that each semester's class has the same set of skills and interest is shortsighted at best. By using a pretest to actually assess the capabilities of the students, the instructor can avoid unnecessary coverage of material in which they are already proficient or provide remediation in prerequisite areas in which they are weak. Of course, this use of pretesting requires an instructor to be flexible in structuring the course content, since it may turn out that the first three weeks of planned material are totally unnecessary, but it allows students to be treated more as individuals and makes instruction more efficient. In modularized, self-paced courses, pretesting can be used to individualize instruction even further by allowing students to skip over modules on familiar material and concentrate on the unfamiliar.

Reprinted with permission from *Engineering Education,* vol. 66, no. 5, pp. 408–410, February 1976.

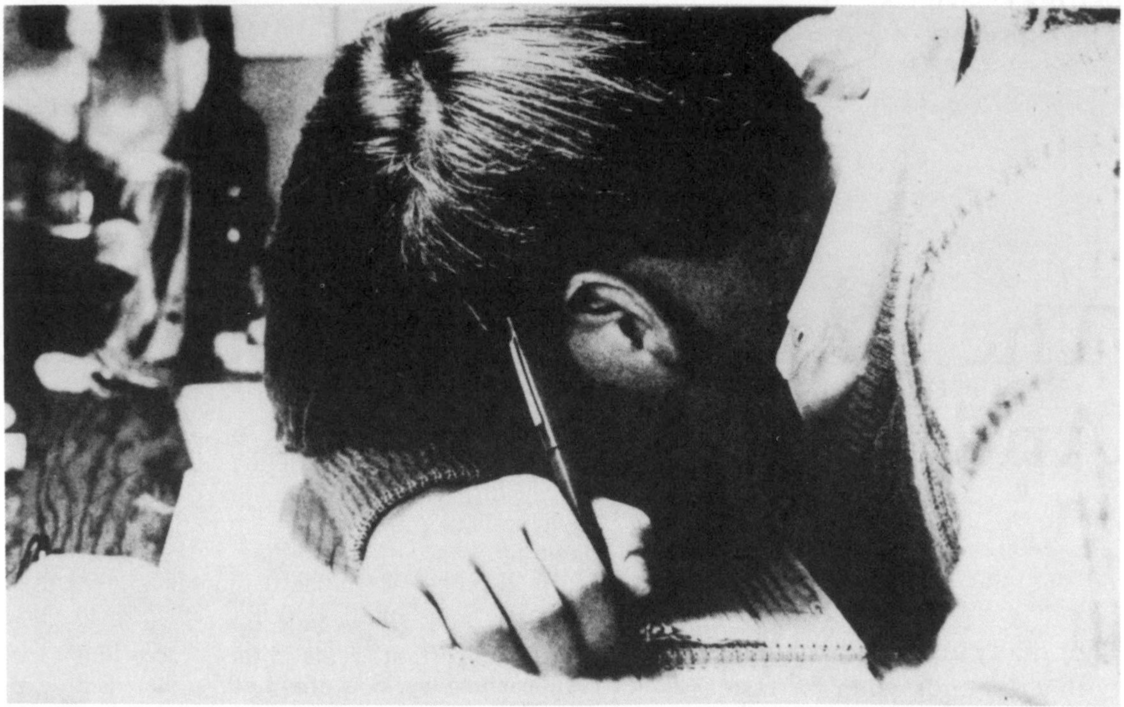

A second diagnostic use of tests calls for frequent, short, non-punitive exams which occur while learning is taking place. The purpose of these exams is to provide the learner with as *rapid and frequent feedback* as possible. Such feedback is an integral part of the learning process. The test gives the learner an opportunity to make the learned response overtly and have it evaluated by another knowledgeable person—a classmate, a proctor or the instructor. Since these tests do not contribute substantially to summative evaluation (the "grade"), they can be administered and taken in an unthreatening, supportive atmosphere in which the instructor and student are viewed as partners in learning rather than as adversaries. Errors on this type of test provide information regarding material that is not fully understood or is being misinterpreted. This information, coming as it does before the summative evaluation, motivates the student to go back and relearn the material in preparation for the "real" exam. And since the diagnostic test points out which areas need the most work, study time can be used with the greatest efficiency. In addition, these diagnostic exams give the student a feeling for the type and level of response expected and minimize the test anxiety which so often hampers performance. It is most important that these tests 1) occur frequently (e.g., once a week), and 2) be viewed by both instructor and student as non-evaluative aids to learning.

The diagnostic use of tests just discussed alerts the *learner* to deficiencies in his understanding, but the same diagnostic tests can be used to alert the *teacher* to weaknesses in the presentation of the material as well. Poor overall student performance on a test or on a particular question could show that the teaching materials covering that point were poorly structured. When an entire class, with only two or three exceptions, fails to grasp a particular concept, it is ridiculous to blame it on "poor student attitude" or "ignorance." The instructor is responsible for locating the source of the problem and revising the teaching presentation so that learning can proceed smoothly. Therefore, the instructor can obtain as much diagnostic feedback from these frequent tests as do the students.

Constructing a Test

When a test is being used as a summative evaluation for the purpose of grade assignment, it deserves to be as carefully planned as the instruction itself. Since the test communicates to the student what types of things must be learned, it should accurately reflect the instructional objectives for the course. Too often, instructional objectives are stated in such terms as, "The student will be able to critically analyze the events of history," while the test questions run more along the lines of, "Who fought the Battle of Waterloo?"

It is not particularly difficult to avoid this kind of problem through planning on the instructor's part. Prior to writing the actual test, the instructor should construct a "test blueprint" as shown in figure 1. The vertical rows represent the major

Levels of Cognitive Complexity	Concepts To Be Covered in Test											
	1	2	3	4	5	6	7	8	9	10	11	12
Knowledge	X			X	X		X		X	X	X	X
Comprehension		X		X		X	X	X	X		X	
Application	X		X		X			X		X		X
Analysis			X			X			X		X	
Synthesis	X											
Evaluation		X				X		X				

Figure 1. A sample test blueprint.

Number of Students in Class: 30 (10 students in a division)

Scores in descending order

100	100	97	95	93	89	85	83	82	79	Top Third of Class (ten students)
78	78	77	72	69	67	63	51	50	50	Middle Third of Class (not used in evaluation)
35	32	30	28	26	25	18	12	10	10	Bottom Third of Class (ten students)

Item Analysis

Question Number	*Number answering correctly*		*Discrimination Ratio*
	Top Third	Bottom Third	
1	9	3	$\frac{9-3}{10} = +0.6$
2	7	7	$\frac{7-7}{10} = 0.0$
3	2	3	$\frac{2-3}{10} = -0.1$
4	1	7	$\frac{1-7}{10} = -0.6$

Figure 2. Computing item discrimination ratios.

concepts being covered in the material to be tested. The horizontal rows represent the levels of the cognitive hierarchy as described by Bloom.[1] The use of this particular hierarchy is not critical. There may be an alternative analysis with which an individual instructor is more comfortable. Most important, the horizontal rows represent increasing levels of cognitive complexity in the handling of a concept. The lowest level (knowledge) involves the mere repetition of facts and definitions, while the highest level (evaluation) involves the critical analysis of theories or other presentations. The instructor then considers at which of these levels the various content concepts should be tested and indicates this in the corresponding square of the matrix. It is important at this point not to become too idealistic or too pessimistic about the levels at which students should be functioning. In a world of geniuses, all concepts could be dealt with at an "evaluation" level. In a normal teaching situation, the levels which are tested should conform to the anticipated abilities of the students. For example, in beginning-level courses the emphasis should be on lower-level skills which will form the basis for future concepts or courses. In upper-division courses students would be expected to perform at the higher levels. Within an individual course the instructor may wish to begin the semester with the lower-level objectives and gradually move toward higher levels near the end of the semester. Attention to such transitions is needed to promote the student's growth within and across courses.*

When the time comes to construct the test, a question appropriate to the level and concept indicated by each X is prepared. This will ensure that the test covers the range of concepts the instructor desires. In writing these questions, the instructor should keep in mind that questions on lower-level objectives can usually be answered more rapidly, and therefore more can be included on the exam. Testing over the higher levels will limit the number of concepts which can be sampled in any one exam. Further details of question construction itself are too numerous to pursue here, but information on this topic is available in a definitive work on testing written by Bloom, Hastings and Madaus.[2]

Once the test is constructed it is helpful to have someone else go over the questions to look for unclear wording. The purpose of a test is to evaluate students' grasp of the material, not their ability to read and interpret questions. "Trick" questions waste both the instructor's and the student's time. As an alternative to external review, the instructor can put the test aside for several days before looking it over again. This may help him to recognize problems in the questions which were not evident during their construction.

After the Test Is Over

For many instructors, concern with a test ends when the last paper is returned. This is most unfortunate, because further analysis of the test scores can provide valuable information about the test questions and instruction. There are several elaborate statistical analyses which can be applied to the test scores and may be available through an institutional bureau such as those which administer large scale achievement tests. The instructor personally can do a simplified analysis which will provide enough information for his or her needs.

The first step in such an analysis is to divide the test papers into three groups on the basis of the overall score (as shown in figure 2). For each question, the number of students in the top third who answered correctly is determined. (For essay or problem questions in which partial credit is given, the instructor must set a limit as to what score constitutes "correct.") The same determination is made for the bottom third of the students. As shown in figure 2, a discrimination ratio is then computed by subtracting the number of students in the bottom division who answered correctly from the number

"The purpose of a test is to evaluate students' grasp of the material, not their ability to read and interpret questions."

*For more on Bloom's hierarchy, see p. 396.

of students in the top division who answered correctly and dividing by the number of students in a division. This discrimination ratio will vary between -1.0 and +1.0. The closer the ratio is to +1.0, the more effectively that question has discriminated the students who apparently know the material (the top third) from those who do not (the bottom third). Questions with high positive ratios (+0.5 or better) can be kept for use in future classes. When the discrimination ratio lies in the negative range, as with Question 4 in the example, the students who should know the material are answering the question incorrectly while the poorer students are choosing the correct answer. One can assume that the question is misleading those who know too much, and should be revised for future use. This can be done by examining the answers given by the top students who responded incorrectly. Is there a similarity among their answers which gives a clue to what is wrong with the question? This information can then be used to improve the structure of the question.

When the discrimination ratio hovers around 0.0, the question is not discriminating among the students very well. In the case of Question 2 in the example, most students are answering correctly, indicating that the question is fairly straightforward and not difficult. The continued use of such a question is a matter of the instructor's personal choice. It is useful to have questions covering a range of difficulty on examinations, particularly when they are to be used diagnostically, and therefore the instructor may wish to keep this question for future use. On questions like number 3, the -0.1 ratio results from the fact that no one did well on it. The instructor should carefully look over that question or the instruction. Such questions aid an instructor in improving his teaching by providing feedback on where improvement is needed.

Questions used on tests plus the discrimination ratio information should be retained. As time passes the instructor will develop a large pool of useful items which can be sampled in future courses, saving time and effort and providing the opportunity to study change in students across semesters.

In Conclusion

The imaginative use of tests in the learning-teaching cycle adds a valuable tool to the teacher's repertoire. They can be used for diagnosis as well as evaluation by both teacher and learner. To be of greatest value, they require careful attention both before and after their use. Used and analyzed appropriately, tests can inform the instructor about deficiencies in the presentation of material which can then be employed to change the teaching strategy and improve its efficiency. By following the suggestions presented here, an instructor can make an immediate improvement in his teaching, and continued use and refinement of these techniques will benefit both teacher and learner alike.

References

1. Bloom, B. S. (ed.), *Taxonomy of Educational Objectives: The Classification of Educational Goals. Handbook I: Cognitive Domain,* McKay, New York, 1956.

2. Bloom, B. S., J. T. Hasting, and G. F. Madaus, *Handbook on Formative and Summative Evaluation of Student Learning,* McGraw-Hill, New York, 1971.

Marilla D. Svinicki
The University of Texas
at Austin

A Test Is Not Necessarily a Test

Let's face it. Evaluation is a fact of academic life. The society outside the ivy-covered walls frequently, however unconsciously, looks to colleges and universities to tell it who the "good" students are. No matter how many studies conclude that academic measures such as grade point average bear only the most tenuous relationship to success in other areas, still things such as good jobs, admission to graduate school and countless other desirables are awarded with a heavy emphasis on the student's grade point average. This places the university professor in a real dilemma. The evaluation measure administered to students for the purpose of assigning grades must be of the highest quality. Those measures must as accurately as possible reflect a student's abilities in that area. The student's performance on those measures might someday determine his or her future.

This is a frightening prospect when one is aware of how most college tests are constructed. In a recent booklet Ohmer Milton and John Edgerly reported some of the abuses of testing they found in a survey of college and university officials. The following is a typical example:

> Students in a senior course were assigned 15 journal articles, the shortest of which was 14 pages long. The only examination question over this considerable volume of material asked students to match the articles' authors with the titles. Challenged by a colleague, the instructor argued he could assume that a student who could do this matching understood the material.[1]

Quite an assumption, that! In my own experience working with faculty I have often had instructors proudly tell me about the "clever" tests they give. For example, one history professor who taught a course on the history of the conquest and settling of the Western Hemisphere has told me that the only question on his final exam (worth at least a third of the students' total course grade) went something like this: *"You are a Spanish explorer who has landed around the Louisiana area in 1500. How would you go about conquering the continent?"* To top it off, the course had been taught totally by lecture and therefore had done nothing to prepare the students for this type of question.

Because the results of testing are used for much more than simply providing feedback to students on their performance, it is incumbent on the instructor to work as hard on constructing good tests as on providing good instruction. There are no simple shortcuts to writing good tests. It is hard work, requiring a lot of time.

A previous article discussed the procedures an instructor could use to plan and use tests appropriately.[2] That discussion dealt mostly with practical concerns of test construction along with how to evaluate the individual test questions for their power to discriminate among students. The present article will be more theoretical, or philosophical, about tests and some of the frequently mentioned concepts related to testing.

The Underlying Statistics

If you read any books or articles on testing, you are bound to run across two crucial measures, *validity* and *reliability*. Just what are these concepts and what do they mean for the classroom teacher?

Both validity and reliability are statistically based on correlations between two independent measures of the same student's performance. (When we say two things are correlated statistically we mean they tend to vary in the same manner.) In terms of student performance this means that students who show excellent performance on one of our measures also show excellent performance on the other measure. The same correspondence of performance holds for students whose performance is poor. This does not mean that good performance on one task *causes* good performance on the second. It is only a statement that they are in some way related, possibly because they are both measures of the same underlying ability. This assumption forms the basis for computations of both validity and reliability. Thus when investigating either of these measures statistically, a teacher will be

Reprinted with permission from *Engineering Education,* vol. 68, no. 5, pp. 410–413, February 1978.

"There are no simple short-cuts to writing good tests. It is hard work, requiring a lot of time."

using a method of computing correlations. One useful and simple method is the Spearman Rank Order procedure.

In this statistical procedure students are ranked on two measures (such as scores on two different exams) and the degree of disparity between the ranks is determined. For example, if all students ranked exactly the same on both tests, there would be no difference (ranked 1 on measure one, ranked 1 on measure two, difference is 0), whereas if the highly ranked students on one measure ranked in the bottom on the second measure, the difference would be great (ranked 1 on measure one, ranked 50 on measure two, difference is 49). See the Computation of Correlations, page 412, for a demonstration of the calculations involved in this procedure. This computation can provide a rough estimate of the correlation between any two sets of scores, for the purpose of evaluating either validity or reliability. Far more sophisticated procedures are available, but this one is easy to compute and most useful for the classroom instructor.

Validity

Validity is a straightforward concept. Measures of validity tell if a test is assessing the knowledge or aptitude the instructor intends it to measure. For example, a test on which the students are required to list the components of a machine is *not* a valid measure of their ability to calculate its efficiency. A far more valid test would provide the students with the necessary data and require them to actually compute the efficiency. Similarly, the ability to *solve* an equation or formula does not mean the student can *apply* the formula appropriately.

Among the several aspects of evaluating the validity of a test you give, three are most useful in the normal teaching situation. First is *content validity*. To assess content validity the instructor determines how well the test samples the array of information for which he is testing. This is not a numerical analysis, but a logical analysis of the test and what it covers. In an earlier article[2] the use of a "test blueprint" (see figure 1) to assist the instructor in writing a test was described. Using this blueprint, the instructor can build a test with good content validity by outlining the concepts and objectives of the unit of study and identifying at which level students are expected to perform. From this a test question may be constructed for each concept at its desired level.

Tests with a one-to-one correspondence between items and course objectives or content as outlined in the blueprint have high content validity. Often, however, it is impossible for instructors to include everything. In those cases they sample from the content universe and use questions that represent the whole. We assume that if we sample correctly, a student's performance on the sample is the same as it would be if we were able to test for everything. The blueprint once again helps the instructor ensure that the sample of questions accurately reflects the entire range of skills desired on each unit.

A numerically rather than conceptually based measure of validity is *predictive validity*. If a test has predictive validity, then student performance on that test will have a high correlation with performance on a later assessment, which presumably is related. For example, a test used to place students in various math classes is based on an assumption of predictive validity, that is, those students who score highly on the test would do well in any math class compared with those who score lower. For this reason they are placed in advanced classes more commensurate with their ability. In the classroom where material is sequenced because concepts build on one another, the instructor can analyze test results "post hoc" for predictive validity. As students move to the next unit, the instructor correlates performance on unit two with the test results from unit one using the statistical procedures described earlier. High correlations indicate that the unit one test does indeed predict which students have the necessary skills for unit two work; and that test is then considered a valid basis for permitting future students to skip unit one and move directly to unit two.

A similar numerically based measure is *concurrent validity* which refers to the degree to which a test is correlated with some other test of

1) To set up a test blueprint, list the concepts to be tested along the top edge of a matrix and the levels of complexity along the left edge. For each concept, check the levels to be tested.

Concepts / *Levels*	1	2	3	4	5	6	7	8	9	10
Knowledge	X			X	X		X		X	X
Comprehension		X		X			X	X	X	X
Application	X		X			X	X	X		
Analysis		X		X	X			X		
Synthesis	X									
Evaluation		X			X	X				

2) Write one question for each X. If that is not possible, at least be sure your test has approximately the same proportion of coverage at each level as is shown on your blueprint.

Figure 1. Constructing a test blueprint.

the same ability or knowledge measured at the same time. But, you say, if we have one test already, why do we need another? Usually it is because the first test is too long or too difficult to administer, or is for some reason not optimal for your use, although it may have very high content validity. This happens quite frequently with intelligence testing. For example, individually administered IQ tests are desirable but impractical if a large number of students are to be tested. The solution is to develop a test that can be administered to many students simultaneously and which is concurrently valid, that is, scores on the group administered test correlate highly with scores on the individually administered tests. Once this correlation is established, the group administered tests can be used as a substitute for the individually administered tests.

Instructors may have similar problems. They may wish to administer individual lab tests, for example, in which each student demonstrates a particular lab procedure, but doing this in a class of 200 would be taxing, to say the least. It therefore would be desirable to develop a written test which could be administered to large numbers of students but which would produce the same evaluation of the students' skills as an individually administered test. To produce such a test, instructors must analyze the concepts involved in the lab demonstration and construct a written test that reflects those concepts. This written test and the individual lab test must then be administered to a large number of students, obtaining a score on each measure for each student. These scores can then be used to compute a correlation similar to the one for predictive validity. If the correlation value obtained is high, the instructor could substitute the written test for the lab test when use of the latter is impractical.

In all three measures of validity just discussed the basic idea is the same: "Does the test measure the underlying ability it purports to measure?" In the cases discussed above, an instructor evaluates the validity of exams by 1) analyzing the content (content validity), 2) examining the accuracy with which the test predicts future related performance (predictive validity), or by 3) comparing test performance to another measure of the same ability (concurrent validity). Discrepancies in any of these areas require the instructor to look closely at the exam to determine what it is about the exam or the instruction that might be causing the difficulty. A good thing for an instructor to do is to engage in the "test blueprint" analysis described earlier to be sure the test reflects the instructional objectives accurately. This activity will have the highest payoff in the end, since it is from this type of careful planning that validity proceeds.

Reliability

Another term frequently associated with testing is "reliability." Reliability is an indicator of the stability with which a test is measuring whatever it is measuring. The scores of the same student on an unreliable test taken at different times will vary widely. This would not be the case for a reliable test. A reliable test measures whatever it measures consistently. There are three ways reliability of a test is evaluated:

1) Internal consistency. This is the method most useful to the classroom instructor. It is based on

Computation of Correlations*

(Spearman Rank Correlation Coefficient: r)

1. Rank the students' scores on each measure (X and Y) individually according to their total score. If ties occur, assign the average rank of the scores. For example, if each of three students has the same top score on one measure, each is ranked as 2.

2. Determine the difference (d_i) for each subject by subtracting the Y rank from the X rank and square this value to obtain d_i^2 for each student. Sum all the d_i^2s to obtain Σd_i^2.

3. If there are no or a small number of ties, compute the correlation coefficient (r) using the formula:

Formula A $$r = 1 - \frac{6\Sigma d_i^2}{N^3 - N}$$ Where N = the number of students

If many ties occur, the formula must be changed to:

Formula B $$r = \frac{\Sigma x^2 + \Sigma y^2 - \Sigma d_i^2}{2\sqrt{\Sigma x^2 \Sigma y^2}}$$

where $\Sigma x^2 = \frac{N^3 - N}{12} - \Sigma T_x$ and $\Sigma T_x = \frac{t_{x1}^3 - t_{x1}^3}{12} + \frac{t_{x2}^3 - t_{x2}^3}{12} \cdots$

and $\Sigma y^2 = \frac{N^3 - N}{12} - \Sigma T_y$ and $\Sigma T_y = \frac{t_{y1}^3 - t_{y1}^3}{12} + \frac{t_{y2}^3 - t_{y2}^3}{12} \cdots$

where t_{x1} = the number of tied scores at rank 1 on measure X and so on

and t_{y1} = the number of ties on Y rank 1 and so on

4. The critical values of r where N is less than 10 are shown in the accompanying table A. When N is greater than 10, the r should be converted to a student's t with degrees of freedom of N-2 by the formula:

$$t = r\sqrt{\frac{N - 2}{1 - r^2}}$$

and the significance compared with those in table B or a comparable Student's t table in any statistics book. If the $|r|$ or $|t|$ value computed by the formulas shown is greater than the value listed in the table, the correlation between the two sets of scores is considered to be *statistically* significant. The values can range from −1.0 to +1.0. Values in the −1.0 to 0.0 range indicate a negative correlation, that is, students scoring high on one measure score low on the other. Values in the 0.0 to +1.0 range indicate a positive correlation, that is, students scoring high on one measure also score high on the other measure. The critical values shown in the two tables indicate what r or t score you must find in order to say that there is less than a 5 percent chance that the correspondence between the two measures is due to chance alone.

the idea that if we were to divide our test into two equivalent tests, the students' performance on these "two" tests would be approximately the same. The instructor can actually divide the test into two halves by matching items of equal difficulty and putting a student's score on one item on one "test" and the score on the matched item on the second "test". Performance on these two split-halves is compared through the same correlation procedure used in the validity tests. A reliable test will show a high degree of correlation between scores on the two halves of the test.

2) Equivalence. Instructors can sometimes produce two actual versions of a single exam and administer the two versions to the same students at approximately the same time. If the exams are reliable, the distribution of student scores on the two exams should be highly correlated. This is not particularly practical for the classroom instructor, but can be used by departments to produce proficiency exams. If one has a test of already demonstrated reliability, other tests can be compared with this "standard" using this method.

3) Stability. Finally, another reliability procedure is to administer equivalent forms of a test to the same student at two different times and to correlate his or her performance on the two versions as before. A reliable test should produce the same approximate distribution of students on both occasions.

An example:

Student	*Rank on X*	*Rank on Y*	d_i	d_i^2
A	1	3	−2	4
B	2.5	5.5	−3	9
C	2.5	1	+1.5	2.25
D	4	2	+2	4
E	5	4	+1	1
F	6.5	5.5	+1	1
G	6.5	7	−.5	.25
H	8	8.5	−.5	.25
I	9.5	10	−.5	.25
J	9.5	8.5	+1	1
				$\Sigma d_i^2 = 23$

Since so many ties occurred, we will use formula B.

$$\Sigma x^2 = \frac{10^3 - 10}{12} - \Sigma T_x \quad \text{where } \Sigma T_x = \frac{2^3 - 2}{12} + \frac{2^3 - 2}{12} + \frac{2^3 - 2}{12}$$

$$\Sigma x^2 = 82.5 - 1.5$$

$$\Sigma x^2 = 81$$

$$\Sigma y^2 = \frac{10^3 - 10}{12} - \Sigma T_y \text{ where } \Sigma T_y = \frac{2^3 - 2}{12} + \frac{2^3 - 2}{12}$$

$$\Sigma y^2 = 82.5 - 1$$

$$\Sigma y^2 = 81.5$$

$$r = \frac{\Sigma x^2 + \Sigma y^2 - \Sigma d_i^2}{2\sqrt{\Sigma x^2 \Sigma y^2}} = \frac{81 + 81.5 - 23}{2\,(81.25)} = \frac{139.5}{162.5} = .86$$

Taking the .86 value computed and the N of 10 students, we find in table A below that we only needed a value of .56. Since .86 is greater than .56, we can conclude that these two measures are correlated in some way so that we could—with some assurance—predict a student's score on one by knowing the other.

Critical values of r for various N's

(Given that the value is in the predicted direction of a positive or negative correlation.)

Table A if N =	then $\|r\|$ =
4	1.00
4	.90
6	.83
7	.71
8	.64
9	.60
10	.56

Table B if N =	then $\|t\|$ =
15	1.77
20	1.73
25	1.71
30	1.70
40	1.68
60	1.67

(For additional t values, consult any table of student's t using N − 2 degrees of freedom and a one-tailed test)

*Seigel, S.[3]

Conclusion

Statistical manipulation of test data by the classroom teacher may or may not always be practical. The concepts of validity and reliability, however, are most important and should be at the back of an instructor's mind whenever he or she is constructing an evaluation instrument. Instructors can, across semesters, use these statistical procedures to produce a file of valid, reliable tests and test items, and use item analysis on each test as described in the article mentioned earlier,[2] thus improving their tests over time. Departments can definitely use the concepts when producing exams for placement or other general purposes. The primary contribution of these concepts is to make instructors sensitive to the need for careful production and analysis tests and to assure that their test procedures accurately reflect the students' performance in their classes.

References

1. Milton, Ohmer and John Edgerly, *The Testing and Grading of Students,* Change Magazine and Educational Change, 1976.

2. Svinicki, Marilla D., "The Test: Uses, Construction and Evaluation," *Engineering Education,* vol. 66, no. 5, February 1976.

3. Siegel, S., *Non-parametric Statistics for the Behavioral Sciences,* McGraw-Hill Book Co., 1956.

Suggestions on the construction of multiple-choice tests

H. T. Hudson
Department of Physics, University of Houston Central Campus, Houston, Texas 77004

Carolyn K. Hudson
Foundations of Education Department, University of Houston Central Campus, Houston, Texas 77004
(Received 20 November 1979; accepted 15 December 1980)

Multiple-choice tests are becoming more and more common in physics classes. In the construction of such tests, attention to certain details will minimize the effects of random guessing. In addition, it has been found that for the majority of students, when a reasonable number of tests are given on a schedule throughout the semester the cumulative grade from such tests can provide essentially the same information as would hand-graded problems.

Classes at the college and university level have been undergoing a transition from the traditional small-sized groups to the giant lecture sections. Like it or not, physics has joined this trend and the introductory level classes have grown and grown, limited only by the physical size of the classroom. Unfortunately, the dark forces that have shaped the destiny of the introductory physics class did not see fit to accompany the compacting of students with commensurate support to the teacher. The end result is a situation where traditional approaches to student-related activities, which depend on one-on-one interactions, are no longer viable.

Figure 1 illustrates the problem. In this figure, student headcount is plotted against teaching assistant support for a number of years for one of us (HTH). It is glaringly obvious that the amount of one-on-one interactions is going to be minimal. The question is, simply, where is the faculty time going to be spent, grading papers or in dialog with students?

For circumstances similar to those illustrated, the multiple-choice test offers some relief. The issue becomes one of how to construct a test that will give the same measure of student performance as would be achieved if the test were free response. To that end, some ideas are presented below that make it possible to remove some of the "multiple-guess" aspect of multiple-choice questions, and that turn the test into a learning device as well as an evaluation instrument.

Physics teachers do have models for multiple-choice tests. For years, faculty have accepted the results of the Graduate Record Examination and have given academic credit based on performance on the Educational Testing Service Advanced Placement Test. However, examination of the procedure by which such tests are constructed is not very useful. For example, in order to construct one of the nationalized tests, some of the more prestigeous teachers are assembled and the questions are written by a committee. In general, several man-hours go into each question. The questions are then used in a number of selected schools. Each question is subjected to item analysis, rewritten if appropriate, and tested again, leading finally to the presentation of the instrument to the intended population.

The teacher faces a somewhat different test development procedure. Each semester this one individual has the sole responsibility of writing each and every examination in the course (which probably amounts to at least three and perhaps four tests). With luck, another member of the faculty may be cajoled into reading the test, but most often the finished product has been created, edited, and proofread by only one person. All questions of any kind of validity (content, construct, etc.) are purely academic. This test and others like it are the basis for the grade.

The suggestions below are directed to that individual who teaches in an environment where the use of multiple-choice tests is necessary. There is no suggestion of academic utopia nor that the description below is necessarily optimum.

Reprinted with permission from *American Journal of Physics*, vol. 49, no. 9, pp. 838–841, September 1981.

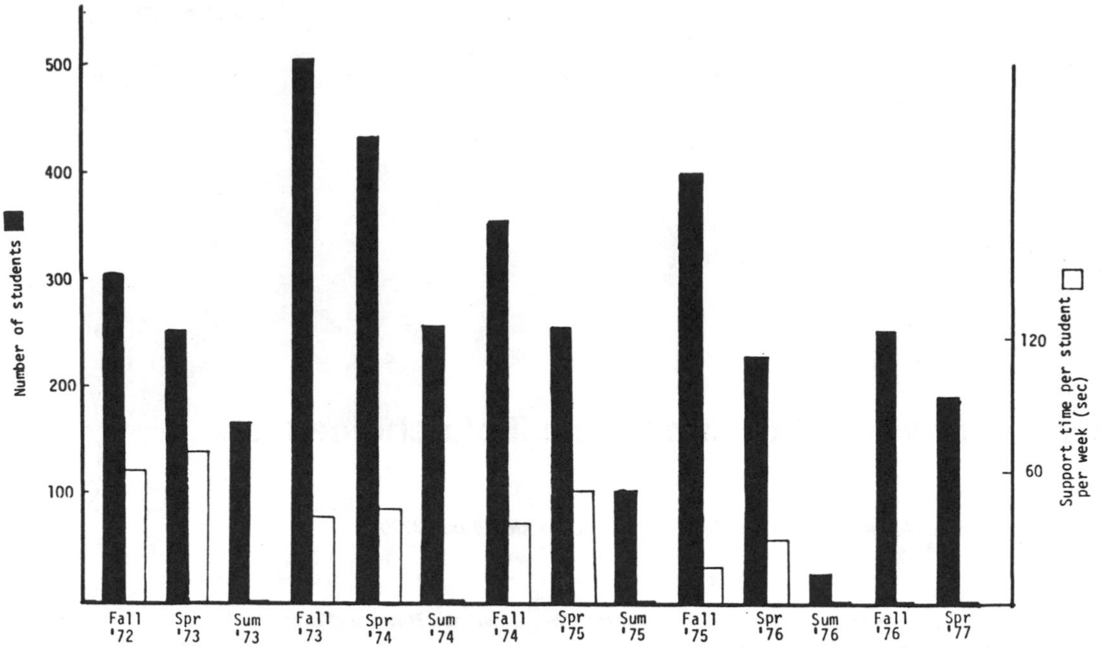

Fig. 1. Student headcount and extra-classroom support is presented for a five-year period.

Rather it is our intent to point out those considerations in test construction that should be approached by positive decision rather than left to chance. The suggestions are made from the philosophical stance that a test should be a learning experience as well as an evaluative instrument.

CONSIDERATIONS ON TEST CONSTRUCTION

(1) *Length of the test.* A test should require the student to think, to reflect, to assimilate, for which time must be allocated. We suggest that a 50-min test of problems should contain 10–12 problems. In order for this number of problems to provide a comprehensive measure of knowledge, the scope of the material must be limited.

(2) *Number of tests in a semester.* The number of tests in a semester will be determined by the total amount of material covered in the course, subject to the condition that each test should comprehensively examine a specific portion of the course. It has been found that four tests per semester (approximately one per month) is a reasonable number. This permits about three questions per chapter.

There is a second reason for giving several tests during the course. Our evidence shows that the correlation between student performance on multiple-choice test and free-response questions increases with the number of tests, both in a temporal sense and in a cumulative sense. This point is discussed later.

(3) *Differential levels of problem difficulty.* Each problem is identified as to how difficult the testmaker judges it to be. A "level number" is assigned to each problem, based on a modification of Bloom's scale,[1] where a level-1 problem requires only a memory operation (i.e., a plug in) and a level-4 problem requires understanding and/or assimilation of several concepts. These levels are, of course, judgement calls on the part of the person constructing the test, but most students seem to find they are reasonable indicators. The levels of difficulty are indicated on the test booklet.

The identification of levels serves two purposes. First, the student can start out with the easiest problems, which contributes to the reduction of anxiety. The second reason for identifying levels is that the instructor can estimate the time needed to work the test. Time is allocated according to: level-1 problems–2 min; level-2 problems–4 min; level-3 problems–6 min; level-4 problems–8 min.

A test made up of two level-4, two level-3, four level-2, and two level-1 problems would require an estimated time of 48 min.

(4) *Provision of formulas are provided on the cover sheet of the examination.* This discourages memorization, eliminates the "cheat sheet" (which too often lives up to its name), and makes a very positive statement concerning the relative importance of "knowing the formula."

(5) *Sequential problems.* It is very tempting to write a sequence of problems where the answer for the first must be used to solve the second. Of course, the attraction of sequential problems (and the least acceptable reason) for using such problems is that it is a quick way to write a test. A more acceptable reason is that such questions may be used to direct student thinking (e.g., a level-1 question may serve as a hint for a level-4 question immediately following). The choice to use such sequencing should be made for academic reasons, with the awareness that the student who misses the first problem will also miss the second, thereby double weighting the first problem.

(6) *Use of symbols.* The use of symbols such as π, ϵ_0, μ, and answer choices indicated by $\sqrt{\ }$, arccos, etc., will require something more than simple manipulation of numbers to match an answer choice.

(7) *Nonrandom incorrect answer choices.* The alternative answer choices should be such that random manipulation of the numbers of the problem will match a choice. This is especially important if students are allowed calculators.

(8) *Avoidance of "1" and "2."* Multiplication or division by 1 results in the same answer. With "2," a student could obtain the correct number by adding 2 + 2 or multiplying 2 × 2. For example, the question "What is the force on a 2-kg mass accelerating at 2 m/sec^2?", could be answered

$4 = 2\ kg + 2\ m/sec^2$ or $4 = (2\ kg) \times (2\ m/sec^2)$. (The units wouldn't match but the student who would add the numbers wouldn't check them anyway.)

(9) *Number of answer choices per question.* A compromise must be made between providing so few choices that the test becomes, in essence, true–false, and providing so many choices that an unreasonable time is required to match the answer. This may be especially critical when symbols are used or the answer is a sum of terms (e.g., vectors). We suggest that five choices is a reasonable number, although answer sheets are available that will accomodate up to ten choices.

(10) *Partial credit mechanism.* One common complaint with multiple-choice questions is "I was able to narrow it down to two, but then I marked the wrong one and got no credit at all." This can be remedied by the simple step of accepting all answer marks and assigning a credit to every mark on the answer sheet. Correct answers count full credit and incorrect answers count −1/4. If the student narrows it down to two choices, and marks both, he receives +3/4 credit (assuming one of the two is correct). Elimination of only one choice as incorrect and marking the other 4 on the answer sheet gives +1/4 credit.

This scheme has the secondary result that guessing is penalized, and it is not uncommon for students to receive a negative score on an examination. Incidentally, the −1/4 credit for wrong answers is the same grading scheme used by the Educational Testing service. Their system differs only in that the score is zero if multiple answers are marked.

(11) *Multiple versions.* If examinations are made in multiple versions, the numbers 6, 8, and 3 are easily misread from a seat or two away. When multiple-version tests are used, the same answer choices appear on each version, the wording or numbers in the problem are different. It has been found that simply rearranging questions between versions is little more than a minor inconvenience to someone seriously intent on cheating.

(12) *Number of correct answers per question.* Avoid more than one correct answer per question. The structure of the examination should be such that it does not distract from the intent of the test questions. If a 10-question test has 12 correct answers, this disturbs students. The distraction is seldom justified.

(13) *Inclusion of free-response question.* It is possible to design answer sheets with a blank section on the sheet.[2] This form may be used for a free-response question that can count a maximum of five times the weight of any other question. The grade is assigned by blanking in up to five bubbles on the answer choice corresponding to the question. For example, if the free-response question is number 10 on the test, and if the student is given 2/5 credit, two of the answer blanks for question 10 would be marked in by the grader before sending the papers to the optical scanner. (Of course, the key should have all answer choices marked for this question.)

(14) *Immediate feedback.* Make it convenient for the student to go over the test. Provide a written solution to the problems of the test on a handout that may be given the student when the test paper is turned in. The student may then take the test booklet outside the classroom and immediately go over the problems missed. This immediate reinforcement makes the test a valuable learning experience.

(15) *Repeated examinations.*[3] If the student has learned from the test, give a second opportunity to demonstrate the new knowledge. Often students are not cognizant of what they do not know until the test. A second chance provides a payoff for review. The repeated examination is made a little harder, is given at a time separate from class time, and is counted as follows: If the second grade is higher, the second grade is the permanent grade. If the second grade is lower, the average is the permanent. Approximately 80% of the students who repeat do better.

ADDITIONAL CONSIDERATION

Given that the usual standard for optimum testing is the free-response problem, the question to be asked is under what circumstances will the multiple-choice test yield results comparable to a hand-graded problem? The Educational Testing Service utilizes both free-response and multiple-choice sections on the Physics Advanced Placement Test. The reported correlation is 0.660.[4] Harke *et al.*[5] compared student performance on multiple-choice questions for a class of 170 students, using 6 free-response questions and 37 multiple-choice questions. They report correlations ranging from 0.44 to 0.73, the correlation increasing with the number of free-response problems.

In an effort to study this issue, 284 students in an introductory, preprofessional physics class (noncalculus) were given a total of 14 free-response problems over the semester. These were worked in a closed-book, proctored, "poptest" situation with knowledge that the results counted toward the final grade. The topics of the questions were taken from material currently under discussion. The problems were each graded on a 0–5 scale, on forms especially designed to streamline the gradebook recording.[2]

Student performance on each of the major course examinations (a total of four, including the final) was correlated (Pearson product-moment) with total poptest scores and, as well, total poptests were correlated with combinations of the multiple-choice examinations. This permitted us to determine how the correlation varied in time (i.e., was there a change over the duration of the course?) and with the number of multiple-choice questions. Each of the regular examinations contained 10 multiple-choice questions and the final contained 25, making a total of 55 multiple-choice problems.

The results are given in Table I. The correlation between hand graded and multiple choice steadily increased over the semester from 0.442 (Test 1) to 0.647 (Final). In addition,

Table I. Pearson product–moment correlation coefficients between hand-graded poptests and multiple-choice problems on examinations.

Test	Test 1	Test 2	Test 3	Test 12	Final	Test 123	Total
Number of problems	10	10	10	20	25	30	55
Correlation coefficient	0.442	0.593	0.616	0.604	0.647	0.679	0.708

the correlation with the sum of Test 1 and Test 2 (Test 12 in the table) was higher than the correlation for either test. The correlation for the sum of three tests (Test 123 in the table) is even higher. The highest correlation was for the grand total of all test grades (55 problems). This points out rather clearly that if multiple-choice testing is to be used, it is important to administer several tests that provide an adequate number of total problems.

As a measure of how meaningful the data of Table I might be, we correlated each multiple-choice test with all the others. Coefficients ranged from a low of 0.58 (Test 1 with Test 3) to a high of 0.78 (Test 3 with Final). The message here is that if an adequate number of multiple choice tests are given, the correlation between hand-graded and total test scores is as good as the intercorrelation between individual tests.

One final point is found from the scattergram of hand-graded problems versus total multiple-choice scores. The number of students who do appreciably better on free response than multiple choice is low. Only five students from this study did that. To the contrary, approximately 10% did appreciably better on multiple choice than hand graded. When these 10% were removed from the sample, the correlation jumped to 0.818. This means that for 90% of the students, the multiple-choice tests gave a reasonable approximation to the same grade as would be obtained with hand grading.

The reason(s) that some students do better on multiple-choice tests than on free-response problems is at this point unresolved. It may be that the presence of the answer choices provides clues or hints as to how to go about solving the problem. Of course, there may be other reasons. Interviews with the better students were of little help in answering this question, because invariably these students followed the procedure of working the problem and then matching their solution with an answer choice. This topic will be the subject of further investigation.

CONCLUSION

Multiple-choice tests can be an effective tool for learning and for evaluation provided certain guidelines are followed in their construction. These include common-sense rules that reduce the tendancy for guessing, and that provide immediate feedback to students after they have completed the test.

Equally important is the fact that only a large number of multiple-choice problems gives a reasonable correlation with hand-graded problems, suggesting that it is critical to administer a series of tests throughout the semester, the more such tests the better.

[1] *Taxonomy of Educational Objectives Handbook 1: Cognative Domain*, edited by B. S. Bloom (McKay, 1956).

[2] H. T. Hudson and R. M. Rottmann, Phys. Teach. **18,** 44 (1980).

[3] D. E. Golden, R. G. Fuller, and D. D. Jensen, Am. J. Phys. **42,** 941–943 (1974).

[4] G. Will Pfeiffenberger and C. R. Modu, Am. J. Phys. **45,** 1066–1069 (1974).

[5] D. J. Harke, J. D. Herron, and R. W. Lefler, Sci. Educ. **56,** 563–565 (1972).

Design of Examinations and Interpretation of Grades

M. W. P. Strandberg
Massachusetts Institute of Technology, Cambridge, Massachusetts
(Received March 10, 1958)

There are two inadequacies of usual teaching methods about which something can be done. The design and evaluation of examinations, as popularly applied, are nonquantitative; and the average student is inadequately prepared for understanding the learning process. Three criteria are presented which may be of use in resolving these difficulties. These criteria are the concept of logical steps, the concept of avenues of approach, and the concept of the nature of examinations.

Of the many situations encountered in teaching, there is one experience that must be universal. At the end of a class session, in which graded examinations have been returned and the instructor has pointed out possible and proper answers to the questions, one or more students linger after the bell has rung and ask, in many different ways, this question: "When you discuss the questions and give the proper answers the quiz seems so simple and straightforward that I have no difficulty following you or even anticipating your arguments. Yet in the examination I am at a total loss about these same things. What is the matter with me? What am I doing wrong? All of the ideas seem so simple." Similar reactions are expressed by students who have listened to a particularly effective lecturer, who has apparently made his subject perfectly clear in all respects, when they try to apply what they have heard to working problems. Again, these students—and there are many good ones among them—express a feeling of frustration that arises from inability to use what appears to be a deep and complete understanding of a subject in anything but the most trivial applications.

The complementary situation is the staff meeting in which an examination is formulated, discussed, and evaluated as to coverage and difficulty. The situation is familiar and, probably, the procedure is universal. The members of the staff who have teaching responsibility for the course voice their opinion about the material that should be covered, and the proposed examination is evaluated by several means: by a discussion of the material covered in the course work and the relevance of the proposed questions to this material; by the formal presentation of appropriate answers prepared by one of the staff; or by having all of the staff "take" the proposed examination in a much shorter time than would be allotted to students. By any, or all, of these means, unanimity (or at least a fair degree of unanimity) is finally reached concerning

Reprinted with permission from *American Journal of Physics,* vol. 26, no. 8, pp. 555–558, November 1958.

the fairness of the examination and its probable difficulties. The students are then subjected to it and, many times, the staff is bewildered by the results. Some good students do poorly, some poor students do very well; many seemingly easy questions are missed completely. Too often, the only consolation left to staff and students alike is that all of the recitation sections (if there are many sections in a course) report equally poor averages, and staff and students are left to a gracious appreciation of the sliding-scale system.

It is far from this writer's intention to find any fault with this system. Indeed, he has lived at ease with this sort of system for some time. Probably, at intervals he was piqued by this irrational chain of events, but not more so than he had come to consider a normal amount of provocation. If there are others who have had less experience in these matters before having conceived a resolution of this unhappy situation, they must take this article as a confession that this writer's awakening has been long overdue and realize that he is now writing in the hope that his experience will be stimulating to others, who have not yet been provoked to action. The writer's experiences, and similar experiences of others, have finally led him to the conclusion that there are two inadequacies of the usual teaching methods about which something can be done:

1. The design and evaluation of examinations, as popularly applied, is nonquantitative.
2. The average student is inadequately prepared for understanding the learning process.

The first point is particularly annoying to a scientist, since the fundamental principle of science is that one understands only what one can measure and describe by numbers. The general attitude that would be evoked by a request for the quantitative design and evaluation of examinations has many facets. The experienced teacher is irritated because through experience he has developed an effectiveness in designing and evaluating examinations. Unfortunately, the transmission of this intuition to others is not readily accomplished. The experienced (and overworked) teacher is repelled when he is asked, for example, "How can the teaching of science be made into a science?" The general attitude is that the problem is too difficult to warrant the effort that would be required to solve it.

The thesis presented here, however, is that the request is neither frivolous, nor arduous, nor trivial. Certainly, a society that will not permit the judging of cows or horses in competition except by the most rigid rules and the most quantitative scoring should not allow the design and evaluation of examinations, particularly examinations in scientific subjects, to be left to skill or artistry, or even to intuition.

This writer does not intend to give an answer to the problem of how examinations can be quantitatively evaluated. His purpose is mainly to stimulate thought and discussion. Some principles have been evolved which may be of use in the evaluation of examinations, but they require experience for their application.

One criterion for the evaluation of a question is the required number of logical steps for obtaining an acceptable solution to the question. We call this the *concept of logical steps.* For example, a quantitative question in physics, commonly referred to by students as "formula pushing," is of the one-logical-step variety. Such a question requires a single deduction from the given situation to the solution of the question. In this kind of question the *given* would be the mass and velocity of a particle, and the *required,* the kinetic energy. The one logical step which must be made to obtain the solution is the connection between the given parameters and the kinetic energy, so that the answer follows the utilization of a single definition or formula, $\frac{1}{2}mv^2$ is equal to the kinetic energy.

The difficulty of a question increases dramatically with the number of logical steps required for its solution. The actual functional dependence of the difficulty on the number of logical steps required for solution is quite possibly an exponential. An example of a two-logical-steps question would be one in which the mass, charge, and velocity of a particle are given and the closest approach of the particle to a second fixed mass m of charge Q is required. The responder must realize the connection between the mass and velocity of the particle, and its initial kinetic energy, and, furthermore, he must realize that the energy of the system is conserved so that the point of closest approach would be that at which all this kinetic energy is traded for potential energy. Multiple-logical-step questions are easily uncovered by perusing the problem section of any physics textbook. The reason that multiple-logical-step questions are often evaluated as simple questions, by staff and students alike, is that any one of the steps required in the solution of the problem is so exceedingly simple. The difficulty of the question arises not from the difficulty of any single step, but the fact that the student is expected to have acquired the ability to foresee the correct path along which he must travel in order to arrive at the desired conclusion. The probability that a student can answer a two-logical-steps question by a random process becomes quite small, while the probability that he will find a correct solution to a multiple-logical-step question by random method is essentially zero.

It is for this reason that the student is abashed at his own inability to work seemingly simple problems. The simplicity of the problem is apparent to him, in that each logical step that is required for the solution of the problem is well known to him, while the hidden difficulty stems from the fact that it is a multiple-logical-step problem. For this reason, a one-hour examination with, possibly, three questions should include a one-logical-step question, a two-logical-steps question, and possibly a three-logical-steps question, in order to adequately grade students on their achievement. This criterion is to be supplemented by other considerations that will be discussed.

A similarly unobstrusive criterion, which, if not properly handled, can lead to as catastrophic consequences as the neglect of the concept of logical steps is the *concept of avenues of approach* to the solution. Stated quite simply, questions can be seemingly quite simple and yet must be evaluated as very difficult, if the student can be expected to know only one convenient avenue of approach to the solution. For example, many problems in mechanics can be solved either by memorizing the equations of motion for a particular situation, or by applying the principles of conservation of energy or momentum, or by the direct integration of Newton's equations. Other questions would test the student's familiarity with a trick method of solution to the problem, or his ability to make a general observation about the system which would simplify the analytical work, or require that the student use a single concept which he is expected to have assimilated; for example, a problem in which the student is required to use the coefficient of restitution. An electrostatic problem that requires the use of images for the solution makes the problem inherently more difficult than a problem that can be solved by several methods, since only one avenue of approach to a solution is available to the student.

Like the concept of logical steps, the concept of avenues of approach to the solution is obvious and, when it is understood, it is almost trivial. Both concepts are useful, just as convenient alphabet and algebra are useful for formulating and solving problems, since they allow a particular situation to be resolved into its component parts.

A third criterion for measuring examinations is the decision about whether the question is quantitative and hence involves the application of concepts, or whether it is qualitative and hence involves the understanding of concepts. This criterion is mentioned here to make the point that, as far as the student is concerned, it is a definite factor in determining the difficulty of a question. This writer has made some simple tests that have convinced him that a class of students can be divided into groups (with some overlapping)—those who find qualitative or quantitative questions easier. An example of a simple experiment for investigating the relative difficulty of qualitative and quantitative questions was a special examination on the atomic model, which I gave to a group of students. A qualitative question asked for a description of the Thomson and Rutherford models of the atom and for a discussion of the relevant experimental evidence for choosing the more valid model. A quantitative problem in alpha-particle scattering was also given. The results showed three quite distinct groups of students: those who understood the problems involved in discussing the Thomson *versus* the Rutherford model and were still incapable of working a quantitative scattering problem; those who were quite adept at working a scattering problem, but who had only a foggy notion of what or who Rutherford or Thomson were; and a few who showed comparable ability in handling both quantitative and qualitative problems. Certainly, this last criterion is more important from the student's than from the instructing staff's point of view, since the questions on an examination are often dictated by the expediency of grading rather than by the desire for developing a standard that is broadened to enable the student to examine himself and his progress in a course of study.

This last point brings us back, essentially, to the relationship between the examination and the grade. It is of the greatest importance that the student have a good understanding of the learning process and an ability to evaluate his grade from the point of view of what it tells him about himself. Certainly, an examination is wasted effort on the part of both staff and student if the student's response is either a feeling of not understanding why he did so poorly on material which he felt he understood, or a feeling of antagonism because he is convinced that he has actually learned more than the examination gives him credit for. Given that the examination has been properly designed to give a graded test to the student, or that it is a test in which the poorest and best students will both be able to show their capabilities, then the student must be educated to make intelligent use of the results of the examination. Possibly, he can be shown that although he knew all of the facts that were needed to answer the question properly, his inability to synthesize the answer from the facts was the particular and important lack that gave him the low grade on a question. Possibly, it would be helpful for a student to realize that a quantitative examination is inherently more difficult for people who have the ability to understand the qualitative or general aspects of a physical situation. Thus, grades in quantitative examinations, which are lower than the student feels he should have, can prove a godsend to him, in the sense that they can be taken as a warning against competing too seriously in an effort or in entering a profession that is most successfully pursued by people with high quantitative ability.

If they can do nothing else at all, logically and precisely designed examinations can make a tremendous contribution to education if they can make a student realize that he did poorly on an examination not because of some frivolous or misunderstood reason, but for the simple reason that although he was able to work the simpler questions, he had neither the synthesizing ability nor the understanding to work the more difficult problems. In like manner, the good student who is able to see, or to be shown, that he is not doing as well as he might, because he has not spent enough effort in constructing a logical edifice and developing facility in the use of a few simple logical steps that are required for understanding his course work, can be prompted to do better work toward developing his own abilities.

The process of education can be made much more efficient if more understanding and responsibility can be obtained from the student. If, in the course of years ahead, American technology is going to be dependent on the education of larger numbers of students by the teaching staffs that must of necessity be only slightly larger than they are today, more efficient teaching methods are going to be required. Larger classes isolate the student more and more from the good offices of the naturally good teacher. There seems to be no question that an investigation and a better understanding of the examination process by the student can develop a more cooperative effort that will allow the students to use their talents more efficiently along lines indicated by a rational and incisive program of examinations.

Part VIII
Evaluation of Teaching

Purpose

EVALUATION of teaching has always been an emotion-laden subject, because it necessarily implies value judgment, and touches many sensitive issues, like academic freedom, personal pride, and career and rewards for faculty. It is also a controversial subject, because the literature on this subject contains some experimental data to support opposing hypotheses, as well as a variety of opinions, both extolling the virtues of course evaluations, and sounding a warning alarm about the apparent reliance on course evaluations. Finally, it is also an easily misunderstood subject, since teaching evaluation takes many forms that are put to many uses—for instance, the college administration's evaluation of an individual's teaching abilities for personnel decisions, an instructor's evaluation of course effectiveness for course improvement, and the students' evaluation of campus courses for disseminating consumer information. Nevertheless, the subject cannot be avoided because the success of the teaching enterprise cannot be judged, nor its operation improved, without the ability to evaluate its effectiveness.

The scope of this part is limited mostly to only one method of teaching evaluation, based on a survey of the students. Although begun mostly in this century, the course, or teaching, or faculty evaluations by students have become almost as permanent a part of the academic scene in the U.S. as blackboards and chalk. In a vast majority of cases, the survey is carried out with the help of a rating form filled out by the students toward the end of the course, indicating their level of satisfaction with various aspects of the course, such as organization, tests, helpfulness of the teacher, etc. Many other methods of teaching evaluation also exist, and some of them are lightly touched on in this part.

The primary purpose of this part of the volume is to increase the reader's awareness of the subject. Such awareness may help a beginning teacher in several ways, such as these:

(a) Anxiety Reduction. Typically, a beginning teacher has no control over the evaluation procedures used on his campus, and his response to evaluation might range from anxiety and dread to curiosity and anticipation. The anxiety can be reduced by understanding the motivation, methods, limitations, and uses of the evaluation.

(b) Proper Interpretation. A beginning teacher may need help in learning proper interpretation of the evaluation results, such as paying more attention to responses in areas where students are a good judge (e.g., the instructor's attitude), and attaching less significance to those where the students are not (e.g., the subject matter competence of the instructor). It is also important to learn to recognize statistical bias (e.g., that the students are generous in their evaluation, on the average), and statistical fluctuations (e.g., that even the highest-rated and award-winning teachers usually receive a few sharply negative comments from a large class).

(c) Instructional Improvement. The beginning teacher may wish to learn how to use the evaluations for improving his course and his teaching skills. He may also have to design his own rating form for this purpose, and can use some guidance in this task.

The Issues

Some of the principal issues brought up in the extensive literature on teaching evaluation, which may be of interest to an engineering educator, are as follows:

- What purposes might a teaching evaluation serve: improvement of instruction, determination of rewards for good teaching, a chance for the students to "blow off steam," etc.?
- Do evaluations measure the students' learning or comfort?
- What is, or is not, measurable by course evaluations?
- What tacit assumptions are inherent in carrying out a course evaluation?
- How do, or should, the various types of evaluations differ from each other?
- By whom should the evaluation forms be designed, administered, and read: faculty, administrators, students, or educational specialists?
- Should the questions on the evaluation forms be precise (or even multiple-choice variety), or should they be open-ended, essay-type?
- What should be the scope of the questions, i.e., should they relate to the course content, or style of presentation, or course organization and materials, or instructor's personality and attitude, or work load and grading?
- How should the results of the evaluations be interpreted?
- What is the reliability (i.e., the degree to which the results are stable and reproducible), and validity (extent to which the results measure what was intended) of the course evaluations?
- What factors influence the results of evaluation (e.g., class size, hour, level of students, whether the course is required or optional, whether the course is within the major field of the student or not, etc.), and what factors influence the evaluation by an individual student (e.g., his course grade, level of maturity, etc.)?
- What alternate means (e.g., survey of alumni, or classroom visitations by senior colleagues) are available for course evaluation?

The Reprinted Papers

The following five papers are reprinted in this part:

1. "The How and Why of Evaluating Teaching," J. A. Centra, *Engineering Education,* vol. 71, no. 3, pp. 205–210, Dec. 1980.
2. "The Role of Faculty Evaluation," W. J. McKeachie, *National Forum (The Phi Kappa Phi Journal),* vol. 63, no. 2, pp. 37–39, Spring 1983.
3. "A Student Evaluation of Teaching Techniques," M. B. Freilich, *Jour. Chemical Education,* vol. 60, no. 3, pp. 218–221, Mar. 1983.
4. "Teaching Effectiveness: Its Measurement," R. C. Wilson, *Engineering Education,* vol. 62, no. 6, pp. 550–552, Mar. 1972.
5. "A Course Evaluation Guide to Improve Instruction," J. C. Lindenlaub and F. S. Oreovicz, *Engineering Education,* vol. 72, no. 5, pp. 356–361, Feb. 1982.

The first paper by Centra, an active worker in the field of teaching evaluation, is an adaptation of his book on the subject. It summarizes the research findings on the six principal methods of evaluating teaching—self-evaluation reports, student ratings, colleague evaluations, alumni ratings, videotaping, and students' performance on achievement tests. For each of these methods, the paper points out the utility and limitations of the information available, and the factors influencing the results of evaluation. The main conclusion from this broad survey is that the various methods of evaluation are complementary rather than duplicative, and that the purpose of evaluation should govern the choice of the evaluation method.

The second paper by McKeachie focusses on one method of evaluation—student rating of instructor and teaching. It addresses four of the more basic questions, from the instructor's point of view, concerning this form of evaluation: why carry out an evaluation, what kind of data can be obtained, how can this data be used for improving instruction, and why should it be acceptable to use the data for two other purposes—personnel decisions concerning the faculty, and students' choice of courses. Once again, this paper is based on a great deal of experimental research work, in which McKeachie has been an active participant for several decades.

Given that student ratings are to be used for evaluation of teaching, a new teacher concerned about his rating, or about its improvement, might well ask the question: what do the students want? Since all teachers have themselves been students, and until recently in the case of new teachers, perhaps the best guide to what the students want is their own intuition. Nevertheless, a comprehensive list of the numerous variables affecting the students' perception of the effectiveness of teaching, and their relative importance in a student's mind, is useful to have, even if it only reinforces one's intuition. The third paper by Freilich provides such a list, backed by a survey of both the students and teachers; it also reminds us that there may be differences between what the teachers and the students think are the variables most helpful in aiding the learning.

One can make other lists of several dozen such variables that can reasonably be expected to enhance learning, and some faculty evaluation forms do indeed contain 80 or 100 items on them. Are all of these variables independent, each requiring a separate question on the rating forms, and individual attention to them from the instructor, or are these merely manifestations of just a few basic factors? The fourth reprinted paper by Wilson reports, among other things, the results of a factor analysis which identifies five essential components of effective teaching. An awareness of these will allow the instructor to place the plethora of "good teaching" advice in a systematic framework. This paper also lists a large number of teaching related activities of college teachers, all of which are an integral part of the teaching task; these should therefore be included in an overall evaluation of one's teaching effectiveness, and it is inappropriate to treat teaching effectiveness as synonymous with classroom effectiveness.

An individual instructor may be interested in carrying out an evaluation of his own course independently of any college-wide surveys, for the purpose of improving the instruction. A good starting point for such an endeavor is the checklist of questions contained in the fifth reprinted paper by Lindenlaub and Oreovicz. The authors suggest the use of this checklist for other purposes as well, such as curriculum planning and peer evaluation of a course.

The Reading Resources

[1] W. J. McKeachie, "Student Rating of Faculty," *AAUP Bulletin,* vol. 55, no. 4, pp. 439–444, Dec. 1969. A clear, concise statement of the purpose of student ratings, factors influencing them, criteria for choosing rating forms, and uses of results, based on empirical data and reasoning; includes a sample, "model" rating form.

[2] W. J. McKeachie, "Student Rating of Faculty: A Reprise," *Academe* (Amer. Assoc. Univ. Profs.), vol. 65, no. 6, pp. 384–397, Oct. 1979. Updates the previous item by summarizing the major research findings of the intervening decade.

[3] F. Costin, W. T. Greenough, and R. J. Menges, "Student Ratings of College Teaching: Reliability, Validity, and Usefulness," *Review of Educational Research,* vol. 41, no. 5, pp. 511–535, 1971. Older but careful review from the educational researchers' point of view.

[4] L. M. Aleamoni and P. Z. Hexner, "A Review of the Research on Student Evaluation and a Report on the Effect of Different Sets of Instructions on Student Course and Instructor Evaluation," *Instructional Science,* vol. 9, pp. 67–84, 1980. A literature survey of the experimental research findings, particularly on the factors influencing the results of teaching evaluation by students.

[5] K. O. Doyle, Jr., *Student Evaluation of Teacher Performance.* Lexington Press, Lexington, MA, 1974. A critical summary of many research findings and a discussion on developing and improving evaluation questionnaires, written in non-technical language.

[6] M. Fishbein, "The Best and the Popular Teachers," *Medical World News,* vol. 14, no. 43, p. 72, 23 Nov. 1973. Describes a hoax (from *The Journal of Medical Education,* 1973) in which a professional actor, masquerading as a researcher, presented a lecture full of doubletalk, neologisms,

non-sequiturs, contradictions, and meaningless references to unrelated topics, and was highly rated by a majority of listeners.

[7] M. J. Rodin, "Can Students Evaluate Good Teaching," *Change,* vol. 5, no. 6, pp. 66–67, 80, Summer 1973. Points out shortcomings of the three methods (test-retest, internal consistency, and agreement among subjects) of determining the reliability of course evaluation forms.

[8] J. E. Stice, "I'm for Evaluation of Teaching, But—," *Engineering Education,* vol. 62, no. 6, pp. 529–532, Mar. 1972. Some practical advice on carrying out teaching evaluation by the teacher himself, by students, by colleagues, and by administrators.

[9] P. Seldin, "Self-Assessment of College Teaching," *Improving College & University Teaching,* vol. 30, no. 2, pp. 70–74, Spring 1972. Summarizes the literature on self-evaluation, its relationship to student ratings, and the distinctions between self-evaluation carried out for personnel decisions and for improving instruction; contains a suggested self-evaluation inventory.

The How and Why Of Evaluating Teaching

John A. Centra
Educational Testing Service

Most teachers and students in engineering schools would prefer not to be evaluated. Since evaluation can be threatening and ill-defined and can sometimes result in unfair judgments, their reluctance is understandable. But unless people are willing to live with decisions made at random or invalidly, evaluation is necessary. The evaluation of teaching at engineering schools may be based on only the subjective judgments of one or two decision makers or on several additional sources of information. There is little doubt about which method results in fairer judgments.

Evidence on teaching effectiveness can be acquired from self-evaluations or self-reports, student ratings, colleague ratings, videotapes of classroom performance, and student achievement. These methods and others can be used not only to help make personnel decisions on promotion, salary or tenure (summative evaluation), but to help individual teachers improve (formative evaluation). Dissonance theory is often cited as the reason to expect assessments to result in improvement. That is, the information or feedback provided by the various evaluation techniques may produce in the teacher some dissonance or dissatisfaction that helps open him or her to change.[1] Whether, in fact, assessment procedures generally produce improvement or change in teachers is one of the issues considered in this article, which is broadly concerned with the several means of evaluating teaching and the strengths and weaknesses of each for formative and summative evaluation. Much of the discussion is based on a series of studies I have been conducting over the past half dozen years.

Self-Evaluation or Self-Reports

How useful are faculty self-evaluations in improving instruction or in helping to make personnel decisions? Robert Burns reminded us that most people do not see themselves as others see them; teachers and the way they see their instruction are apparently no exception. The research evidence clearly indicates that most teachers do not view their teaching as their students, their colleagues or the administrators at their colleges view it. Blackburn and Clark, for example, reported little agreement between faculty self-ratings of overall teaching effectiveness and ratings by students, colleagues or administrators.[2] These last three groups, however, did agree substantially on how they rated teachers at their institution.

A study I completed a few years ago investigated self-ratings of faculty members and student ratings on 21 items dealing with instructional practices.[3] The sample consisted of 343 teachers from five colleges. In addition to the general lack of agreement between self-ratings and student evaluations (a median correlation of .21 for the items), there was also a tendency for the teachers to give themselves better ratings than their students did. Discrepancies were most notable in student-teacher interaction, course objectives and the instructor's openness to other viewpoints. These aspects of instruction would seem to be ones on which many teachers could profit from other sources of information. It should be pointed out that although about a third of the teachers generally rated themselves considerably higher than students did, a few—5 or 6 percent—gave themselves much lower ratings.

Self-ratings, then, would probably not be very useful in making personnel decisions; as an aid to instructional improvement, they might best be used in conjunction with student ratings or other evaluations to highlight discrepancies for the individual instructor.

Although self-ratings have questionable value for summative purposes and only limited value for formative purposes, some kind of *self-report* of teaching and other professional activities can be critical to either purpose. The self-report gives faculty members the opportunity to describe their responsibilities and accomplishments over the year. Included in the description should be indications of responsibilities and performance in teaching, student advising, research and scholarship, community service, college service and whatever other areas may be appropriate. The teacher's report should, whenever possible, include illustrative materials and evidence of accomplishments, especially when it

Reprinted with permission from *Engineering Education,* vol. 71, no. 3, pp. 205–210, Dec. 1980.

is to be used in personnel decisions. For teaching, these materials might include the objectives and syllabus of the course, the methods and materials used in instruction, assignments and examinations, and, if possible, evidence of student learning or accomplishments in the course, such as course projects, term papers, and pretest and posttest results. The self-report should be one of the first steps in the evaluation process and, in a sense, could provide one basis for judgments by colleagues, promotion committees and chairmen.

The self-report or some kind of self-analysis of one's teaching can also be useful for instructional improvement. Faculty development practitioners, for example, might use it in advising teachers or in deciding on appropriate development activities. Teachers might be encouraged to identify aspects of their teaching that they hope to change or improve. Used in this way, the self-report is similar to what has come to be referred to as a growth contract or individualized development plan.

Student Ratings

Although student ratings are increasingly being considered in personnel decisions, they have been used primarily to improve instruction. Research results suggest that such use can benefit some teachers. This, at least, was the major conclusion of a study I conducted at five colleges a few years ago[4] and which was corroborated recently.[5] The study had an experimental design in which random groups of teachers received their rating results, while those in the control groups did not. Although there were no overall differences between the experimental group and the control group, teachers whose self-evaluations were considerably higher than their student ratings made some adjustments in their teaching within as little as half a semester after receiving the rating results. Over a longer period, a wider variety of teachers made some positive changes.

Teachers who change as a result of student ratings are probably those who place a fairly high value on collective student opinion and who know how to go about making changes. Undoubtedly, some teachers merely write off student judgment as unreliable or unworthy, and for these individuals changes are unlikely even though they may be called for. No doubt the kinds of items included in a rating form and the way in which results are reported and interpreted are critical. The typically generous ratings give many faculty members an inflated view of their teaching. For example, only 11 percent of a national sample of 852 engineering classes were rated below average by students.[6] Student generosity in ratings is not unique to them as a group; colleagues are even more likely to rate each other kindly.[7] Because of this bias some kind of normative or comparative data is helpful in interpreting student evaluations of teaching. Comparing themselves with other relevant groups, such as teachers in the engineering school, or members of their department, or perhaps a national sample of engineering teachers, can give faculty members a better understanding of their ratings than simply looking at a mean scale or item score.

> **"Oral directions, especially if the teacher makes a subtle appeal to generosity, might have a sizable effect on ratings."**

In sum, properly designed student rating programs probably can have some modest effects on improving instruction. But the ratings are not a panacea for teaching ills; they should be accompanied by a variety of teaching improvement activities in the engineering school.

Different Circumstances, Different Ratings?

When student ratings are used to make personnel decisions, the means of collecting and interpreting the data become particularly important. In administering the forms, for example, we might expect that students would be more lenient or generous if they were informed that the ratings would be used for tenure, promotion or salary considerations. Conversely, we might expect students to be more frank and possibly more severe in their ratings and criticisms if they understood that the results would be used for improving the course or instruction; such information, they might logically assume, could lead to needed changes. At least two research studies have shown, however, that the differences in ratings under the two circumstances were only slight.[8] Students who had received different *written* instructions regarding the purpose of the ratings did not generally rate teachers differently. One possible explanation for the findings is that many students simply do not read the instructions on student rating forms very carefully. Oral directions, especially if the teacher makes a subtle appeal to generosity, might have a sizable effect on ratings. Some standardized procedures for the administration of forms would therefore be advisable if the results are to be used in personnel decisions. These procedures might include the requirement that a staff member or student distribute, collect and place the questionnaires in a sealed envelope and that the teacher not be present during the administration.

When using the ratings for summative purposes, one especially needs to know whether the responses are influenced by specific characteristics of the students, teachers or courses that have little to do with actual teacher performance and course effectiveness. In general, research indicates that most extraneous variables have a relatively weak relationship to ratings; for example, the ability level, sex, or class level of students in a class generally do not produce significant differences in ratings (for example, see Costin, Greenough and Menges,[9] and Centra and Creech[10]). A few course characteristics, however, can apparently influence ratings somewhat, and they should be kept in mind in interpreting results. Class size is one of these. Very small classes—that is, those with fewer than 15 or so students—usually get especially high ratings.[10] The reasons for the high ratings could be more a matter of the situation than the teacher; students may enjoy very small classes because they allow more individual attention and interaction with a teacher. In larger classes, class size does not seem to have a highly significant relationship to ratings. All in all, then, ratings from classes with fewer than 15 or so students, which generally compose a small percentage of courses taught at engineering schools, need to be interpreted cautiously. Responses from larger classes should also be viewed cautiously if a small, nonrepresentative sample of students filled out the forms.

Students also tend to give higher ratings to courses that are major requirements or electives, as compared with courses they take to fulfill a college requirement. How a course fits into the college curriculum might therefore be taken into account in interpreting ratings of a teacher's overall effectiveness. Probably what should be done to reduce possible bias, no matter how small, is to look at ratings over time and from whatever courses a teacher offers. A set of ratings would probably be a fairer indication of how the teacher has affected students than the ratings for a single course; it would allow trends to be noted, as well.

Also to be considered in the use of student ratings are the questions asked. Some items are especially appropriate for summative evaluation. The forms usually contain a variety of items dealing with specific instructional practices and may have one or two items assessing the overall effectiveness of the teacher and the course. The former includes, for example, student views of how well the teacher organized and presented material, how helpful the teacher was to students, and how difficult or stimulating the course was. (Forms should also include space for instructors to add their own items, as well as provisions for students to make additional written comments.) Research indicates that the overall rating items tend to be more highly related to measures of student learning than are the specific items.[11, 12] That is, when students' ratings are correlated with their end-of-course examination performances, the global ratings of teacher effectiveness are better estimations of how much students have learned (correlations at about .50) than are ratings of the specific behavior of the instructor.

Global ratings may be more valid estimates of students' academic achievement, because they are not tied to a specific instructional style. Some teaching methods probably work well for some, but not all, teachers. Every teacher, for example, does not have to develop close relationships with students to facilitate learning in his or her course. For some, these relationships are part of their teaching style and may well contribute to their effectiveness. But other teachers may use different practices that account for their success. So if one assumes that ratings should be at least moderately related to student learning in order to be used in personnel decisions, and if teachers ought to be allowed to develop their own teaching style, then the global ratings are more defensible than the ratings of specific practices. Ratings of specifics, however, would be more useful in instructional improvement because they can more readily suggest changes for teachers.

In summary, if student ratings are used in making personnel decisions, there should be a sufficient and representative number of students responding for each class, and the forms should be administered and collected in a prescribed and systematic manner. In addition, global or overall ratings rather than ratings of specific practices or behaviors ought to be used, and ratings for several courses over a period of time would probably provide the best basis for making judgments.

Colleague Evaluations

What part should colleagues play in faculty evaluation? Although other teachers definitely can provide invaluable judgments about a faculty member's research and scholarship, exactly how they might contribute to the evaluation of teaching effectiveness is less certain. One point of view is that colleagues' ratings of classroom instruction would provide a more professional and valid assessment than students provide. The evidence, however, indicates that ratings based *primarily* on classroom observations would generally not be reliable enough to use in making decisions on tenure and promotion—at least not unless faculty members invested much more time in visitations or training sessions.[7]

The preceding conclusion was reached in a study conducted at a college that had set up a reasonable classroom visitation schedule for colleagues. Each teacher was observed and rated twice by three colleagues, for a total of six separate ratings. Colleagues could be expected to base their ratings on actual classroom observations, rather than on teaching reputations, because the institution was in its first year of operation. Colleagues rated specific instructional practices that could be observed during a class visit (for example, whether the instructor used class time well, the extent to which the instructor used examples or illustrations for clarification). Some global items were also included, such as the overall effectiveness of the teacher and the quality of the lecture, the class discussions and the textbook. As a special aid to visiting scholars, each teacher was encouraged to prepare and discuss with them a course description, including objectives and planned methods of instruction.

Colleagues were generous in their ratings: 94 percent of the teachers were rated excellent or good. The ratings were far less reliable than student ratings, mainly because of this positive bias and because student ratings are usually based on a full term of observation by more raters. If more colleagues visited each teacher, their average responses would be more reliable, just as achievement test scores become more reliable when the number of test items is increased. But such an effort would require several visits to each class by at least a dozen colleagues, a time investment that many faculty members would be unwilling or unable to make.

Under ordinary circumstances, then, colleague evaluations of classroom teaching performance are not reliable enough to use in personnel decisions. For purposes of both summative and instructional improvement, however, there are several important aspects of teaching that colleagues seem able to judge. Among these are the instructor's qualifications and his knowledge of a subject, the course syllabus and objectives, the reading lists and materials employed in instruction, and the assignments and examinations. Colleagues might also be able to judge concrete evidence of student achievement in a course, such as term papers, course projects, and possibly pretest and posttest results. These aspects, which should be part of the instructor's self-report (described earlier), could be assessed best by colleagues within the same discipline (including the department chairman) and without necessarily setting foot inside the instructor's classroom.

The reliability of colleague ratings is less of an issue when the purpose of evaluation is instructional improvement. Visitations followed by an informal chat could be worthwhile. Visitors might, in fact, learn a great deal about their own instruction as well. In addition, simply encouraging a free exchange of ideas among colleagues might help improve instruction. For example, departments might arrange weekly informal sessions among members to discuss various methods of approaching a particular unit in a course. Another possibility is more joint teaching. Other informal and nonthreatening methods could also be employed instead of formal visitation ratings, which—aside from the problems already discussed—can adversely affect morale.

Alumni Ratings

Information on teaching or course effectiveness can also come from alumni. Some people argue, in fact, that alumni views are more valid than those of current students in a course, because alumni are more mature and have gained a perspective that students lack. One often hears of teachers who think their students do not appreciate them until the students have gotten out into the world. The evidence from two published studies, however, clearly indicates that current students and alumni agree substantially on effective and ineffective teachers.[13, 14] These studies compared ratings by alumni who had been out of college for five or ten years to the ratings of students who had just completed the course. Given the generally high agreement between alumni and student ratings (correlation of .75 for overall teacher effectiveness), there seems to be little need to obtain both alumni opinions and current student ratings. Since alumni ratings are more difficult to gather regularly, student ratings would probably suffice. Moreover, in making personnel decisions one may not be able to wait until enough of a new instructor's students have become alumni.

Alumni opinions are also of questionable value in helping to improve the performance of individual teachers, because most alumni would have difficulty recalling the kind of specific information useful to instructors. Surveys of alumni, however, may provide useful information for adjusting the curriculum and environment of a college. Alumni views of the utility of particular courses or experiences, for example, are likely to be of greater benefit than their ratings of the instructional procedures of particular teachers.

Use of Videotaping

In the past few years a number of colleges and universities have made available facilities for faculty members to observe their teaching on tape. Established mostly for instructional improvement, the use of videotaping can help faculty gain a unique awareness of their teaching. Some teachers may be able to view their classes alone and notice ways in which they can improve; for other teachers, however, the video replay has greater impact if someone—perhaps a faculty development practitioner or "master" teacher on the faculty—points out particular practices or behaviors that could be improved.

Videotapes of instruction might also be helpful in making promotion and tenure decisions. Teachers could have the option of submitting selected tapes of their classes as part of their self-report and of having their classroom performance judged by senior colleagues, administrators and, possibly, promotion committees.

Assessment of Student Learning

Some people argue that the best way to judge how effective a teacher has been is to assess how much students have learned. The proponents of this approach say that the emphasis should not be on what the teacher does but on whether appropriate learning has taken place. Of course, an end-of-course assessment is only a measure of short-term learning and is therefore less ideal than an assessment of long-term learning. But long-term achievement—performance that might be assessed several years later—is not easily measured and even harder to attribute to the impact of an instructor or course. So, in a way, an end-of-course assessment of what students know—and this should include knowledge in both the cognitive and affective domains—is the best estimate of the long-term effects.

Situations in which an evaluation of student learning could be used in making personnel decisions seem limited. Student achievement measures, for example, could not provide a common yardstick for comparing teachers in different fields or even in different courses within the same field. Comparisons might be possible in multiple-section courses, but to ensure fairness at least two conditions should be met: students should be assigned to sections on a random basis (in an attempt to distribute evenly the brightest or most highly motivated students), and a common examination that is unknown to the teachers should be used (to avoid teaching to the test). Such experimental conditions would be impossible to arrange on a continuing basis.

The institutions or departments most likely to use assessments of student learning for summative evaluation are those that employ a systems approach to instruction (also known as criterion-referenced measurement, or mastery learning, among other things). The approach requires teachers to define specific instructional objectives and levels of performance for students; then the extent to which a teacher's students attain these goals determines the teacher's effectiveness. Some faculty members, however, may not be as effective as their students' performance suggests. For example, by formulating easily attained, low-level objectives or by teaching only to the test questions, teachers can achieve good passing rates; in truth, though, they would have shortchanged their students. It must be remembered that a test represents only a sample—indeed, a very small sample—of the subject matter in a course. Or, teachers might manage to bring students to a desirable level of achievement (by stressing memorization rather than understanding, for example), but fail to motivate them for further learning; indeed they may have dampened any further interest. Judgments by colleagues or students of the kind discussed earlier in this article might provide a check on these occurrences. Thus, even with mastery learning or other systems approaches to instruction, assessing the effects on students is not a foolproof basis for making summative evaluations of teachers.

"Teachers might manage to bring students to a desirable level of achievement but fail to motivate them for further learning; indeed they may have dampened any further interest."

When information about student learning is to be used only by the instructor for instructional improvement, the teacher will be less threatened and less concerned about how the results will appear to promotion committees. Pretesting students at the outset of a course and testing them periodically as the course progresses is a type of formative evaluation that allows teachers to modify their instruction for individual and class needs. Many teachers, unfortunately, assess student learning only to award grades. Conveying the results of such tests to individual students also promotes their learning. Thus, the periodic assessment of student learning is beneficial to both students and teacher; it is critical not only to the so-called systems approaches to instruction but to the more traditional methods of instruction as well.

Conclusions

The evidence clearly indicates that no one method of evaluating teaching is infallible for making personnel decisions. Each source is subject to contamination, whether it be possible bias, poor reliability or limited objectives. And, of course, each shortcoming becomes especially important when the results are to be used in making decisions about people.

How then does an institution make fair judgments about teaching performance? Only by using several of the methods as a system of checks and balances, so that the limitations of one method are balanced by the strengths of another. It may well be that different methods are appropriate at certain types of colleges because of the special circumstances found there. A case in point would be the use of student learning measures at institutions where a systems approach to instruction is common. But even in these instances, using additional sources of information would result in more equitable decisions about people.

The first step in acquiring evidence for personnel decisions should be a self-report of teaching and other professional activities. Illustrative materials and evidence of accomplishments—including, when possible, information on student achievement—should be a major part of the self-report. Colleagues might then play a principal role in judging such aspects of teaching as the course syllabi, reading lists and instructional materials, examinations and any evidence of student learning. Student ratings of a course's value and the teacher's effectiveness, as well as information on student learning, ought to be included as evidence, too, if proper guidelines have been established for collecting and interpreting the results. Looking at patterns of ratings or student learning across courses and over time should be part of the evaluation guidelines; being aware of alternate explanations for results is also necessary. For example, did the teacher set easily attainable objectives or teach to a test? Were especially low student ratings caused by course characteristics over which the instructor had no control?

For instructional improvement, guidelines are less critical, but each of the methods of evaluation can be beneficial. Unfortunately, however, maximum use is seldom made of the results. Faculty members who most need to improve their teaching seem least willing to participate in instructional improvement practices on their campuses. For instance, results from a faculty development survey indicated that at 40 percent of the institutions, very few of the instructors who were judged inadequate as teachers had been involved in development activities, whereas good teachers who wanted to get better appeared to participate the most. Given the fact that participation in most development activities is voluntary, we should not be especially surprised that these good teachers compose the major clientele; after all, they are frequently the most interested in teaching. Ultimately, however, ways must be found to reach more of the faculty members who need to improve.

References

1. Festinger, L., *A Theory of Cognitive Dissonance,* Row, Peterson, Evanston, Ill., 1957.
2. Blackburn, R.T. and M.J. Clark, "An Assessment of Faculty Performance: Some Correlates Between Administrator, Colleague, Student and Self-Ratings," *Sociology of Education,* vol. 48, no. 2, 1975, pp. 242-256.
3. Centra, J.A., "Self-Ratings of College Teachers: A Comparison with Student Ratings," *Journal of Educational Measurement,* vol. 10, no. 4, 1973, pp. 287-295.
4. Centra, J.A., "The Effectiveness of Student Feedback in Modifying College Instruction," *Journal of Educational Psychology,* vol. 65, no. 3, 1973, pp. 395-401.
5. Marsh, H.W., H. Fleiner and C.S. Thomas, "Validity and Usefulness of Student Evaluations of Instructional Quality," *Journal of Educational Psychology,* vol. 67, no. 6, 1975, pp. 833-839.
6. Educational Testing Service, *Comparative Data Guide for the Student Instructional Report* (1975-1976), College and University Programs, Princeton, N.J., 1975.
7. Centra, J.A., "Colleagues as Raters of Classroom Instruction," *Journal of Higher Education,* vol. 46, no. 1, 1975, pp. 327-337.
8. See for example, Centra, J.A., "The Influence of Different Directions on Student Ratings of Instruction," *Journal of Educational Measurement,* vol. 13, no. 4, Winter 1976, pp. 277-282.
9. Costin, F., W.T. Greenough and R.J. Menges, "Student Ratings of College Teaching: Reliability, Validity, Usefulness," *Review of Educational Research,* vol. 41, no. 5, 1971, pp. 511-535.
10. Centra, J.A. and F.R. Creech, "The Relationship Between Student, Teacher, and Course Characteristics and Student Ratings of Teacher Effectiveness," PR-76-1, Educational Testing Service, Princeton, N.J., 1976.
11. Sullivan, A.M. and G.R. Skanes, "Validity of Student Evaluation of Teaching and the Characteristics of Successful Instructors," *Journal of Eduational Psychology,* vol. 66, no. 4, 1974, pp. 584-590.
12. Centra, J.A., "Student Ratings of Instruction and Their Relationship to Student Learning," *Research Bulletin 76-6,* Educational Testing Service, Princeton, N.J., 1976.
13. Druckers, A.J. and H.H. Remmers, "Do Alumni and Students Differ in Their Attitudes Toward Instructors?", *Journal of Educational Psychology,* vol. 42, no. 3, 1951, pp. 128-143.
14. Centra, J.A., "The Relationship Between Student and Alumni Ratings of Teachers," *Eduational and Psychological Measurement,* vol. 34, no. 2, 1974, pp. 321-326.
15. Centra, J.A., "Faculty Development Practices in U.S. Colleges and Universities," PR-76-30, Eduational Testing Service, Princeton, N.J., 1976.

The Role of Faculty Evaluation in Enhancing College Teaching

Wilbert J. McKeachie

Faculty evaluation is currently a hot topic in higher education. I think the current interest stems from a number of changes in our society and particularly in higher education. As institutions have grown larger, we have become more bureaucratic. As we become more bureaucratic, we put in more layers of authority. And as we get more bureaucratic, we try to get more systematic about things. Issues are handled much more in contractual, legal, sometimes adversarial ways, rather than informally. We rely upon formal evaluation procedures rather than impressionistic ones.

A second factor promoting change came as one residue of the student uprisings during the 60s and 70s: Students expressed increased concern about the quality of teaching—a concern which was undoubtedly a valid one and was accompanied by pressures from alumni, parents, the public, and legislators for accountability—for proving that we're really accomplishing some of the things that we say are our goals.

Each of these constituencies probably has a somewhat oversimplified view of what's involved in faculty evaluation. For example, those who wish to have evaluation to improve teaching often presuppose different models of evaluation. This is apparent in what they say: "We will evaluate all faculty members, give feedback to them, and they will become better teachers." In some cases students have had an alternative model: "We evaluate the faculty. When the administration sees which professors are bad, they will fire them."

Neither one of these goals is realistic. In this paper I propose to discuss topics illustrating the fact that teacher evaluation is somewhat more complex.

The topics I wish to address are:

1. Why evaluate?
2. What kind of data can we gather?
3. How can the data be used for improving teaching?
4. How can the conflicting uses of evaluation data be reconciled?

Why do we evaluate?

It is important to be explicit about our purposes because we often assume that one method of evaluation is going to meet all purposes.

Generally, there are three purposes for which evaluations are undertaken. One of them is to help an individual faculty member improve. A second one is to make some sort of personnel decision rewarding individuals who are good teachers or getting rid of those who are not good teachers. A third reason is to help students choose courses more wisely.

What kind of data can we get?

I would argue that the ultimate criterion of effective teaching is evidence of impact upon student learning. What we are concerned about in evaluating teaching is not what the teacher has done, but what has happened to the students. This statement has some important consequences because it puts into focus the idea that as long as students learn, it does not matter how the teacher has gone about getting that learning to occur.

The term "evaluation" implies value judgments about what goals are most important. Which is more important?

1) that students in a course *develop a desire for learning and curiosity* about the subject matter so that they keep on trying to learn? or
2) that they *acquire skills in thinking* and in structuring the material so that they are able to understand what they read or what they observe? or
3) that they *master the knowledge* of what is known in this subject matter?

In evaluating teaching, someone or some group has to make a decision about which of these goals are most important, because relatively few of us are equally effective in achieving all the educational goals important for higher education.

So *student achievement* is one criterion by which we can measure teaching effectiveness. But there are some difficulties with student achievement data. The major problem is how do you compare student achievement in mathematics with student achievement in psychology? Here we have a group of mathematics test papers; there we have a group of psychology test papers. Which of the groups was taught better? We have no good basis for equating so many units of gains in mathematics with so many units of gains in psychology. We thus tend to avoid data on student achievement in evaluating teaching. Whenever we compare the teaching of two teachers teaching different courses, we should remember that we are comparing apples and oranges even though we have numerical ratings which appear to be directly comparable. Simply obtaining a mean rating of 2.1 for one teacher and 2.2 for the other does not mean that their teaching has magically become directly comparable.

Who is competent to judge the relative values of the many different kinds of teachers who make up the faculty? If then our ultimate criterion for teaching effectiveness is impact upon student learning, if we are to make judgments about whether or not student learning has been greater or less than might be normally expected, the judgment must be made by individuals who have expertise and experience in the areas of teaching involved. It becomes almost self-evident that such judgments can seldom be made in very precise, fine-grained terms. A frequent mistake in faculty evaluation is to try to put each faculty member on a numerical scale; e.g., one person rates 2.5; another, 2.4; and another, 1.7. Computing mean student ratings to decimals or counting pages of publications gives promotion committees a false sense of objectivity.

If we have inadequate information about student learning, what other useful data can we get? The most common source is *student ratings*. Often student ratings are criticized as being invalid. Faculty members say, "Students can't really know what good teaching is." The research evidence on validity is now quite clear. Cohen located studies of validity involving 68 courses and found an overall validity coefficient of .4 between mean student ratings and mean achievement.[1] This is much higher than we would expect and is probably very reassuring to those who are concerned about whether or not

WILBERT J. McKEACHIE is director of the Center for Research on Learning and Teaching at the University of Michigan.

Reprinted with permission from *National Forum,* vol. 63, no. 2, pp. 37–39, Spring 1983.

students can really judge when they are learning. While it represents only 16 percent of the variance, it is probably substantially more than you are likely to get by any alternative method of evaluating teaching.

Student ratings probably are the best single source of information about whether or not students are learning from a teacher, other than measures of actual achievement. Students are in class almost every day and they know what's going on. They are the ones we are trying to affect, and they have some sense of whether they are learning.

A third method for evaluating teaching is *peer judgment*. Often, peers judge teaching without visitation. Faculty members are inclined to think that anyone who talks well and is a nice person is probably a good teacher. Peer ratings do correlate fairly well with student ratings. The correlation is probably due to the fact that peer ratings are likely to be based upon hearsay of complaints from students, since students with complaints are more likely to speak to an advisor or department chair than students with commendations for their teacher. Since only complainers are heard and satisfied students are not, peer ratings are not something to be highly commended.

"But," you say, "there is no bias when colleagues base their ratings upon actual observation of classes. Surely this is the ideal method of evaluating teaching."

Unfortunately this does not seem to be the case. We cannot be very good judges of one another's teaching on the basis of one or two class visits. Such a visit creates an artificial situation. In an article written for the *Journal of Higher Education*, John Centra reports research findings which show peer observations to have low reliability.[2]

On the other hand, peer visitation may be a useful basis for consultation about teaching. What is useful in evaluating teaching does depend upon your purposes. Peer visitation is probably an effective method for improving one's teaching but not useful for evaluating teaching for promotion.

Nonetheless, in evaluating teaching for promotion, peers have an important role to play. Peers are the best judges of the content and goals of a course. Students trust administrators to staff courses with teachers competent in the discipline. Students thus do not make discriminating judgments about their teachers' subject-matter competence. If you have doubts about a faculty member's scholarly competence for teaching a course, peers can evaluate student examinations, student papers, syllabi, or reading lists to see whether these are appropriate for the goals of that particular course in this particular program. Thus the combination of peer judgment with respect to the content and organization of the course and student judgments about what the impact has been on them is preferable to either one taken in isolation.

There are other techniques of evaluation. The Universities of Illinois, Texas, and Washington and other institutions have been using group interviews. In the group-interview technique, a team of colleagues or a representative from the center having to do with instructional improvement comes into a class and interviews the class or a subgroup of the class about what's going on in the class and how they evaluate the teaching. The advantage of the group interview is that it typically enables more probing and specific questions about what the actual problems are or about what the teacher is doing that makes that teacher particularly effective.

". . . the third condition of improvement from feedback is that you have some alternative ways of behaving."

How can these methods be used for improving the teaching by individual faculty members?

Usually our implicit theory of change has been that we will get data, we will give the data to the teachers, and they will get better. More simply put, the theory is "Feedback (knowledge of results) produces improved performance." Unfortunately things are not that simple. Feedback results in improvement only when at least three conditions are satisfied:

First, you learn if you get *new* information from the feedback. And you may not learn—you may not improve—if you already know what the feedback is telling you. One of the problems with many forms of evaluations of teaching (and research as well) is that you don't learn anything new: You already knew the information provided by the evaluation.

Another problem, particularly with student-rating forms, I think, is that sometimes you get more information than you can assimilate. In his book, *Gestalt Psychology*, David Katz called this phenomenon "mental dazzle."[3] In general, you can see things better if there is more light, but if the light is too intense you may be dazzled and not be able to see things at all. Katz suggested that if you increase the amount of information beyond a certain point, people are dazzled and unable to comprehend. That sometimes happens in student-rating forms. Committees developing a rating form often come up with a list of things that could be problems in teaching and may develop a list of a hundred items. They then prune the form down to 60 items and administer the form, and give the faculty member a computer printout with the results—means, standard deviations, and distributions of 60 items. For most of us, it is simply impossible to assimilate a computer printout with about 60 different items on it; so we ignore the whole thing.

A *second* condition of improvement as a result of feedback is *motivation*. Even though you gain information about things that are different from what you had thought, you are not going to improve unless you are motivated to improve. One may get information that the students do not like the way one lectures. If you feel that you are doing the right thing or if you don't respect the students' opinions or if you simply don't want to take the trouble to do differently, you aren't going to change.

Such lack of motivation is probably relatively rare; more often the problem is not lack of motivation, but the fact that student ratings may have negative effects upon motivation. I have found that the biggest problem of people who are ineffective is that they know that they are ineffective and have a great deal of anxiety about going into the classroom. One of the reasons that they do not like student ratings or other methods of evaluation is that they know they are going to get negative information, which is going to make them even more uncomfortable in the classroom. To the degree that student ratings or information leads a teacher to think, "I'm in a situation where people don't like me," that situation is likely to be even more difficult to cope with.

A *third* condition of feedback centers on the question, What do I do? One can find information suggesting that changes are indicated; one can be motivated to change; yet, if one does not know how to change, one is not going to improve. So the third condition of improvement from feedback is that you have some *alternative ways of behaving*.

What can we do then to bring about these three conditions of providing information, securing motivation, and suggesting alternative approaches? If we are using student ratings, one aid is to give faculty members a chance to choose their own items. One of the problems with standard rating forms for student ratings is that if a form is to be used throughout the college, the items have to be so general that they are not likely to be very applicable to any individual course. They are likely to be so general one does not know what to do if one gets a low rating. So to the degree that one is able to ask questions that one wants to know about—that an instructor can do something about—one is more likely to learn and improve. The Purdue Cafeteria System and most of the major systems of collecting student ratings provide opportunities for faculty members to choose their own questions.

A second important element is consultation. There have now been several studies which suggest that student ratings of teaching produce some positive changes in teaching but also that the change is much more likely to occur if the teacher receives consultation as well.[4]

What does this consultant do? The consultant often first calls attention to information that the teacher might not notice.

A second thing that the consultant can do is to give encouragement. When we get evaluative comments, we tend to overemphasize negative comments. One of the functions of the consultant is to remind us that everyone gets some negative evaluations and to point to the positive aspects of the ratings along with the possibility that some changes should be made.

The third thing the consultant might do is to suggest alternatives. When I act as a consultant to my own teaching assistants, the scenario might go like this: "You are concerned about the fact that students feel that you're not very well organized. Have you thought about anything that you could do?" The teaching assistant says, "I could write the main points on the board as they occur in the discussion." I say, "That's good. Or you could provide a handout that lists some of the main questions that are going to be involved in the discussion."

Using Evaluative Data for Personnel Decisions

With respect to personnel uses of evaluative data on teaching performance, one gets into a bind when one is trying to use the same evidence for improvement and personnel purposes. For purposes of improvement, we want to know the areas where there is room for improvement; so we choose student-rating items that are about things where there are some doubts. If that information is also going to be used to judge whether or not a faculty member gets tenure, is the faculty member going to choose items where he or she may be doing poorly?

To the degree that evaluative information is going to be used for personnel purposes, we also increase the individual's anxiety. I have already suggested that one of the problems in ineffective teaching is teacher anxiety. A faculty member worrying about promotion is probably not going to take risks in trying new things, especially when performance in these new areas will be evaluated.

A third problem in using evaluative data for personnel purposes is that a heavy emphasis upon evaluation is likely to convert a situation which should be intrinsically satisfying into a situation where one is working for promotions and merit increases—much as students sometimes work for grades, rather than for what they can learn. An increasing body of evidence in psychology suggests that when one puts emphasis on extrinsic rewards and punishments, one is likely to kill curiosity and intrinsic interest in an activity. In the job market today, assistant professors who enter the academic profession with some sense of enjoyment of research and enjoyment of teaching may begin to get so concerned about getting tenure that they feel they have to turn out extra publications; they have to count pages; they have to get good student ratings. What happens when tenure is achieved? All too often research and teaching are no longer fun, and administrators wonder why it is that these promising young faculty members are no longer producing the way that they were before they got tenure. Essentially what we want when giving tenure is not to reward someone for what they have done before, but to make a prediction of what they are going to do for the rest of their lives. The very process of evaluation may distort our ability to predict what a person finds basic satisfaction in doing.

"Peer ratings do correlate fairly well with students ratings . . . . probably due to the fact that peer ratings are likely to be based upon hearsay. . ."

Using Evaluative Data for Student Choice of Courses

Gathering evaluative data can help students make wiser choices of courses. Here again there is a potential conflict. Typically, when students say that they need to have information in order to choose their courses more effectively, they want to publish the results. On large campuses where students may not know anyone else who is taking or has taken a course, it seems legitimate for the students to get as much information as they can to judge whether that course is a good choice in their curricula. But part of the motivation for a published evaluation is more threatening. Often what the students really want to do is to expose the lousy teachers and make them so embarrassed that they will get better. The problem here is exacerbated in some student booklets by the desires of the publishers of the booklet to make it interesting reading. They will caricature a particular professor's approach and quote the most cutting comment about the professor, rather than the more typical reactions.

Publicity has a negative impact upon particularly poor teachers—the opposite effect from what students would have liked. It is bad enough to go into class and not feel very comfortable with the students to start with; it is far worse to go in and feel that the whole campus knows that students dislike my course.

In working with our students on their publications, I suggest that they get information that is descriptive of the course, rather than evaluative. For many students, the important things to know about courses are questions about such things as the nature of the assignments and the degree to which the class is structured or unstructured. Course difficulty is a reasonable item. Surprisingly, there does not seem to be a strong relationship between course difficulty and ratings. As reported in the *Journal of Educational Psychology* recently, students rate difficult courses highly and will choose highly rated difficult courses over easier courses rated less highly.[5]

To the degree that the student booklet works and students are choosing courses for valid reasons, faculty members do have a chance to get a group of students who are there because they have made a choice. We know that a very strong element in motivation for all of us is a sense of personal choice and personal control. So student use of evaluative information may also contribute to effective teaching by giving us a group of students whom it is easier to teach.

Evaluation is a Means, Not an End.

My final point is that evaluation is not an end in itself. I frequently find a college or a university which has devised a very systematic plan of evaluation; yet the system does not work because it was worked out by a group of experts or administrators, and faculty members do not trust it. It is much more important that the faculty have confidence in what's going on than that the evaluation plan meet standards that evaluation specialists might impose. Our ultimate goal is student learning. In some cases less evaluation makes for more student learning. Let us not let evaluation take preeminence over our ultimate goal of helping students become better educated. ■

Notes

1. P. A. Cohen, "Student Ratings of Instruction and Student Achievement: A Meta-analysis of Multisection Validity Studies," *Review of Educational Research* 51 (1981): 281-309.

2. J. A. Centra, "Colleagues as Raters of Classroom Instruction," *Journal of Higher Education* 46 (1975): 327-337.

3. D. Katz, *Gestalt Psychology*. (New York. Ronald Press, 1950).

4. P. A. Cohen, "Effectiveness of Student-Rating Feedback for Improving College Instruction: A Meta-Analysis of Findings," *Research in Higher Education* 13 (1980): 321-341.

5. J. Coleman and W. J. McKeachie, "Effects of Instructor/Course Evaluations on Student Course Selection," *Journal of Educational Psychology* 73 (1981): 224-226.

A Student Evaluation of Teaching Techniques

"None of Them is Unimportant"

Mark B. Freilich
Math/Science Division, Junior College of Albany, Albany, NY 12208

As Robert Travers points out in "Criteria of Good Teaching" (*1*), it did not become necessary to formally evaluate teacher effectiveness until students were *required* to go to school (either as a result of compulsory education laws regarding elementary through secondary school pupils or the compulsion of the market place in regard to higher education). The effective teachers in early Greece were easy to recognize: they were the ones (such as Socrates) around whom those who wanted to learn would gather for discussion and argument. When the University of Paris opened in the tenth century, the effective teachers were the only ones who were paid—the students paid fees directly to the professor and only those teachers who could attract enough students proved able to support themselves.

Obviously, the situation is very different today and, under the dual protections of tenure and faculty unions, the teacher finds himself/herself in a much more secure situation. However, while the teacher may have gained a great deal, the student, the ultimate consumer of the teacher's product, may have lost a great deal. Recently, an increasing number of our students have recognized that they *are* consumers and have a right to demand a quality product. This, combined with greater administrative pressures to ensure that shrinking budgets are not used to support poor teaching (poor teaching is very bad from a public relations standpoint!), has led to a phenomenal growth in research into effective teaching and the use of various methods of evaluating individual teacher effectiveness. (At this point, let me emphasize that this paper is neither in defense of nor an attack upon the practice of student evaluations of faculty.)

Only a few years ago, Richard Meeth felt compelled to write that "in spite of the numerous teaching improvement centers, research on teaching is nominal and incidental in most colleges and universities" (*2*). While this may have been true then, it is certainly less so now. A bibliographical search performed in late 1979 using ERIC (*3*) on certain aspects of teacher effectiveness turned up one reference in 1966, two in 1967, nine in 1968, and *44 pages* of references (with an average of almost three references per page) in 1977.

Sadly, upon reading the literature on improving instruction, we encounter a great deal of conflicting advice. For example, we are told that the use of behavioral objectives (*4*) help students learn, but we are warned that this may "discourage independent thinking on the part of students" (*5*). We are told that frequent testing is beneficial (*6*), but students complain that they are being tested so frequently across all of their courses that they have little time to digest the material. As a result, we begin to wonder which of the many things we could do are likely to be *the most important.* Until recently, this information has either been lacking or merely the chaff of other studies (*7*).

Attempts have been made to correlate teacher effectiveness with non-instructional activities (such as participation on committees, work as a consultant, etc.) (*8*). One study by Martin (*9*) investigated the correlation between the evaluations different sections of a large course gave of the lecturer and the final exam averages of those sections. Several studies have dealt with teacher characteristics (*10–19*) and some (*11, 13, 18*) have indicated that certain general qualities are more important than specific techniques. These studies of teacher characteristics and the more general analyses of student evaluations of teachers (*20–24*) give us a basis for discerning what helps students learn. They show that different types of students respond favorably to different teaching formats (*19*) and that faculty responses to student evaluations have "contributed to the improvement of college teaching" (*24*).

Readers of THIS JOURNAL may remember the papers presented in the *Symposium: Student Evaluation of Teaching* in March 1974 (*25*). Whereas the general tone of the symposium was somewhat critical of the state of the art,[1] no participant questioned the importance of the actual items used in written questionnaires. In order to clarify student responses to the items in such evaluations, Larsen (*25e*) took six often-used items (such as "the teacher is always well prepared") and asked students to describe the properties which would lead to favorable and unfavorable responses to those items.

To extend this type of analysis and in an effort to get some idea of which of the variables are most important in instruction, 107 students in a general chemistry course for freshman engineering students at Purdue University and 106 students in a general chemistry course for liberal arts/science students at California State University, Hayward, were given a list of the 28 items duplicated in Table 1 and were asked to evaluate them. The students were told to choose the five items which were most important in helping them learn and also the five least important, though it was emphasized that none of the items was necessarily unimportant. In addition to choosing the five most and least important items, the students were asked to rank the items in order of importance. The surveys were conducted in the spring, with the students nearing completion of the year-long general chemistry sequence. All students had been exposed to at least two different lecturers and/or recitation leaders. Additionally, several faculty members and a few experienced graduate teaching assistants (for a total of 23 teachers) in a wide variety of disciplines were asked to evaluate these items.

The raw data were treated as follows: each time an item was chosen as most important it was assigned a value of +5; if second most important, +4, down to +1 for the fifth most important item. When an item was chosen as the least important of all, it was assigned a value of −5, etc., to an assignment of −1 for the fifth least important (or, alternatively, the twenty-fourth most important) item. Thus, if all of the students thought that connecting the material to be learned with their own future well being (item *b*) was *the most important* item, it would accumulate 213 × +5 or +1065 points. The correspondingly least important aid to learning could accumulate a maximum total of −1065 points. An item to which the students were relatively indifferent would accumulate 0 points as would one which was ranked equally high and low by the respondents.

Results

Of the questionnaires which were distributed to the stu-

[1] In contrast, see reference (*24*), which lends strong support to the reliability and validity of students evaluations.

Reprinted with permission from *Journal of Chemical Education,* vol. 60, no. 3, pp. 218–221, March 1983.

dents, 192 were evaluated, and 21 were discarded because they were not completed properly. Table 2 lists all 28 items in order of decreasing importance as indicated by the total accumulated points of each item. Two separate listings are given: one for the results of the student survey and one for the results of the teacher survey. (More detailed data, including separate data for the engineering and the liberal arts/science students, may be obtained by writing to the author.)

For the large fraction of readers of THIS JOURNAL who were active in higher education during the late sixties and early seventies, either as students or as faculty, some of the results of the survey may come as a surprise. However, the general

Table 1. Items Thought to Help Students Learn

Students learn best when:

(a) They understand why the subject is being taught
(b) They can connect the material with their own future well being
(c) They feel they have some say in their own education
(d) They are allowed, within limits, to follow their own interests
(e) Individual initiative is not severely suppressed
(f) They are told what is to be learned and what is expected of them
(g) A certain schedule is established and adhered to reasonably well
(h) They are provided with study problems, of varying degrees of difficulty, on new material and the opportunity for review of these problems is soon after exposure to the material. [Problem solving at different levels of Bloom's taxonomy][a,b]
(i) They are quizzed frequently
(j) They are provided with prompt feedback on their achievements (promptly graded homeworks and exams, for example)
(k) There is visual as well as verbal reinforcement [Demonstrations]
(l) There are reading assignments as well as lecture
(m) They are told the goals and given learning objectives[c]
(n) They are given concrete and varied examples
(o) Abstract concepts are attached to models
(p) There are connections among the concepts and principles they are asked to learn
(q) New ideas can be related to already established ones
(r) New information is presented in a logical progression
(s) They are given practical (working) rules as well as theoretical concepts
(t) Distractive mannerisms of the teacher are held to a minimum
(u) The teacher is able to adapt the stress he/she puts on an item to the importance of that item
(v) They (the students) feel free to ask questions or challenge the ideas of the instructor (either before, after, or during class)
(w) They feel confident that the instructor knows his/her material
(x) They perceive the instructor as a person who is interested in them and their welfare
(y) They perceive the instructor as being well organized (as indeed he/she must be)
(z) They try to explain to or teach one another
(aa) They can "hash out" ideas in small groups
(bb) They are allowed to test their ideas for themselves (to experiment wilth their ideas)

[a] The phrases in parentheses appear as they were seen by the students. The phrases in brackets are either verbal explanations given to the students who asked for such or are intended to clarify the item for the reader.
[b] Bloom, Benjamin S., (*Editor*), "Taxonomy of Educational Objectives, Cognitive Domain," David McKay Co., Inc., New York, **1956.**
[c] Though only the author, as lecturer or as recitation class leader, provided the class with objectives, approximately 75% of all of the respondents in the survey were in the author's class at the time the surveys were conducted.

Table 2. Items In Order of Relative Importance[a]

Student Evaluation

Students learn best when
1) New information is presented in a logical progression. [*r*]
2) They are given concrete and varied examples. [*n*]
3) They are provided with study problems . . . [*h*]
3) They are provided with prompt feedback on their achievements . . . [*j*]
3) They feel confident the instructor knows the material. [*w*]
6) They feel free to challenge the instructor and to ask questions. [*v*]
6) They are told what to learn and what is expected of them. [*f*]
8) There are connections among the concepts and principles they are asked to learn. [*p*]
8) They can connect the material with their own future well being. [*b*]
10) New ideas can be related to already established ones. [*q*]
11) They understand why the subject is being taught. [*a*]
12) They perceive the instructor as being well organized. [*y*]
13) They perceive the instructor as a person who is interested in them. [*x*]
14) There is visual as well as verbal reinforcement. [*k*]
*[b]
15) Individual initiative is not severely suppressed. [*e*]
15) The teacher is able to adapt the stress put on an item to the importance of that item. [*u*]
15) Abstract concepts are attached to models. [*o*]
18) There are reading assignments as well as lecture. [*l*]
18) They are allowed, within limits, to follow their own interests. [*d*]
18) They are told the goals and given learning objectives. [*m*]
21) They feel they have some say in their own education. [*c*]
22) A certain schedule is established and adhered to. [*g*]
23) They try to explain to or to teach one another. [*z*]
24) They are allowed to test their ideas for themselves. [*bb*]
24) They are given practical (working) rules as well as theoretical concepts. [*s*]
26) They are quizzed frequently. [*i*]
27) They can "hash out" ideas in small groups. [*aa*]
28) Distractive mannerisms of the teacher are held to a minimum. [*t*]

Teacher Evaluation

Students learn best when
1) They feel confident the instructor knows the material. [*w*]
1) They are told what to learn and what is expected of them. [*f*]
3) They are provided with prompt feedback on their achievements . . . [*j*]
4) They are provided with study problems . . . [*h*]
5) They perceive the instructor as well organized. [*y*]
5) They perceive the instructor as a person who is interested in them. [*x*]
7) A certain schedule is established and followed. [*g*]
7) New information is presented in a logical progression. [*r*]
7) New ideas can be related to already established ones. [*q*]
10) They are given concrete and varied examples. [*n*]
11) There are connections among the concepts and principles they are asked to learn. [*p*]
11) They are allowed, within limits, to follow their own interests. [*d*]
*[b]
13) They feel free to challenge the instructor and to ask questions. [*v*]
13) They can connect the material with their own future well being. [*b*]
13) They understand why the subject is being taught. [*a*]
13) They are given practical (working) rules as well as theoretical concepts. [*s*]
17) There is visual as well as verbal reinforcement. [*k*]
17) Individual initiative is not severely suppressed. [*e*]
17) The teacher is able to adapt the stress put on an item to the importance of that item. [*u*]
17) They are told the goals and given learning objectives. [*m*]
17) Abstract concepts are attached to models. [*o*]
17) They are allowed to test their ideas for themselves. [*bb*]
23) They feel they have some say in their own education. [*c*]
23) They can "hash out" ideas in small groups. [*aa*]
25) They are quizzed frequently. [*i*]
26) There are reading assignments in addition to lecture. [*l*]
27) They try to explain to or to teach one another. [*z*]
28) Distractive mannerisms of the teacher are held to a minimum. [*t*]

[a] Items given the same numerical rank had total point accumulations which were insignificantly different.
[b] Items above this point had positive total accumulations. Items below this point had negative total accumulations. Items clustered about this point had statistically insignificant differences in total accumulations.

tone of the results does lend support to those who claim that today's students are rather practical, goal (job) oriented, and willing to look to the faculty for, or actually demand from the faculty, rather strict guidance. It seems that, within a given course, students attach a much lesser importance to giving free rein to individual initiative (item *e*), allowing them to follow their own interests (item *d*), or to allowing them to feel that they have some say in their own education (item *c*) than to having confidence in the instructor (items *v*, *w*, *x*, and *y*) and in the way the course is structured (items *f*, *h*, *j*, *n*, and *r*). (Each of the items *c*, *d*, and *e* mentioned above actually accumulated significant negative totals.) The rather significant dependence upon teacher control of learning lends a great deal of support to Dudley Herron's suggestion (*26*) that students need a reason for learning; that the teacher must "establish a need to know" on the part of the students.

Both teachers and students agree that it is important to have confidence that the teacher knows the material (*w*), that there are study problems followed by a prompt review (*h*), that there is prompt feedback in general (*j*) and that the students must be told what is expected of them (*f*). However, by far the most important item to the students (it was placed in the top five by 67 students and in the bottom five by only 4; only one item was placed in the bottom five by fewer students) was that new information be presented in a logical progression (*r*). While this was important to the teachers, it only tied for seventh place! The teachers attached considerable importance to the interest an instructor displays toward his/her students, ranking item *x* fifth, whereas the students ranked this item thirteenth. Apparently, the mechanical workings of a course are more important than the personal ones as long as the teacher displays some degree of interest in the students.

The students also thought it extremely important that they be provided with concrete and varied examples (*n*). While we all recognize this as critically important in a course such as chemistry, it is interesting to note that the cross-section of teachers included in this study ranked that item tenth. Obviously, other faculty must be made aware of the importance of including examples in their presentations and we, the science faculty, must take cognizance of the fact that approaches which are essentially built-in facets of our subject are not second nature to the college community at large. As a matter of fact, the teaching of general chemistry (and chemistry in general) lends itself very well to most of the items thought of as most important by the students.

Some interesting results were hidden as a consequence of listing the items in the order of total accumulations. For example, item *b*, connecting the material with their own future well being (perhaps meaning a good grade in the course?) was rated as *the most important* item by more students than any other item, yet it ranked only eighth overall. Obviously, this is extremely important to a significant fraction of any given group of students. Surprisingly, this same item was ranked thirteenth overall by the faculty and four other items (confidence in the instructor's knowlege, instructor is interested in students, students are told what is expected of them and students are given opportunities for practice) received more "first place votes" from faculty. Of all the items, the need for prompt feedback (item *j*) was second in appearance as one of the five most important items to the students (the need for a logical progression was first) and appeared most frequently in the faculty rankings (66/192 students and 12/23 teachers). Most disconcerting is that 32 percent of the students listed the need for examples (*n*) among the five most important, but only three of the twenty-three teachers did so.

A surprising finding was the very low importance assigned by both faculty and students to frequent quizzing (item *i*). (However, as mentioned earlier, studies (*6*, *11*) have shown testing to be important.) With the exception of item *t*, this item received more total "votes" from the students than any other item (a total of 94, 24 as being among the five most important, 70 (!) as being among the five least important). Clearly, a significant fraction of chemistry teachers regard frequent quizzing as a spur to learning general chemistry. It is therefore important that we recognize the existence of this attitude on the part of our students and other faculty and that we make it clear to the students just why they are quizzed frequently. It was also surprising to discover the low importance that students attach to the adherence to an established schedule (item *g*). We teachers certainly thought it was rather important!

Students *can* learn well by discussing material with one another, by trying to teach one another or by being given the opportunity to try out their own ideas (*27*). Thus, it was somewhat disconcerting to see the very low importance assigned by the students to explaining to one another (item *z*, ranked twenty-third overall), to discussion in small groups (item *aa*, ranked twenty-seventh overall) and to the students' testing of their own ideas (item *bb*, ranked twenty-fourth overall). However, as indicated by Steiner (*27*), these rankings may have resulted from a combination of a lack of confidence in themselves apparently experienced by many underclassmen and the limited experience most students have had in trying to learn in this manner.

Apparently, an instructor's distractive mannerisms (Who? Me?!) are no hindrance to learning. This item, *t*, was ranked as one of the five least important items far more frequently than any other by both the students (92 times) and the instructors (13 times). It was one of only two items (*aa* being the other) to average across the *entire* group of students as one of the five least important items. Considering any meterstick used in this survey, the students and teachers agree in ranking this item as least important.

Particularly for faculty teaching chemistry, the annual rite of choosing a textbook need no longer be regarded with such trepidation; simply choose the one which is least expensive, the one with the best selection of easy problems for you to solve, or the one with the nicest pictures and diagrams. Only sixteen students ranked item *l* (reading assignments in addition to lecture) as one of the five most important and nearly twice that many placed it among the five least important. (To avoid confusion about the meaning of item *l*, I surveyed a random sample of students to ascertain that they did indeed regard the textbook as a reading assignment in addition to lecture. They did *not* regard the faculty member's or a graduate assistant's lecture notes in the library as such.) To those who think the choice of a particular textbook (or perhaps the assignment of readings in THIS JOURNAL or *SciQuest*) *is* important, it is worth noting that none of the teachers assigned item *l* as one of the five most important and nearly one-third placed it among the five least important.

In closing, it is worth noting that each and every item was ranked as one of the five most important by at least five students and as one of the five least important by at least three students. One student, in not filling out the survey form properly, wrote "None of them are (sic) unimportant." (Two students regarded the need for prompt feedback as least important of all, and yet this item was only one short of receiving more rankings in the top five than any other!) It is also significant that many of the higher rated items tend to be those which lead to the accumulation of facts and that the lowest rated ones (items *z*, *bb*, and *aa*) are those which approximate the problem solving situations which one finds upon entering the "real world." When we ask our students what helps them learn best, are we actually asking them what best helps them to memorize, not think?

Acknowledgment

I would like to thank Professors J. Dudley Herron and John Feldhusen, both of Purdue University, for their suggestions on this manuscript. I am grateful to the students and faculty of Purdue University and to the students of California State

University, Hayward, who participated in this survey and to Professors Milton Fuller and John De Vries, both of CSUH, for allowing me time during their lecture periods to conduct parts of the survey.

Literature Cited

(1) Travers, R. M. W., "Handbook of Teacher Evaluation," Millman, J., (*Editor*). Sage Publications, Beverly Hills, CA, **1981**, Chapter 2.
(2) Meeth, L. Richard, *Change*, **8**, 3 (1976).
(3) Educational Resources Information Center, National Institute of Education, U.S. Dept. of Education, Washington, D.C.
(4) Herron, J. D., *J. Res. Sci. Teach.*, **8**, 385 (1971).
(5) Haight, G. P., *Change*, **8**, 4 (1976).
(6) Martin, R. R., and Srikameswaran, K., J. CHEM. EDUC., **51**, 485 (1974).
(7) In this vein, see Meeth in reference (2) above.
(8) McCullagh, R. D., and Roy, M. R., *J. Exp. Educ.*, **44**, 61 (1975).
(9) Martin, R. R., J. CHEM. EDUC., **56**, 461 (1979).
(10) Gadzella, B. M., *Coll. and Univ.*, **44**, 89 (1968).
(11) Hampton, E., *Imp. Col. Univ. Teach.*, **19**, 248 (1971).
(12) Hanke, J. E., and Houston, S. R., *Education*, **92**, 97 (1972).
(13) Haslett, B. J., *J. Exp. Educ.*, **44**, 4 (1976).
(14) Mannan, G., and Traicoff, E. M., *Imp. Col. Univ. Teach.*, **24**, 98 (1976).
(15) Mueller, R. H., Roach, P. J., and Malone, J. A., *Psych. in the Schools*, **8**, 161 (1971).
(16) Quick, A. F., and Wolfe, A. D., *Imp. Col. Univ. Teach.*, **15**, 133 (1967).
(17) Sandefur, J. T., and Adams, R. A., *J. Teach. Educ.*, **27**, 71 (1976).
(18) Sherman, B. R., and Blackburn, R. T., *J. Educ. Psy.*, **67**, 124 (1975).
(19) Andrews, J. D. W., *Res. High. Educ.*, **13**, 321 (1980).
(20) Ory, J. C., Brandenburg, D. C., and Piper, D. M., *Res. High. Educ.*, **12**, (1980).
(21) Jones, W., and Somers, P., *J. Exp. Educ.*, **44**, 44 (1976).
(22) Pohlman, J. T., *J. Educ. Meas.*, **12**, 49 (1975).
(23) Baum, P., and Brown, Wm., *Res. High. Educ.*, **13**, 233 (1980).
(24) Cohen, P. A., *Res. High. Educ.*, **13**, 321 (1980).
(25) Siebring, B. R., J. CHEM. EDUC., **51**, (a)150, (b)152, (c)155, (d)161, (e)163 (1974).
(26) Herron, J. D., J. CHEM. EDUC., **55**, 190 (1978).
(27) See, for example, Steiner, R. P., J. CHEM. EDUC., **57**, 433 (1980).

Robert C. Wilson

Teaching Effectiveness: Its Measurement

The most common conception of college and university teaching among legislators and the general public is that of a professor standing at the front of a classroom lecturing to a group of students for six to twelve hours a week. Those who are more familiar with academic life recognize that classroom teaching is only the most visible part of teaching, that classroom teaching is based upon a great deal of much less visible activity and that much of teaching takes place outside the classroom.

There have been a number of workload studies of college teachers which indicate that they average between 50 and 60 hours of work per week. Nevertheless, the very damaging public image of the college teacher's six to 12 hour week exists and has contributed greatly to the hostile and punitive actions which several state legislatures have taken against public higher education.

Perhaps teachers, too, share some blame for having contributed to this image. Even though there are many activities which teachers must engage in, they generally do not talk about them when discussing teaching assignments and workload distribution. Because classroom hours are convenient units to think in terms of, teachers and administrators tend to talk about them as though they were equivalent units and also the most meaningful units to use in distributing the teaching resources of college campuses. The lack of real equivalence receives some acknowledgment, however, in the way assignments are made. The number of courses calling for different preparations is sometimes taken into account, and workload credit may be given for the supervision of theses and independent study.

Ways must be found to assign and report teaching workloads which reflect the real breadth of direct and indirect teaching activities. The following is a sample of some teaching related activities engaged in by many college teachers.

1) *Classroom teaching activities*—lecturing, leading discussions, suggesting reading references, making assignments.

2) *Preparatory classroom activities*—reading assigned books, preparing notes, constructing reading lists, devising assignments, preparing laboratory demonstrations, securing equipment such as audio-visual aids.

3) *Associated housekeeping activities*—preparing attendance rosters, making problem sets, preparing quizzes and examinations, reading and grading quizzes and examinations, reading and evaluating term papers, evaluating class projects.

4) *Course planning activities*—reconsidering the needs and interests of students, the state of the field, and the condition of society, reviewing possible textbooks, planning course sequences.

5) *Out-of-class teaching activities*—talking with students about classroom discussions, clarifying assignments, helping students plan and prepare term papers or projects, holding paper conferences or examination conferences, discussing intellectual matters with students, helping students learn how to study, supervising independent study.

6) *Advising and counseling activities*—discussing students' vocational aims and plans, advising about academic programs, discussing students' personal problems, gathering relevant information from other faculty or administrators, acting to help students with difficulties, writing letters of recommendation.

7) *Student extracurricular activities*—advising student organizations, attending student social functions, discussing campus issues with student groups.

8) *Keeping one's knowledge up-to-date*—reading books and professional journals in one's specialty, reading in related fields, reading about general cultural developments, attending professional meetings, corresponding with colleagues elsewhere, writing for books, articles, and papers.

9) *Keeping informed about campus issues which affect students and teaching*—discussing issues with colleagues in one's own department as well as with colleagues in other departments, meeting with members of various committees and with administrators, reading school newspapers, reading memos, reading position papers and planning documents.

10) *Departmental governance activities*—attending department meetings, serving on department committees, writing memos and proposals concerned with the educational program and the students and resources involved.

11) *Division, college, or university governance activities*—attending meetings, serving on committees, writing memos, proposals and position papers.

12) *Graduate education activities*—selecting stu-

Reprinted with permission from *Engineering Education,* vol. 62, no. 6, pp. 550–552, March 1972.

dents from applicants, recommending financial assistance for students, preparing, administering and evaluating graduate examinations, and serving on thesis committees.

The above, although lengthy, is not an exhaustive list of activities in which college teachers engage. Not all faculty perform all of these activities. Each activity listed, however, is directly or indirectly related to effectiveness in teaching when it is broadly conceived. Clearly, teaching involves a complex of activities, only a few of which occur within the familiar confines of the classroom.

What is Effective Teaching?

Considering the diversity of activities which may be considered part of a professor's teaching duties, how shall one go about measuring teaching effectiveness (the goodness or badness of performance of these duties)? What shall the criteria of good performance or bad performance be? What kinds of evidence of performance are acceptable?

First, it is apparent that "effective" is a value term, a social judgment rather than a description. This means that the same teaching activity may be judged differently depending upon the personal values and social position of the persons making the judgment.

It is also apparent that people differ about the kinds of data or information they are willing to accept as evidence of teaching effectiveness. Most of the controversy about teaching effectiveness and how to measure it boils down to disagreements about the adequacy or acceptability of different sources of information. The question, "How do you know you are *really* measuring teaching effectiveness?" should be translated into three declarative statements. "I do not find the data you propose as evidence of effectiveness acceptable. I think you should use some other kind of data. Ask me what I think you should use." It may be useful, at this point, to list some of the criteria or sources of information which have been proposed as indices of effective teaching.

1) *Teachers' own judgments about how effective or ineffective they are.* This criterion has the virtue of allowing the person closest to the situation, the teacher himself, to assess his effectiveness. Since he is largely responsible for designing and teaching courses, he may be in the best position to assess progress toward his own objectives. The major limitations of this approach lie in the fact that many teachers don't know how effective their own teaching is; they may be so involved in the teaching process that they cannot assess their effectiveness accurately. However, this is one source of information which can be used in conjunction with others.

2) *Judgments of effectiveness on the basis of student achievement.* On the surface, this criterion appears to be the most directly relevant and valid of all, for is not the proof of good teaching found in the actual achievement of students? This is the kind of criterion which most people would like to use if possible. However, there are several problems in its use. First, people disagree as to what kinds of achievement should be measured; second, there are real difficulties in developing achievement tests which are sufficiently sensitive to give reliable measures of change; third, there are many important kinds of change in students which are not easily measured with the kinds of tests we presently have available.

3) *Judgment of effectiveness by alumni.* It is sometimes suggested that students don't really appreciate what they have learned from a teacher until they have been out in the real world for a while. A few studies have been made of alumni judgments, and the agreement with student judgments is generally high. Major limitations of this approach arise from faculty turnover and also the fact that the junior faculty were probably not on campus when the alumni were.

4) *Judgments of teaching effectiveness by colleagues.* While colleagues are often involved in such judgments, the evidence they use is generally indirect. A few institutions have made use of classroom visitation with some success. But it requires a school or department in which there is a good deal of mutual confidence and trust. It is also a time-consuming and expensive method, particularly when the number of teachers involved is large.

5) *Student judgment of teaching effectiveness.* After 50 years of research on the use of student evaluations of teaching, this approach seems to be coming into more common use. There now exist a number of measuring instruments for obtaining reliable data from students. In our own studies, we have attempted to separate the evaluation function from the description function. That is, we have attempted to use students as classroom observers to describe what a teacher does, reserving the evaluation function as a later stage in which some individual or group of individuals interprets the descriptive data obtained and combines it with data from other sources.

Our studies at the Davis campus of the University of California were focused on student descriptions of effective teachers and colleague descriptions of effective teachers.[1] Our objective was to contribute to the improvement of teaching at the university by characterizing effective performance and providing a satisfactory basis for the evaluation of teaching.

Three surveys were conducted in 1967 and one in 1968. First, 338 students (including 60 graduate students) identified the "best" and "worst" teachers they had had in the previous year, and answered 158 questions about the teaching of each. Second, 119 of the faculty identified the best and worst teachers among their colleagues, and answered 103 questions about the teaching and other academic activities of each. Third, 162 of the faculty reported how often they had performed various academic pursuits in stated time periods. Finally, a validation survey was made: fifty-one classes were selected to include, in about equal numbers, those of instructors previously identified as best teachers by three or more students or colleagues, those of instructors previously identified as worst teachers, and classes of other instructors not previously named as either best or worst. The 1,015 students in these classes answered many questions about the teaching of their respective instructors.

The principal results follow:

1) There is excellent agreement among students, and between faculty and students, about the effectiveness of given teachers.

2) Best and worst teachers engage in the same professional activities and allocate their time among academic pursuits in about the same ways. The mere performance of activities associated with teaching does not assure that the instruction is effective.

3) After performing an item-analysis, 85 items emerged that characterized best teachers as perceived by students, and 54 items that characterized best teachers as perceived by colleagues. All items statistically discriminate best from worst teachers with a high level of significance.

4) A factor analysis of the items characterizing best teachers as perceived by students produced five scales, or components of effective performance. These may be summarized as follows:

a) *Analytic/Synthetic Approach*
Has command of the subject, presents material in an analytic way, contrasts points of view, discusses current developments, and relates topics to other areas of knowledge.

b) *Organization/Clarity of Presentation*
Makes himself clear, states objectives, summarizes major points, presents material in an organized manner, and provides emphasis.

c) *Instructor-Group Interaction*
Is sensitive to the response of the class, encourages student participation, and welcomes questions and discussion.

d) *Instructor-Individual Student Interaction*
Is available to and friendly toward students, is interested in students as individuals, is himself respected as a person and is valued for advice not directly related to the course.

e) *Dynamism/Enthusiasm*
Enjoys teaching, is enthusiastic about his subject, makes the course exciting, and has self-confidence.

5) A factor analysis of the items characterizing best teachers as perceived by colleagues produced five scales of components which may be summarized as follows:

a) *Research Activity and Recognition*
Is well known and highly regarded for his scholarly activity, publications, and research. Confers with colleagues about research and keeps abreast of recent developments in his field.

b) *Intellectual Breadth*
Has broad knowledge both within and beyond his field. Is sought out by students and colleagues for information and academic advice.

c) *Participation in the Academic Community*
Attends and participates in campus lectures, social functions, and student-oriented activities. Maintains a congenial relationship with colleagues.

d) *Relations with Students*
Maintains an informal and congenial relationship with students beyond the classroom. Is consistently available to students for consultation about personal and academic concerns.

e) *Concern for Teaching*
Expresses concern for teaching and consults with colleagues about issues related to teaching.

The majority of the colleague scales are mainly concerned with activities that take place outside the classroom. This is because the majority of our respondents had not observed the person they nominated in the classroom. However, when colleagues do observe in the classroom, scales quite similar to the student scales are used.

6) In general, as most other investigators have found, student ratings of teachers showed only negligible correlation with the academic rank of the instructor, class level, number of courses previously taken in the same department, class size, required versus optional course, course in major or not, sex of the respondent, class level of the respondent, grade-point average, and expected grade in course.

In conclusion, it would seem that both student and colleague descriptions of teachers can provide useful information for a variety of purposes. Such information is probably most useful when it is obtained for the purpose of improving teaching and is incorporated as part of a larger institutional program for the improvement of teaching.

Copies of the above-mentioned questionnaires or permission to use the questionnaires may be obtained by writing to the author at the Center for Research and Development in Higher Education at the University of California, Berkeley.

References

1. Hildebrand, M., Wilson, R. C., & Dienst, E. R., *Evaluating University Teaching*, Berkeley: University of California, Center for Research and Development in Higher Education, 1971.

A COURSE EVALUATION GUIDE TO IMPROVE INSTRUCTION

John C. Lindenlaub
and
Frank S. Oreovicz
Purdue University

Finding an appropriate mechanism to evaluate college teaching has been and continues to be a difficult problem. Student evaluation is probably the most prevalent mechanism today. While studies have shown that student evaluations of college teaching are substantially correlated with other types of evaluations,[1-8] we cannot deny that other people—peers, administrators, outside visitors—can bring important dimensions to the evaluation process. For one thing, these people can assess the feasibility of carrying out improvements in the context of available resources.

The evaluation of college courses is a sensitive subject. Some faculty members feel that any such evaluation invades their private turf. Yet to be realistic, if we expect and desire rewards for good teaching, we must assist in—and insist on—evaluation. Recognizing the sensitivity of the area and its close connection to the concept of academic freedom is crucial in designing and implementing an evaluation method.

There are several good reasons for gathering data on the effectiveness of college teaching. They include providing information to guide faculty development and self improvement, promotion and tenure decisions, and merit raise considerations. These are compelling reasons, especially in today's world of accountability. Still another reason why one might want to evaluate courses is to improve them and the instructional process. That was the reason for developing the Course Evaluation Guide presented here.

Description of the Course Evaluation Guide

The Course Evaluation Guide was designed to help identify specific areas in which courses might be improved. The guide is intended to supplement the student evaluation now conducted on many campuses.* The guide serves to make faculty members more conscious of course design, and to point out systematically the factors that influence course design and are also under the instructor's control.

The guide examines a course from three perspectives: course characteristics (how it is taught), subject matter (what is taught), and assessment (how results are measured). The final section of the Guide provides an outline for summarizing results and recommendations. The guide appears on pp. 357-359.

Underlying the guide is the idea that all courses are good but that even the very best of courses can be improved in some areas. The guide is therefore not intended to identify poor courses but to assess strengths and weaknesses, and to help those in charge of a course plan for improvement.

The guide can be used for curriculum planning, by supervisors of multi-section courses, for soliciting peer evaluation and by individual instructors for self-evaluation. To help communicate a clear understanding of how the guide is to be used in a particular situation, notes are offered here on its use for the four

*At Purdue a CAFETERIA Course and Instructor Evaluation System is administered through the Center for Instructional Services, 402 ENAD, Purdue University, West Lafayette, Ind. 47907.

Reprinted with permission from *Engineering Education,* vol. 72, no. 5, pp. 356–361, February 1982.

"Underlying the guide is the idea that all courses are good but that even the very best courses can be improved."

purposes just mentioned. A comparison of these four sets of instructions will reveal a little overlap among them, which we have retained for the sake of making the context clear in each case.

Use of the Course Evaluation Guide for CURRICULUM PLANNING

The course evaluation guide can assist faculty groups in assessing courses to develop a mechanism for curriculum planning. It may be used as a supplement to student evaluation, such as CAFETERIA. The guide, however, focuses particularly on those areas more readily assessed by professional educators and subject specialists than by students. The guide emphasizes those factors over which faculty and instructors have control, so that the results of the review might provide useful guidelines for initiating change. The word "change" is to be understood in all its constructive implications—for example, incorporating recent instructional innovations—and not merely as change for the sake of change. There may also be compelling reasons to change because the subject area itself has undergone growth in recent years. A good way to see whether it is necessary, or in what way it may be accomplished, is through a systematic evaluation. This course evaluation guide aims to fulfill that need.

The items in Section II, Subject Matter, and those questions under Section III concerning the assessment of student workload are probably of more immediate interest to curriculum planners; but the items under Section I concerning course characteristics apply to whole curricula just as they apply to individual courses.

In evaluating a course, reviewers should keep in mind the factors to be retained as well as those that need to be changed. This is important for several reasons, not the least of which is that a more balanced view is achieved and the person in charge of the course is more likely to accept conclusions when made in this more balanced manner. Positive suggestions about areas needing improvement strengthen the review and help the instructor in implementing needed changes. In recommending changes, reviewers need to remember that their suggestions should fall within the realm of possibility, that is, that they should be possible under the constraints within which the course is offered. For example, it is easy to suggest that a certain course could be taught more effectively under a computer-managed format, but such a recommendation should be within the capabilities and resources of the department or the institution.

Use of the Course Evaluation Guide for COURSE SUPERVISORS

The course evaluation guide may be used by faculty members who are responsible for multi-division courses and who supervise the associated staff. Intended to provide information to supplement student evaluation, such as CAFETERIA, the guide focuses on areas more readily assessed by professional educators and subject specialists than by students.

Just as technology grows, expands, and changes, so does the area of instruction. Innovation is just as likely and as beneficial in this field as it is in scientific research. This is particularly true for multi-division, large courses. A good way to see whether or not change is necessary, or in what way it may be accomplished, is through serious evaluation of the course as it presently stands. This course evaluation guide has been designed to fulfill that need.

Course supervisors will find the first section of the guide, Course Characteristics, particularly useful, since it probes areas that may be affected by the establishment of course operation and management procedures. One subsection deals with instructional personnel.

While the importance of subject matter content, Section II of the guide, should not be minimized, very often subject matter is the concern of a wider group of faculty than just the course supervisor. This is because of the interrelationship of courses stemming from prerequisite structures.

The course supervisor will also be interested in parts of Section III, Assessment, especially those questions concerning the assessment of student workload and student performance.

On completing the course evaluation guide, the course supervisor may wish to seek comments from colleagues or members of the appropriate curriculum committee. If questions arise concerning the feasibility of certain types of changes having to do with the use of media or instructional formats, he or she may wish to consult with staff members of the school's instructional development center or similar group.

Use of the Course Evaluation Guide for SELF EVALUATION

The course evaluation guide may be used by faculty to assess their own courses. It provides a systematic mechanism for identifying strengths and weaknesses in a course. Innovation is just as likely and as beneficial in instruction as it is in scientific

"The heart of the guide's approach is a no-fault theory of review or evaluation."

research. A good way to see whether or not change is necessary, or in what way it may be accomplished, is through serious self evaluation.

The course instructor should answer as many of the questions in the guide as possible and try to find one or two areas where it is clear that improvements can be made. The identification of areas where improvements could be made should not be interpreted negatively. It is reasonable to assume that all courses are good, but it is also true that all courses can be improved.

On completing the course evaluation, an instructor may wish to discuss the results with a neutral party, such as personnel at the school's instructional development center.

The items listed under Section I, Course Characteristics, are more under the control of the instructor than those in Section II, Subject Matter. The latter is often determined in part by prerequisite and succeeding course requirements and generally involves an element of curriculum planning. Section III, Assessment, should be of interest to both the course instructor and those interested in curriculum planning.

Use of the Course Evaluation Guide for PEER EVALUATION

The course evaluation guide may be used by faculty to assess their own courses, but it also provides a mechanism for soliciting peer evaluation of courses.

The guide may be used by an individual professor to assess his or her course. When used in this mode, the guide helps the professor find one or two areas where it is clear that improvements can be made. Used in this manner, the guide becomes a self-evaluation instrument.

When the guide is used to obtain peer evaluation of a course, it serves the same function for teaching as peer review of journal articles serves in a faculty member's research efforts. In both areas the aim is to improve the product. Peers can judge how current a course and instructional materials are, and they can offer advice on the syllabus, assignments, handouts and so on, which are important to any course. Furthermore, someone who has had similar experience can help update the materials and is aware of criteria for assessing student performance. This aspect of the guide is designed to provide constructive suggestions for course improvement.

Approach to Peer Evaluation

The heart of the guide's approach is a no-fault theory of review or evaluation. It is not intended that the instructor feel that the worst is being looked for; rather, the opposite is maintained: all courses are good, but all courses can be improved.

Before something can be improved it must be examined in all its aspects, with strengths and weaknesses identified. As in any endeavor, some things are fixed, but the assessment process is begun with the assumption that ideal circumstances exist and that unlimited resources are available.

In assessing strengths and weaknesses, one must examine all aspects of a course. The instructor is very much a part of this overall picture and an interview with him or her, as well as with the teaching staff and the students, is necessary. In considering the materials of the course, the following items should be considered: content outline, schedule, objectives, methods for assessing student performance, textbook and references, strategy for course evaluation and change, instructional methods employed, personnel, student population, assessment of student workload, and feedback mechanisms (to students).

An examination of all these aspects will obviously vary from course to course, for not all things apply to all situations. But one way to clarify this series is to have the person in charge of the course rank-order the applicable features.

Hints for the professor in charge of the course being evaluated: The instructor can facilitate the evaluation by having the necessary materials ready. These materials might include, apart from the general considerations listed above, one-page descriptions of the course—its goals and objectives—textbook, course prerequisites, course topics and time allocated to each set of homework assignments, sets of tests and examinations, and so on.

Hints for the reviewer: It is important that the reviewer keep in mind the underlying assumptions stated earlier—that improvement of the course is the aim of the evaluation.

In evaluating the course, the reviewer must consider the factors to be retained as well as those that need to be changed. This is important for several reasons, not the least of which is that a more balanced view is achieved this way, and the reviewee is more likely to accept conclusions when made in this more balanced manner. Suggestions about areas needing improvement will further strengthen the review and help the instructor implement needed changes. It is important to remember that any suggestions made should fall within the realm of the feasible, that they be possible under the constraints with which the instructor is operating.

Course Evaluation Guide

This evaluation guide examines a course from three perspectives:

- *Course characteristics—how is the course taught?*
- *Subject matter—what is taught?*
- *Assessment—how are results measured?*

The fourth section of the guide provides an outline for summarizing results and recommendations.

The guide may be used by individual instructors, course supervisors, or faculty groups for purposes ranging from self-evaluation to curriculum planning.

I–Course Characteristics

A. Objectives

1. Are course objectives presented to the student:
   at the beginning of the course? ____ yes ____ no
   at the beginning of each class? ____ yes ____ no
   at the beginning of each major section (chapter)? ____ yes ____ no
   at any other time? ____ yes ____ no. Explain:
2. How are objectives transmitted to the student?
   ____ written handout form ____ verbally
   ____ "osmosis" ____ written on blackboard
   ____ on overhead transparency ____ other. Explain:
3. Do the objectives state what the student will be able to do upon completion of:
   a well–defined block of material? ____ yes ____ no
   the entire course? ____ yes ____ no
4. Does the text or other student materials specify how the student will know if he or she has met the objectives? ____ yes ____ no

Instructional objectives can be classified into different levels. A convenient scheme for many engineering courses is to classify objectives as dealing with facts, closed-end problems (single answer problems) or open-ended problems (many correct answers, design problems).

5. Have course objectives been categorized according to the fact, closed-end, open-end (or similar) scale? ____ yes ____ no
   Scale used:
6. Are the objectives ____ global or ____ specific?
7. How are objectives reviewed?
8. What approach or mechanism is used to see if course objectives match the prerequisite requirements of succeeding courses?

B. Instructional Methods Employed

1. How is the course scheduled? ____ 3 lectures/week
   ____ lecture/recitation ____ lecture/lab ____ other. Explain:
2. Do you feel this is the optimum way to schedule the class? ____ yes ____ no
3. How are classes utilized (estimate percentages where possible)?
   Lecture:
   ____ information transfer ____ expansion of text ideas ____ examples ____ other. Explain:
   Recitation:
   ____ work example problems ____ information transfer ____ group projects ____ other. Explain:
4. Would there be a better way to schedule this class? If so, what?
5. What mechanisms are used to accommodate individual learner differences? (Check appropriate items)
   ____ office hours ____ help sessions ____ supplementary materials ____ files of previous exams/homework ____ tutoring ____ optional course formats (lecture self-study) ____ other. Explain:

C. Mechanisms for Providing Feedback to Students

1. How can your students tell if they are doing well in the course?
2. What mechanisms are used to provide feedback to the students? (Check appropriate items)
   ____ homework grades ____ lab reports critiques
   ____ test grades ____ discussing homework in class
   ____ allowing time for questions and answers in class
   ____ other. Explain:

D. General Course Characteristics

1. Is the student given an overview of the entire course before the first class? ____ yes ____ no
2. Does the student receive feedback after tests, lab experiments, exercises, etc.? ____ yes ____ no
3. Is the feedback keyed to appropriate parts of textual material so that students may easily return to the proper section for remedial work? ____ yes ____ no
4. Is the course learner-centered (i.e., does it require the student to interact frequently by answering questions, participating in exercises case work, etc.)? ____ yes ____ no
   Comments:
5. Is there a reasonable balance between "explaining, exercising, and evaluating"? ____ yes ____ no
   Comments:
6. Are all instructions and test items clearly worded? ____ yes ____ no

E. Student Population

1. Usual level of students taking this course (sophomore, junior, etc.):
2. Usual major(s) of students taking this course (ME, CE, etc.):

This guide was prepared by the authors on November 30, 1979.

3. Describe the kind of background expected or required:

   Concepts needed:

   Prerequisite courses:

4. How is student background assessed?

   ____ stating assumed prerequisites ____ administering a pretest ____ not at all ____ other. Explain:

F. Instructional Personnel

1. The course is usually staffed by:

   ____ professors ____ GTA's ____ combination (give details):

2. Describe staff supervision and/or interaction with people responsible for related courses:

3. How long has the staff been associated with the course?

4. What orientation and/or training is provided for new staff?

5. What mechanism is used to provide in-service staff development?

G. Use of Media

1. What media are used? ____ chalk board

   ____ overhead projector ____ video tapes

   ____ audio tapes ____ movies ____ other (list):

2. Does use of media support/complement objectives of the course? ____ yes ____ no. Comments:

3. Are the visual portions (slides, video, etc.) clear and understandable? ____ yes ____ no. Comments:

4. Is the audio understandable? ____ yes ____ no

   Comments:

5. Does the narrator's voice, presence and appearance add to or detract from the effectiveness of the course? ____ add ____ detract

6. Do the audiovisual presentations logically coordinate with other course materials? ____ yes ____ no

H. Other Materials and Structure Characteristics

Comment on any other materials and structure characteristics that you believe pertinent to the evaluation of this course.

II–Subject Matter

A. Content Outline

1. Does a current content outline exist for the course? ____ yes ____ no

2. How is the content outline organized?

   ____ major topics only ____ major and sub-topics

   ____ by week ____ by day ____ other. Explain:

3. Does the outline correlate with text (or notes) used? ____ yes ____ no

4. Is content up-to-date? ____ yes ____ no. Comment:

5. Does content overlap with other courses in curriculum? ____ yes ____ no

   Comment:

6. Are topics sequenced in a "logical" order? ____ yes ____ no

B. Relation to Other Courses

1. What courses serve as prerequisites to this course?

2. Do the prerequisites cover all prerequisite concepts? ____ yes ____ no. Comment:

3. Are the official prerequisite courses "really needed"? ____ yes ____ no. Comment:

4. For which course(s) does this course serve as a prerequisite?

5. Does this course adequately cover the necessary prerequisite topics and concepts needed by these courses? ____ yes ____ no. Comment:

C. Schedule

1. Does a day-by-day or week-by-week schedule exist? ____ yes ____ no

2. Is the pace reasonable? ____ yes ____ no. Comment:

3. On the average, how many new concepts are introduced each week?

4. Are there opportunities for students to practice and/or apply new concepts? ____ yes ____ no

5. How many tests are given per semester? ____

6. How many review periods are scheduled? ____

7. Is time allowed for discussion of current literature? ____ yes ____ no

8. Is there any other flexibility in the schedule? ____ yes ____ no. Explain:

9. On the average, how many homework problems are assigned? each day ____ each week ____

D. Text and/or References

1. What text (notes) is specified for the course?

2. Are the notes or the text used as a text? ____ or ____ reference?

3. Are the notes or text written for a newcomer to the field? ____ yes ____ no

4. Are students given a list of references? ____ yes ____ no

5. What books and materials are placed on reserve in the library?

III–Assessment

A. Assessing Student Work Load

1. On the average, how many pages of text are assigned per week?

2. Describe types of outside assignments:

   Reading

   Closed-end homework problems (i.e., single answer)

   Open-end homework problems (i.e., many correct answers)

   Individual projects

   Group projects

   Lab reports

   Other

3. Frequency of each type of assignment.
4. Estimate time it would take instructor to do assignments.
5. Estimate time it would take students to do assignments.
6. Basis of this estimate:
   Do you ask students?

B. Assessing Student Performance

1. Exams
   Frequency:
   Length (minutes):
   Type of questions used:
   ____ multiple choice ____ closed-end
   ____ problem ____ open-end
   Are questions based on course objectives?
   ____ yes ____ no
2. Is homework graded? ____ yes ____ no
3. Are quizzes used to evaluate students?
   ____ yes ____ no
4. Are projects assigned? ____ yes ____ no
   If yes, % of final grade ____. Explain evaluation criteria:

C. Course Evaluation and Change

1. What methods are employed to evaluate the course?
   ____ CAFETERIA ____ staff input
   ____ content review ____ teaching strategy review
   ____ other:
2. How frequently are these methods employed?
3. Extent to which examination results are used to assess teaching effectiveness:
   Is an analysis made of student performance of each test question? ____ yes ____ no
   Are test questions keyed to objectives?
   ____ yes ____ no
4. What mechanism is used to evaluate appropriateness of the course with respect to the rest of the curriculum?

IV–Summary and Recommendations

A. Sort out the priorities of the course.
B. List the strengths of the course.
C. List the weaknesses of the course.
D. List the constraints (within which the instructor is operating).
E. Taking into account the constraints, what changes do you recommend?

Once the reviewer completes a first draft of the review, he or she should arrange a follow-up meeting with the instructor to discuss findings and recommendations. From such a meeting both individuals can assess the amount of time the instructor can devote to improving the course, as well as decide on how to proceed. Because both are professionals in the field, such a conference can be beneficial to both.

Finally, a checklist for the reviewer to follow might look something like this:

- ☐ Sort out the priorities of the course
- ☐ List the strengths
- ☐ List the weaknesses
- ☐ List the constraints
- ☐ Recommend changes taking the constraints into account.

Conclusion

The Course Evaluation Guide has not been tested and validated in any scientific sense. Its purpose is not to obtain precise quantitative measures of the various attributes of a course. It is intended rather to serve as a *guide*—to point the way towards improving instruction. Perhaps it can also serve as a starting point for developing a similar instrument for use in other engineering education environments.

References

1. Aleamoni, L.M., "Typical Faculty Concerns About Student Evaluation of Instruction," *Symposium on Methods of Improving University Teaching*, The Technion Institute of Technology, Haifa, Israel, 1974.
2. Drucker, A.J. and H.H. Remmers, "Do Alumni and Students Differ in Their Attitudes Toward Instructors?" *Journal of Educational Psychology*, 1951, vol. 42, pp. 129-143.
3. Hildebrand, M., R.C. Wilson, and E.R. Dienst, *Evaluating University Teaching*, U. of California Press, Berkeley, 1971.
4. Kulik, J.A. and W.J. McKeachie, "The Evaluation of Teachers in Higher Education," in F.N. Kerlinger (ed.), *Review of Research in Education*, vol. 3, Peacock, Itasca, Ill., 1975.
5. Marsh, H.W., "The Validity of Students' Evaluations: Classroom Evaluation of Instructors Independently Nominated as Best and Worst Teachers by Graduating Seniors," *American Educational Research Journal*, vol. 14, 1977, pp. 441-447.
6. McKeachie, W.J., "Student Ratings of Faculty," *AAUP Bulletin*, 1969, vol. 55, pp. 439-444.
7. McKeachie, W.J., Y–G. Lin, and W. Mann, "Student Ratings of Teacher Effectiveness: Validity Studies," *American Educational Research Journal*, 1971, vol. 8, pp. 435-445.
8. McKeachie, W.J., "Student Ratings of Faculty: a Reprise," *AAUP Bulletin*, 1979, vol. 65, pp. 384-397.

The work reported was supported by the National Science Foundation under grant SER 77-05922. The views expressed are those of the authors and not necessarily those of NSF or the Purdue Research Foundation.

Part IX
Other Academic Activities

Scope and Purpose

As every beginning teacher rapidly finds out, classroom teaching is not the only (or even the major) activity of a member of the faculty in an engineering college. A professor does (and is usually expected to do) many other things, some of which are in direct support of the classroom teaching, while others are only indirectly related. The article by Wilson, reprinted in Part VIII of this volume, contains a long list of activities, more or less directly related to teaching, in which a teacher normally gets involved. In addition, depending on his interests, available opportunities, extent of time or commitment required, perceived benefits, and time pressures, a college teacher may also engage in some of the following activities:

- Advising students.
- Writing textbooks.
- Writing book reviews.
- Reviewing book manuscripts for book publishers.
- Serving as a consultant.
- Serving as an expert witness.
- Working as an inventor, or developing patentable ideas.
- Serving as an officer of local or national professional societies.
- Serving on technical committees of volunteer groups, professional societies, or local, state, or federal government.
- Serving as a reviewer of manuscripts for professional journals.
- Serving as a reviewer of research proposals submitted to sponsoring agencies for funding.
- Serving on university committees.
- Carrying out administrative assignments (such as admission or transfer credit evaluation) within the university.
- Serving as lecturer in short courses or extension courses.
- Carrying out research on various aspects of engineering education.

The motivations behind engaging in each of these activities are different. Some, such as student advising, may be a required part of the responsibility as a faculty member at one's institution. Others, such as consulting, have financial benefits, in addition to the opportunity for research, for collaboration with professionals, and for state-of-the-art work. Many of the activities listed above have professional benefits, ranging from developing contacts and receiving the recognition of peers, to deriving personal satisfaction and pursuing one's interest. Even the administrative/committee work, often considered a necessary evil, may be psychologically beneficial to an individual, in that it gives him a sense of control over his environment. Even if the beginning teacher feels that a lack of time prevents him from engaging in most of these activities, he may wish to be aware of them. The purpose of this part is to bring about this awareness for some of the more common professional activities.

The Issues

The literature concerning these professional activities is of two kinds. First, there are "how-to" guides on many of these activities, which can help the uninitiated get started. Second, there are discussions in the literature on issues such as:

- To what extent is each professional activity in support of the educational, or the overall, mission of the university?
- Does heavy involvement in outside activities significantly influence the institutional commitment and involvement of the faculty member?
- Can some of the outside activities of a faculty member create a potential conflict of interest with his university duties?
- In what manner does the university benefit from the various professional activities of the faculty—national visibility, enrichment of the classroom work by real-life experience, ability to attract and retain highly active individuals as faculty, etc.?
- How should the performance of an individual in these activities be evaluated and accounted for in his overall performance evaluation?
- Does the university adequately reward the individuals for their contributions to the university through such auxiliary activities?

The Reprinted Papers

The scope of the reprinted papers in this chapter is limited to six of the activities mentioned earlier: advising, thesis supervision, book reviewing, book writing, consulting, and educational research. In addition, the reprinted articles are selected to be mostly of the "how-to" variety; some discussions of the issues of fairness, rewards, etc., are contained in the reading resources cited at the end of this chapter. The following papers are reprinted in this part:

1. "The Professor as a Counselor," P.C. Wankat, *Engineering Education,* vol. 71, no. 2, pp. 153–158, Nov. 1981.
2. "Responsibilities of the Thesis Advisor," N. R. Scott, *American Soc. Agricultural Eng. Winter Meeting,* Chicago, IL, Dec. 1978, Paper 78-5517, 18 pp.
3. "Book Reviews of Scientific and Technical Books," M. S. Gupta, *IEEE Trans. Professional Communication,* vol. PC-26, no. 3, pp. 117–120, Sept. 1983.
4. "Writing a College Textbook," D. G. Newman, *Engineering Education,* vol. 71, no. 4, pp. 279–282, Jan. 1981.
5. "Getting Started in Consulting," A. S. Diamond,

Chemical Engineering, vol. 82, no. 19, pp. 139–140, 142, 144, Sept. 1975.

6. "Research in Engineering Education: An Overview," W. K. LeBold, *Engineering Education,* vol. 70, no. 5, pp. 406–409, 422, Feb. 1980.

The first paper by Wankat on student advising has some good advice for the faculty advisor. Student advising encompasses many different aspects. At the factual or information transfer level, a student may need help in course selection, meeting graduation requirements, or following some other university rules or procedures. Or, he may need counseling on professional and career matters, such as choice of specialization and employer, or perhaps assistance in helping him define his career goals. Finally, a student may need counseling on personal problems, or study habits, or learning strategy for a given discipline. Obviously, it is difficult to produce a "how-to" guide for faculty advisors covering all of these situations. Many of the mechanical parts of advising have sometimes been relegated to computers [3], and this is desirable if it leads to a more meaningful use of the advising time for a deeper interaction between the student and the advisor. Wankat's article emphasizes the interpersonal interaction aspect of advising.

The article on interaction analysis by Root, reprinted in Part IV of this volume, is also pertinent from an advisor's point of view. It shows how an advisor can select his responses so as to promote growth in the student.

The article on thesis supervision, by Scott, is a careful and detailed statement of the role of a thesis supervisor in the various stages of the thesis work of a student. It begins by asking some basic questions, like what is the purpose of research, and what are the conditions necessary for discovery, and uses their answers to deduce some guidelines. In addition to making the duties of a thesis supervisor explicit, the article also considers the attributes desirable in a thesis supervisor from a student's point of view.

A teacher may review a book for many different reasons: as a consultant to a textbook publisher, to publish a book review in a journal (although these are often invited by an editor rather than contributed by the reviewer), to consider a book for classroom use, and to examine existing books when considering the writing of a new book. The next paper reprinted here is the only "how-to" guide available on the subject of reviewing books, is rather comprehensive, and can serve two purposes. For a book reviewer, this article can serve as a prompting aid and a guide. For a prospective book author, this article will serve as a reminder of the reader's viewpoint and expectations.

Writing a book, especially a textbook, is a very time consuming task, not only because good writing takes time, but because there is a gestation period for good ideas to form. One author's rule of thumb in book writing is: Make a realistic estimate of the amount of time it will take you to write the book under consideration, multiply it by five, and you have a rough estimate of how long it will actually take. A beginning teacher should therefore undertake this task with caution, and with careful consideration of its time demands and career implications. In addition, it is well to remember that the number of books actually published is a fraction of the books that get started by well-intentioned authors who have a serious interest, and that a large fraction of the published books were never actually "finished"; the authors just gave up further improvement at some point in time (usually determined by the publisher's deadline). The paper reprinted here, by Newman, provides some helpful guidelines to authors, but there is much else a prospective author can learn from other experienced authors, from editors of publishing companies, and from the writers' manuals published by major publishing houses. The writing of advanced and professional books is sufficiently different from textbook writing that different considerations must be taken into account. The references below include some guides for authors of such books.

Consulting, like academia, is a professional field by itself, and there are numerous books and journals (such as *Consulting Engineer*) devoted to it. Since few new engineering teachers can find the time to engage in this activity except for a small fraction of their time, many of the guidelines for consultants are not directly applicable to them. The reprinted article by Diamond is a brief, but broad, introduction of the demands of this activity. More detailed guidelines can be found in some of the items included in the list of reading resources.

Very few engineering educators (perhaps no more than a few percent nationwide) ever get involved in a formal research program or project in engineering education. This appears to be a result of four factors: (a) the extensive time requirement for initiation and execution of educational research, (b) the small perceived benefit of the research to the individual's career, (c) the difficulty of obtaining support for carrying out the work, and (d) ignorance about educational research. The first three are valid reasons, worthy of careful consideration; the fourth is unnecessary, and the brief introduction here is intended as a first step in mitigating it. Roughly speaking, the educational research carried out by an engineering teacher may be of two kinds: empirical research on some aspect of teaching, or characterization of some aspect of the educational system. The empirical research on teaching may be aimed at determining the effectiveness of something new or different, such as a new curriculum, or instructional material, or teaching method, that the teacher has tried. It is clear that classroom teachers are in an ideal position to experiment with new strategies, and carry out this kind of research. There are numerous problems in providing for a control, for example, when there is only one section of the course being taught, or when the instructor is highly and contageously enthusiastic about his innovative idea. Nevertheless, the value of this kind of experimentation cannot be denied. The systematic characterization work involves determining some trait or characteristic of the students, the faculty, the curricula, or some other element of the educational enterprise. The work typically requires data collection, and its goal is to provide some hard information on which decisions and recommendations can be based. The data will often be collected at one institution, or in one geographical region, which will determine the amount, and generalizability, of the work. The reprinted paper by LeBold gives an overview of mostly the second kind of the educational research. It provides many examples of past work in its long list of citations.

The Reading Resources

A. Advising.

[1] E. R. Hines, "Academic Advising as Teaching," *Improving College and University Teaching,* vol. 29, no. 4, pp. 174–175, Fall 1981. Considers academic advisement from the institutional policy point of view; suggests viewing advising as a form of non-classroom teaching; recommends including it in the determination of the rewards (e.g., release time).

[2] J. Choma, Jr. and D. J. Walukas, "Advising Function in School of Engineering," *IEEE Trans. Education,* vol. E-13, no. 3, pp. 213–214, Sept. 1968. Makes eleven recommendations for administering a successful advising program; some of these relate to administrative and policy matters, and others to the duties of an advisor.

[3] A. M. Ali and D. Gates, "Computer Program to Counsel Engineering Students," *Engineering Education,* vol. 62, no. 7, pp. 823–824, Apr. 1972. Describes the capabilities and makeup of a computer program that is used to display an up-to-date record of a student's standing in a program, and the requirements that remain to be completed.

B. Thesis Supervision.

[4] P. A. Vesilind, "Attributes of a Thesis Advisor," *Jour. Engineering Education,* vol. 58, no. 2, pp. 151–152, Oct. 1967. A graduate student's viewpoint of the attributes desirable in an ideal thesis advisor, including professional competence, interest and involvement in the subject of the thesis, willingness to criticize and candidly appraise the work, and recognition of the student as an individual.

C. Book Reviewing.

[5] A. Sekey, "Personal Views on Book Reviews," *Canadian Electrical Engineering Jour.,* vol. 6, no. 3, pp. 35–36, July 1981. Some suggested guidelines for book reviewers.

D. Book Writing.

[6] R. C. Brinker, "Should I Write a Textbook," *Jour. of Engineering Education,* vol. 52, no. 8, pp. 513–521, Apr. 1962. Practical information on the non-writing aspects, e.g., book costs, marketing, publisher acceptance, and contracts; the quantitative cost figures are badly outdated, and off by about a factor of five, but many other observations are still valid.

[7] M. P. Rosenthal, "How To Write a Technical Book and Get It Published," *IEEE Trans. Engineering Writing and Speech,* vol. EWS-7, no. 2, pp. 11–20, Sept. 1964. Details of the mechanics of professional/technical book writing, along with a long list of publishers, which is somewhat dated.

[8] O. C. Wells, "How to Write a Book," *Physics Today,* vol. 34, no. 6, pp. 9, 78, June 1981. Also in *IEEE Trans. Professional Communications,* vol. PC-25, no. 3, pp. 131–132, Sept. 1982. Suggestions for authors of advanced and professional technical books, dealing with bibliography, reference collection, citations, material compilation, organization, etc., based on the author's own experience of writing one book.

[9] *The McGraw-Hill Author's Book,* McGraw-Hill Book Co., New York, 1968, and *A Guide for Wiley Authors,* John Wiley & Sons, New York, 1973. Authors' manuals from two of the major publishers of engineering books in the U.S.

E. Consulting.

[10] C. V. Patton, "Consulting by Faculty Members," *AAUP Bulletin,* vol. 66, no. 4, pp. 181–185, May 1980. A summary of the findings of two major surveys; the data fail to support the hypothesis that those who do paid consulting work teach less, or that they are less active in academic work, based on several different measures of academic activity.

[11] A. Wildavsky, "Debate over Faculty Consulting," *Change,* vol. 10, no. 6, pp. 13–14, June-July 1978. Some arguments in favor of consulting by faculty.

[12] R. Aggarwal, "Faculty Members as Consultants: A Policy Perspective," *Jour. College and Univ. Personnel Assoc.,* vol. 32, no. 2, pp. 17–20, Summer 1981. Pros and cons of outside consulting by the faculty.

[13] S. W. Golomb, "Faculty Consulting: Should It Be Curtailed," *National Forum (The Phi Kappa Phi Journal),* vol. 69, no. 4, pp. 34–37, Fall 1979. Strongly defends faculty consulting, with arguments based on economics, technological currency of faculty, technology transfer, research inspiration, and benefits to the institution and the community it serves.

[14] H. Holz, *How to Succeed as an Independent Consultant.* John Wiley, New York, 1982. A practical guide on various aspects of running a consulting business, including fees and contracts, reports and presentations, etc.

[15] M. A. Neighbors, "You Can Become a Consultant," *Measurement & Control,* vol. 16, no. 8, pp. 309–311, Aug. 1983. Answers some of the commonly asked questions for beginners.

[16] J. A. McQuillan, "Forensic Engineering: CE as an Expert Witness," *Consulting Engineer,* vol. 62, no. 1, pp. 48–50, Jan. 1984. A simple guide to the legal procedures, along with suggestions for a consulting engineer serving as an expert witness.

[17] R. A. Connor, Jr. and J. P. Davidson, *Marketing Your Consulting and Professional Services.* John Wiley, New York, 1985. A detailed, pragmatic source of tips and ideas for someone with a serious commitment to consulting.

F. Educational Research.

[18] N. L. Gage, Ed., *Handbook of Research on Teaching.* American Educational Research Assoc., Rand McNally, Chicago, 1963; and R. M. W. Travers, Ed., *Second Handbook of Research on Teaching.* American Educational Research Assoc., Rand McNally, Chicago, 1973. Useful as an introduction to the professional literature on the nature of research problems, methods of investigation, and technical matters.

[19] *Engineering Education,* vol. 70, no. 5, Feb. 1980. Special Issue on Educational Research. Articles including establishing a climate for educational research, and locating funding to support it.

The Professor As Counselor

What do you say to a student in a distress? How do you respond to a student's anger? Crisis intervention techniques, such as learning to listen without being judgmental, can help any professor be a more effective counselor.

Phillip C. Wankat
Purdue University

As well as teaching and doing research, engineering professors occasionally counsel distressed students. As pointed out by Ericksen,[1] counseling is a ubiquitous part of a professor's life. Counseling can range from academic guidance on what electives to take, to advice on graduate schools, to counseling on personal problems. This process takes place both formally and informally, and conversations can easily change from academic to personal topics if the professor is receptive. Like many of the other functions an engineering professor performs, counseling, particularly personal counseling, is not something the professor has been trained to do.

Professors need effective listening and counseling skills to help students. Because feeling accepted and understood is a basic need,[2] it is important to really listen to students and let them know you understand. It takes skill to discover how people feel and to let them know you understand their feelings. Fortunately, these skills can be learned, and will be discussed in this article as part of the first step in a counseling technique based on crisis intervention theory.

Counseling and Crisis Intervention Methods

Many theories and methods of counseling and helping people have been developed. Crisis intervention theory was developed into a technique for training volunteers for telephone and walk-in crisis intervention, and is based on considerable practical experience and research.[3-5] The method is simple to learn, easy to use and effective. Crisis intervention and the counseling methods it is based on are not therapy, however, and should not be used as a substitute for therapy. The intervention and counseling methods discussed here are based on the premise that the person-in-crisis is usually able to function satisfactorily, but his coping mechanisms are not working in this instance. These methods will help bring the person back to normal functioning as quickly as possible. The term "crisis" is interpreted broadly to include everything from wondering what elective to take to thinking of suicide.

Several similar methods are discussed by Delworth,[3] Edwards,[4] and McGee.[5] We will discuss the ABC model,[4,6] since it is easy to use and easy to remember. This three-step system consists of:

- A—Acquiring information and rapport;
- B—Boiling down the problem;
- C—Coping (helping the person to cope).

Each step does not require the same amount of time: Step A usually takes more than half of the time involved. This step, which demands active listening, is the one engineering educators will be least familiar with. Steps B and C, which can be considered problem solving, will be more familiar (and, at first, comfortable) to most engineering teachers and to most of the students they talk to.

A—Acquiring Information and Rapport

Before you can help a student cope with a problem, you need to gain rapport and find out what the problem is. Gaining rapport and obtaining the necessary information take place together, as the student is shown that his situation is understood. Understanding, in this case,

Reprinted with permission from *Engineering Education,* vol. 71, no. 2, pp. 153–158, November 1980.

means knowing both how he feels and what the facts are. If a personal relationship and trust do not already exist, they must be developed to at least some degree.

Developing a relationship and gaining trust normally take a long time, but in the stress of a crisis, a person will probably take risks in order to gain relief. Trust, rapport and a relationship can all be developed much more quickly than usual. As a professor, you probably already have a certain degree of rapport with your students, at least as far as technical, academic and career-related problems are concerned.

Learn to Listen

To be a "good listener" the professor must be an "active listener." A variety of techniques are used in the active listening process.[7,8] The process requires that the listener be non-judgmental, since nothing turns off an upset student faster than hearing what he feels to be arbitrary judgments. For example, suppose a student has just told you that she is considering dropping out of school. You can respond,

"That's the dumbest thing I've ever heard of," or,

"What the hell do you want to do that for?" or,

"What do your parents think?" or,

"That's a serious step. Tell me about it."

The first response is your judgment and may shut off communication. The second response will probably be interpreted as being judgmental. The third puts you on thin ice. If the student's parents are screaming at her *not* to quit, she might assume you agree with them even if you don't. The last response is neutral, and it doesn't put the student on the defensive. It acknowledges the seriousness of her problem and indicates your interest.

The listener must be sincere. Faking interest in the student will not work. The professor should also demonstrate unconditional positive regard for the student—the attitude that the student is an O.K. person, even if some of his behavior is undesirable. If you do not wish to talk to students, this attitude will be telegraphed to them, and gaining rapport will be extremely difficult.

Use Empathy

With this attitudinal background the active listening process then centers on empathy.[7,8] Empathy is letting the person know you understand how it is to be in his shoes, but at the same time being clear that you are *not* in his shoes. Part of this process is to find out what the student's world is like. A student may be under enormous pressures that you are unaware of, and his actions may be controlled by internal restrictions and messages that are irrational. Without some understanding of what his world is like, it will be impossible to empathize and gain rapport. When you have reached the point of accurate empathy, the other will feel heard and understood. Naturally, this builds the rapport that allows you to proceed to the problem-solving steps.

The other parts of the active listening process provide the foundation for empathy. Use open-ended questions to encourage a student to explore his problem areas. Closed questions, which require only yes or no answers, or a rapid-fire series of questions can prevent exploration. In active listening, the listener must provide feedback as to what he has heard. Thus, restate and paraphrase what the student has said. It also helps to summarize the student's conversation, since this helps both of you clarify what he is saying.

Accurate empathy requires a focus on feelings, which may be alien to the professor's normal method of responding. You need to reflect the student's feelings both by reflecting on his statements and by uncritically labeling his feelings. A student may be angry, but unwilling to admit it until you label his anger. Labeling feelings can be particularly important if the anger is directed at you, since the student is then allowed to vent his feelings. This is important, because problem solving is not possible until emotions have been discharged. To help someone vent strong feelings, you need to match the level of his emotions and be non-judgmental.

Accepting feelings and being non-judgmental does not mean you agree with a person or feel his emotions are justified. Instead, it is a statement that this is the way someone feels and you accept his feelings, even though you may not agree with them. It is best not to commit yourself to his side, since you do not have all the facts, and agreeing will usually not help to resolve the crisis. If asked, "What do you think?", refocus on the student. This can usually be done by asking what he thinks, or by stating that you need more information.

Know When to Be Silent

The active listener directs the process by accenting certain points made by the student, or by asking questions. Although some direction is desirable, the professor should avoid over-direction. Without some exploration and apparently aimless wandering, the most important problem may be missed. The student must be given time to explore and think. The resulting silences may be uncomfortable for you, and you may have a strong desire to say something—anything. Usually, it is best to allow the student to think when he wants to and to let him break the silence when he is ready. A monitoring of the student's nonverbal behavior helps tell if the silence is being used to work through some phase of the problem. For instance, a student may stare with fierce concentration at the ceiling or hold his hand in front of him in a stop signal.

Letting a student know you are listening will encourage him to talk. Maintaining direct eye contact, nodding and gesturing at appropriate times, saying "uh huh," and keeping an attentive body position will show that you are listening. When you feel that you have acquired sufficient rapport and information, you can move on to clarifying the problem. You must first be sure that enough "venting" has taken place, so that the student is ready to move on. More than half of the time in dealing with the crisis probably should be spent on Step A; it is best not to rush it.

"Since professors are familiar with being authorities who present the correct way to do something, giving advice is an easy trap for them to fall into."

B—Boiling Down the Problem

This step helps clarify what the most immediately pressing problem is. It is comparable to the "defining the problem" step in problem solving. The student-in-crisis may be so disoriented that he cannot focus on what is distressing him. Thus, the professor wants to pick out the problem causing most of the current stress from a host of problems that may be bothering the student. The problem may be obvious to the professor, but not to the student. Once the professor thinks he knows what this problem is, he formulates a clear, concise statement of it and presents this hypothesis to the student. For example, you might say, "You're so upset about fighting with your Dad that you can't concentrate on studying." If the problem statement does not seem to fit, then return to Step A and get more information (both on feelings and facts). If the problem statements fits, then move into the coping stage.

Since there may be several problems, it is important to spend enough time acquiring information and rapport to identify the problem causing the current crisis. It may not be the first problem mentioned. Also, the student must have sufficient time to vent his emotions so that he can rationally consider what the underlying problem is. Failure to pinpoint the problem on the first trial is not critical.

What do you do if you strongly disagree with the student's interpretation of the problem? In the long run, it is best to be honest about your misgivings. Then, if the student stays with his interpretation, you can either move on to Step C and help him cope with his definition of the problem or try to formulate another problem statement you can both agree on. Since you have gained rapport and trust, the student should be able to listen to your objections. If he cannot hear you, you may have to state that you disagree and then end the conversation. Even if you cannot help the student cope, you have served the useful purpose of listening and helping him express his feelings.

C—Coping

Once a clear problem statement has been defined, one can help a student to cope. Remember that the crisis intervention procedure is based on the premise that students are usually able to cope, but their normal coping mechanisms are not working. The key is to help them cope with a problem, not to solve it for them. This step can be considered the problem-solving step.

Giving advice does *not* help someone cope. Advice should not be given for three reasons:[4,6]

1) The student may refuse to take the advice and be discouraged from receiving help. Advice can shut the door to a possibly helpful interaction.

2) The student may try the advice and it may not work. He can then blame this failure on you and not take responsibility for his actions.

3) The student may take the advice and it may work. His perception that he cannot cope on his own is then reinforced, and a dependency relationship may be fostered.

Avoid giving advice; this is particularly important to the relationship. Many students are overly anxious to take advice, but it does not help them become mature, independent people, capable of solving their own problems. Since professors are familiar with being authorities who present the correct way to do something, giving advice is a very easy trap for them to fall into.

If you do not give advice, what do you do? You help students develop possible solutions to their problem. First, find out what has worked in the past. Have they tried that previously successful coping mechanism, or forgotten to use it? In a crisis, a student is often disorientated and cannot use the techniques that have worked in the past. When asked to search for what he has done before, he may find his own solution to the problem.

Find out what he *has* tried for this crisis and what happened when he tried it. This review may suggest modifications or new approaches. It may also show that he is in the process of solving the problem but has not given himself enough time for the approach to work. A thorough exploration of what he has tried also helps to guard against what Eric Berne called a game of "Yes, but," in which the student uses that rejoinder to reject all proposed approaches.[9]

What alternatives are possible? With some encouragement the student may think up new possibilities. With your different experience and viewpoint, you can also generate alternatives. You may know of resources that he is not aware of, such as an emergency loan fund in the dean of students' office, aptitude and interest testing, counseling at a counseling office or the student health center, pregnancy information, and available part-time jobs through the financial aid office. Most campuses have a variety of services available, and many students are surprisingly ignorant of them. Of course, in order to generate alternatives you have to know what is available on campus or in the community. When presenting alternatives, you must present them as such and not as advice that should be followed.

The Action Plan

Once the possible actions have been explored, the professor provides assistance in choosing what alternatives to try and in developing a plan of action. A student may find this decision very simple or may be unable to make what seems a simple decision. Since the crisis is his problem, it is imperative in most cases that the solution be his. Help and encouragement are not the same as telling him what to do. The action plan should be as specific as possible. Remember that to a person partially immobilized by a crisis, tasks may appear very difficult. A specific plan will be easier to follow and harder to ignore.

Once a plan of action has been developed it may be appropriate to obtain a commitment to return to you for follow-up. The appropriateness of following up will depend on the nature of the problem and your relationship with the student. If follow-up is appropriate, a definite time and date should be set. A record of the coping plan will help both of you determine if the plan has been successfully implemented. When the follow-up session is held, start at Step A to regain rapport and determine what the current situation is.

The ABC process can lead students from an initial feeling of hopelessness and despair to an understanding of what the underlying problem is, and then to a rational plan of action. Since this process returns the students to their usual coping mechanisms, remarkable behavioral changes can occur in a short time.

Learning Crisis Intervention Techniques

One does not have to be a trained counselor to be a good listener or

Example: An Angry Student

Bob comes into your office to complain about his laboratory grade.

BOB (*controlling himself*): Why did I get a B in lab? I think I deserve an A, and I want to see the grades.

You show him the grades:

PROFESSOR: As you can see, you got low grades because your lab partners didn't do well.
BOB: That's not fair, I'm being punished for my lab partners' laziness.
PROFESSOR: You think the grade isn't fair because you were graded as a team and not on your personal contribution.
BOB: Yeah, it's not fair. My lab partners screw around all semester and get B's because I worked hard all semester, and I also got a B. It's not fair.
PROFESSOR: I can see you're really angry about this.
BOB: Damn right I'm angry! I've been screwed out of a grade I deserve. If I'd had different lab partners I'd have gotten an A.
PROFESSOR: You feel that assigning group grades for projects has screwed you?

Note that the professor does not attack Bob, and he does not become defensive. Instead, he first presents Bob with the information he has demanded and then focuses on Bob's feelings. These feelings are accepted, but the professor never compromises his position that the laboratory projects must be graded as team efforts. Once Bob has calmed down, the grading policy can be discussed. Bob may even have some good suggestions for how the grading can be improved. There is probably no solution to this problem that will totally satisfy Bob. The advantage of actively listening to him, however, is that he will feel understood and will not walk around with a chip on his shoulder next semester.

Example: A Student Unsure About Engineering

Nancy is a sophomore who has talked to you previously about the engineering curriculum and what engineers do. Now, towards the end of the semester, she comes back with doubts about her choice of career.

NANCY: Do you have a few minutes?
PROFESSOR: Sure, come in and have a seat.
NANCY: Well . . . I'm just not sure I want to stay in engineering.
PROFESSOR: Why don't you tell me about it?
NANCY: I don't know what I want to do. The other students seem so sure they want to be engineers, but I don't know if it's right for me.
PROFESSOR: Your grades look O.K. What seems to be the problem?
NANCY: I can handle the courses all right. They're a lot of work, but I can do them. I'm just not sure it's worth it.
PROFESSOR: You think you might enjoy something else more?
NANCY: I think being an engineering might be . . . well . . . dull. All they do is work with numbers, and a lot of engineers aren't interested in people.
PROFESSOR: You like to work with people.
NANCY: Yes, but I don't know what else I'd like to do.
PROFESSOR: You sound pretty unsure of your future.
NANCY: I really am. I can do the engineering, but I don't know if I want to. And I don't know what else to go into.
PROFESSOR: What other fields have you thought of?
NANCY: Psychology sounds interesting, and maybe history. I always enjoyed history in high school, but I don't think I can make a living at it. I really don't know much about psychology.
PROFESSOR: So you're worried about earning a living, but you want your work to be something you'll enjoy.

After more discussion the problem is somewhat clearer and you move on to defining the problem and looking for solutions:

PROFESSOR (*summarizing/Step B*): The problem is you don't want to drop out of engineering now, but you want to try a lot of things to see what you like. What kinds of things have you done to explore what you enjoy doing?

NANCY: I took an aptitude test before coming to school. It said I was good at science and math.

PROFESSOR: How about other things, like summer jobs?

NANCY: I've worked three summers at a "Y" camp. I really enjoyed working with the kids. It was just great!

PROFESSOR: You seem to really like working with people.

NANCY: Oh, I really do. . . .

PROFESSOR: Have you ever taken one of the interest tests at the dean of students' office?

NANCY: No. What do they do?

After explaining the purpose of interest tests, other possibilities are explored:

PROFESSOR: Several times you've mentioned you are not sure you'd enjoy an engineer's job. You know engineers do a variety of jobs. Have you talked much to the speakers who come in to present seminars on their work?

NANCY: No, I've always been busy with school work. I guess I never saw the purpose of those seminars. Maybe I should go next semester.

PROFESSOR: It might help. I've also noted you have some electives to take then. Have you thought about using these to explore different fields?

NANCY: That's something I wanted to talk to you about. Can I . . .?

After exploring other alternatives, a tentative action plan can be developed:

PROFESSOR: I have a class at 10:30. Let's see if we can summarize some of the alternatives and come up with a tentative plan.

NANCY: Well, I've decided to stay in engineering for now, but to look around at other areas, too. I'll take Psych 100 and History 120 as electives next semester. They should give me an idea of what psychology and history are like, and I can use them as electives in almost anything. Also, I'm going to make a point of talking to the engineers who give seminars. Maybe I'd enjoy engineering after all. Let's see, what else did we talk about . . .?

Although Nancy's problem is not solved, she has come up with a rational plan that should help her find herself. Note that the professor has not made a biased argument for engineering as a career, but has helped Nancy consider alternatives and come up with a plan that should help her decide for herself.

to use crisis intervention techniques effectively. In fact, research shows that trained telephone volunteers are as effective or more effective than professionals in crisis intervention, and, not surprisingly, that trained volunteers or professionals are more effective than untrained lay personnel.[10] Effectiveness decreases in time, however, unless some sort of refresher course is taken.[11]

The best way to learn these skills is a combined didactic/experiential approach. For engineers, as mentioned, probably the hardest to learn are the listening skills used in Step A, where it is necessary to focus on feelings and gain rapport. Listening skills can be learned through courses[12] or with the aid of books[7] and articles[8]. Courses or reading must be followed by practice. Role play is a useful training method.

The boiling down (statement of the problem) and coping (problem solving) steps are more natural for engineers. Thus, less practice is needed for these steps. A pitfall for engineering professors is trying to get to these steps too quickly. A conscious effort must be made to slow down and spend sufficient time acquiring information and rapport.

The ABC approach also works on relatively minor crises that occur at home or at school. These crises can give you an opportunity to practice the approach and to resolve these problems more effectively. Listening skills can also be used in situations that may not normally be classified as crises, such as in talking to a graduate student who is frustrated because his research is not progressing.

The main thing is to practice, so that listening skills and crisis intervention skills become second nature and can be used as needed.

References

1. Ericksen, S.C., *Motivation for Learning*, University of Michigan Press, Ann Arbor, 1974, pp. 122-124.

2. Many sources could be cited. For example, Maslow, A.H., *Motivation and Personality*, Harper and Row, NY, 1954.

3. Delworth, V., E.H. Rudow, and J. Taub (eds.), *Crisis Center/Hotline: A Guidebook to Beginning and Operating*, C.C. Thomas, Springfield, 1973, chapters 3 and 10.

4. Edwards, R.V., *Crisis Intervention and How It Works*, C.C. Thomas, Springfield, 1977.

5. McGee, R.K., *Crisis Intervention in the Community*, University Park Press, Baltimore, 1974, chapter 12.

6. Wankat, P.C., "Helping Other Engineers Cope with a Crisis," *Chemical Engineering*, vol. 86, no. 5, Feb. 26, 1979, p. 127.

7. Hackney, H., and S. Nye, *Counselling Strategies and Objectives*, Prentice-Hall, Englewood Cliffs, N.J., 1973.

8. Root, G., and D. Scott, "The Interpersonal Dimensions of Teaching," *Engineering Education*, vol. 66, no. 2, Nov. 1975, pp. 184-188. (Has a good list of references).

9. Berne, Eric, *Games People Play*, Grove Press, New York, 1966.

10. O'Donnell, J.M., and K. George, "The Use of Volunteers in a Community Mental Health Center Emergency and Reception Service: A Comparative Study of Professional and Lay Telephone Counseling," *Community Mental Health Journal*, vol. 13, no. 3, 1977.

11. D'Augelli, A.R., and M. Levy, "The Verbal Helping Skills of Trained and Untrained Human Service Paraprofessionals," *American Journal of Community Psychology*, vol. 6, no. 23, 1978.

12. For example, AIChE has a short course on "Managing Human Interactions," ACS has a short course on "Effective Management Techniques," and ASEE has presented a "Workshop in Transactional Analysis and Engineering Education." All of these courses cover some aspects of active listening skills. Other societies offer similar courses.

RESPONSIBILITIES OF THE THESIS ADVISOR

N. R. Scott, Member ASAE

INTRODUCTION

The preparation of a thesis is a universal requirement for the Ph.D. (Doctor of Philosophy) degree and is usually required for the M.S. (Master of Science) degree. The thesis advisor plays a key role in the total process from inception of the research to the culmination of the research in the form of an "approved" thesis. Because of this important role, it is fitting that the role of the advisor, as well as attributes of an advisor, be discussed in a Symposium on Graduate Education. For me, personally, the preparation of this paper has encouraged me to review an extensive body of literature and has forced me to integrate my ideas and experiences as a thesis advisor into a concise and organized presentation.

THESIS DEFINED

From a brief historical review one finds that the word "thesis" has been used to mean different things. Almack (1930) states that originally it meant a proposition which the student proposed to defend and to maintain. Today we view the thesis as a coherent report of research, in which both process and the results are given. Almack (1930) states,

> "The chief thing to remember is that a thesis is a report of the process and results of research, extending from a central proposition, hypothesis, or problem to a definite generalization growing out of facts."

Koeford (1964) has presented a composite definition from the major dictionaries to define a thesis as:

> "A dissertation which embodies the results of original inquiry and research, and which strictly substantiates a specific view, especially the solution to a problem."

The significant words in this definition are "original inquiry and research" and "strictly substantiates a specific view".

FUNCTION OF THE THESIS ADVISOR

The function of the advisor, according to Almack (1930), is chiefly to see that the thesis meets university requirements and measures up to university standards. The responsibility for planning, executing and organizing the thesis rests with the student (Almack, 1930). Almack (1930) writes,

"In order that the advisor may know the progress being made, the student should consult with him/her, from time to time, and particularly when: (1) the problem has been selected and stated, (2) when the literature of the subject has been reviewed and the bibliography prepared, (3) when a plan of work has been outlined, (4) when the results have been arrived at, and (5) when the thesis has been written in preliminary form. The appearance of unexpected difficulties or discoveries warrant other conferences".

My own personal mode of advising is one of a much more active involvement than the rather passive role described by Almack. Support for the need of an active participation with the graduate student is given by Katz and Hartnett (1976) who have documented their detailed exploration of the experiences of graduate students. They have concluded:

"Graduate student relations with members of the faculty is regarded by most graduate students as the single most important aspect of the quality of their graduate experience, unfortunately, many also report it is the single most disappointing aspect of their graduate experience".

Certainly, the opportunity for a strong working relationship between the student and his/her advisor exists and it is a trememdous shame when this relationship does not develop.

When a solid working relationship exists between student and advisor, the reward is an exciting search together in co-discovery of research knowledge. To emphasize this cooperative pursuit of "original research to substantiate a specific view", I have tried to capture the essence of the nature and meaning of this process by a modification of a quotation by Whitehead (1959), where my insertions are underlined:

"The justification for <u>thesis research</u> is that it preserves the connection between knowledge and the zest for life, by uniting the young and the old in the imaginative consideration of learning. <u>Thesis research</u> imparts information, but it imparts it imaginatively.... This atmosphere of excitment, arising from imaginative consideration, transforms knowledge. A fact is no longer a bare fact: it is invested with all its possibilities. It is no longer a burden on the memory: it is energizing as the poet of our dreams, and as the architect of our purposes."

Thus, my purpose in this paper is to develop further the role of the thesis advisor as one which yields a high quality thesis amid the development of mutual respect and friendship between student and thesis advisor.

ROLE OF THESIS ADVISOR IN SELECTION OF RESEARCH TOPIC

Although the selection of the research topic is a subject of another paper of this symposium, I do wish to expand on several of the criteria with respect to the interaction of the student and advisor in problem selection. The Committee on Research and Graduate Studies of the Engineering Research Council (1965) proposed that project research to be used for an advanced degree must allow the graduate student to have full responsibility for his/her work including the freedom to make mistakes. If the circumstances of the project or obligations to a project do not permit this independence and freedom, then the topic is inappropriate for a thesis. Thus, the advisor must not force graduate students to do their thesis research on his/her grant projects, unless the project offers the appropriate ingredients. On the other hand, the advisor can play an important role in helping the student to define a problem that is soluble in a reasonable time frame.

The Right Question. Krebs (1968) refers to the importance of asking the right question in choosing a research topic, by avoiding those which can be solved easily and concentrating on those which are really worthwhile. Paul Weiss (1945) wrote:

> "The primary aim of research must not just be more facts and more facts, but more facts of strategic value".

While research has the principal aim of increasing our knowledge and understanding, it should lead to the clarification of the problem or deeper insight, or to the integration of previously unrelated concepts.

Causey (1968) writes that,

> "The world of science is most interested in surprising new discoveries".

Surprising discoveries are not expected on the basis of present knowledge, but have a double value. They are a gift and a challenge because they add to our knowledge and reveal gaps in understanding (Causey, 1968).

No simple formula exists to select a worthwhile, soluble problem which provides a "surprise". However, fortunate indeed is the graduate student who has a thesis advisor who has learned the importance of asking the right questions and who challenges the student to do likewise.

Certainly, a very influential factor in this process is the close association with a group of researchers. Krebs (1968) writes that association with a leading teacher (researcher) automatically leads to association with other outstanding students because great teachers attract good people. Students learn as much from fellow students as from their seniors (Krebs, 1968).

A concern which frequently arises for the agricultural engineer in research selection is the one of the practical versus scientific. In reality the problem is not: Should the research be practical or scientific? It is really a question of balance between the two, and how does the balance depend upon the student's own talents, professional tastes, and personal and professional goals (Du Bridge, 1968)? It is clear that practical technologies become obsolete rapidly, while basic principles and theories have a longer life. Therefore, the thesis advisor must be flexible in order to challenge the student to select the right question for "worthwhile discoveries" as a part of a process of education of a human being for participation in life.

Rules of Research. In terms of selection of a research topic, it seems to me that the graduate student is well advised to be aware of three rules of research as given by Causey (1968). The advisor should challenge the student to select a topic which will: (1) maximize chances of being surprised, (2) require an explanation of what one understands the least, and (3) require that only the fittest theories survive (Law of the Jungle).

The first rule states that the chances of discovering something unexpected are better if one is the first to explore a subject. Also, new techniques and inventions could be utilized to develop dramatic surprises. The literature is full of discoveries which were made with the telescope, microscope, x-rays, scanning electron microscope, and ultrasonic sensor.

The second rule reiterates the suggestion that the student work on the most perplexing problems. They are not solved overnight, but a surprising discovery presents the challenge to explain it with existing theories first. If this cannot be accomplished, then one must invent new theories to explain the phenomenon and to test them.

The third rule, the Law of the Jungle, is based on the survival of the fittest. Thus, the theory should be severely tested. An important element of graduate education is to develop in the student a courage to test his/her theories in the "jungle". Feynman (1965) states,

> "One of the ways of stopping science would be only to do experiments in the region where you know the law. But, experimenters search most diligently, and with greatest effort, in exactly those places where it seems most likely that we can prove our theories wrong. In other words, we are trying to prove ourselves wrong as quickly as possible, because only in that way can we find progress".

The thesis advisor should encourage and challenge the student to ask the right question and to select a topic worthy of the student's intellectual abilities. To do otherwise will rob the student of the pride in accomplishing the complex and the thrill of making surprising discoveries.

ROLE OF THESIS ADVISOR IN EXECUTION OF RESEARCH PLAN

Once the topic of research has been selected, the attention turns to executing the plan of research. The advisor, in my opinion, plays an active role by conveying a standard of excellence, helping to develop a climate for discovery, requiring a research plan, being available for frequent and regular consultations, helping the student to develop a time table as a guide in the conduct of the research and lastly, reviewing and approving the thesis.

Standard of Research. Although this has been alluded to earlier, I want to emphasize the importance that the advisor can play in developing in the student his/her standard of excellence. Krebs (1968), in discussing the importance of an outstanding teacher (researcher), quotes Warburg (also a Nobel laureate):

> "The most important event in the career of a young scientist is the personal contact with the great scientists of his time".

As Krebs (1968) traced the scientific "genealogy" from Berthollet to Krebs, he points out that,

> "The association between teacher and pupil was close and prolonged, extending to the mature stage of the pupil, to what we would now call postgraduate and postdoctoral levels. It was not a matter of attending a course of lectures, but of researching together over a period of years".

This student-teacher genealogy emphasizes the fact that "distinction breeds distinction, or, in other words, distinction develops if nutured by distinction (Krebs, 1968).

While most of us are not likely to reach the distinction of Krebs or the other great scientists in his scientific genealogy we must not forget the significant influence we have on our students. Students, and we as well, measure everything in terms of comparisons and how can a student learn the high standards of research unless they see this excellence in their advisor? Krebs (1968) writes:

> "Mediocre people may appear big to themselves (and others) if surrounded by small circumstances. By the same token, big people feel dwarfed in

in the company of giants, and this is a most useful feeling."

Krebs (1968) summarizes what he learned in particular from Warburg (his teacher):

> "I would say he was to me an example of asking the right kind of questions, of forging new tools for tackling the chosen problems, of being ruthless in self-criticism and of taking pains in verifying facts, of expressing results and ideas clearly and concisely and of altogether focusing his life on true values".

Were any of us to have our graduate students make this type of statement, we would certainly have fulfilled our responsibilities as thesis advisor!

Conditions of Discovery. Based on studies of scientific and scholarly discovery, particularly related to training students for productive thinking, Katz (1976) states conditions for discovery (development of the mind).

1. Attention and recognition from professors. This is particularly crucial as the student begins graduate study because it sets the pattern of developing professional identity.

2. Regular exchange of ideas. This includes interactions with both peers and professors. To emphasize this point I quote from Loewi (1954) (himself a Nobel laureate) who said about the leading physiologists of the 19th century:

   "They shared to the highest degree the qualities of contagious enthusiasm, broad mindedness and imagination, humility and deep devotion to their pupils. These are qualities which in themselves suffice to attract outstanding students ... Besides the art of experimenting and observing, the pupils learned the ways of thinking required by science. They learned how to select the object to be explored, how to interpret and evaluate the results obtained, and how to integrate them into the whole body of knowledge. In this way students were not only made familiar with methods and facts, but were imbued with the general scientific spirit which shapes the pattern of the true scholar and investigator".

3. The capacity of working with others. The quotation of Loewi above says it all in terms of the value of collaborative search with both peers and the advisor.

4. Cultivation of the imagination. This condition permits the development of new ideas and fresh conceptualizations that bring newness to the research. Katz (1976) suggests that imagination can be aided by (a) exposure to other disciplines, (b) artistic activities,

(c) practical applications of what one knows, and (d) encouragement of impulse in fantasy, free association, and spontaneous activity.

5. Sequence of learning that is reasonably harmonious with one's own curiosity. I paraphase this as the "right time". Requirements or pressures, if at the wrong time or in inappropriate areas, can interfere drastically with the desire to develop intellectual autonomy in the student.

6. Establishing a broad knowledge base. Pasteur wrote:
"In fields of observation, chance favors only the prepared mind" (Berelson, 1965).
The broader the knowledge base, the greater the opportunities to integrate seemingly detached concepts into a specific solution.

7. Reasonably secure financial support during graduate studies and for the future. While the financial support of graduate students is typically very minimal, it is increasingly important, in light of other financially attractive options, to develop reasonable levels of financial support. It is clear that the zest for learning and autonomy of discovery must be a key interest on the part of the graduate student or no student would find graduate assistantships reasonable in comparison with todays opportunities in agricultural engineering! On the other hand, the opportunities for the M.S. and Ph.D. (particularly) graduates in agricultural engineering are extremely favorable, both in terms of available positions and salary.

In my opinion, the thesis advisor plays an absolutely paramount role in the development of these conditions of discovery. The advisor can help make thesis research an exciting and imaginative quest of surprises or pure hell! In my mind, his/her responsibility is to make it the former.

Research Proposal (Prospectus). The thesis advisor should require that the student submit a research proposal to him/her and the special graduate committee when the student has reached a state of confidence about a meaningful, intellectual achievement of his own (Koefod, 1964). At this point, the student is ready to move from perception and comprehension of critical questions to the presentation of a resolution of the problem. Allen (1973) suggests that the proposal delineates the specific area of research by stating the purpose, scope, methodology, overall organization, and limitations of the proposed study. I also require that the proposal include a review of relevant literature and an indication of the expected contribution of the research.

The benefits of a well-prepared research proposal are great for both the student and advisor. The well-conceived proposal will help the student: (a) think out the critical questions, (b) lay the foundation for the research, (c) isolate pending problems and suggest action before wasting inordinate amounts of time, (d) serve as a "map" for the research, and (e) force a thinking through of the whole process, indicating the need for an integrated approach (Allen, 1973). The proposal should contain the hypotheses stated concisely. At Cornell, it is usual for the student to defend the proposal at the time of admission to candidacy for the Ph.D. degree. Certainly, if the proposal is successfully defended at this time, the student can begin his/her research with a feeling that the topic is appropriate and the committee understands what is planned. While only the beginning, the proposal does serve to reduce disasters of misunderstanding later.

Time Schedule. An often difficult and challenging aspect of graduate research is the problem of developing a realistic time schedule for the thesis. A well-organized, self-disciplined student will conduct his/her thesis work with a minimum of delays and a timetable can be a tool for efficient development of the thesis (Cox, 1964). The advisor should challenge the student to develop a timetable in order to have a chance to discuss realistically the tasks involved.

Timetables are likely to vary in form and approach due to particular tasks and personal habits. For example, De Nevers (1965) suggests that the thesis can be subjected to a critical-path method (CPM) which would include the principal tasks of:

1. Select topic and thesis advisor.
2. Have doctoral committee appointed.
3. Prepare budget and obtain approval.
4. Search the literature for pertinent prior work.
5. Develop the necessary theory.
6. Design equipment and order necessary components.

7. Build equipment.
8. Debug equipment.
9. Obtain experimental data.
10. Analyze data, revise theoretical models as needed. Obtain more data as suggested by this analysis.
11. Write and submit thesis.

The critical path has some obvious steps - 1, 3, 6, 7, 8, 9, 10, and 11, while the other items can be fitted into the process. Preferably, the design of equipment should follow the literature search, but may begin after a preliminary literature search. Certainly, the student can write up the introduction, theory, and literature review when time permits on the critical path.

Cox (1964) presents a planning approach using the bar chart method, given in Figures 1 and 2, for an experimental and an analytical thesis, respectively. No technique such as these can be used successfully without an appreciation of the need for flexibility. The timetable is not an end in itself and only to the extent that the timetable or CPM, or other techniques help students to think about the problem and possible routes of solution are they of value.

Thesis Approval. The final step in the process is the approval of the thesis as a scientific product. The right criteria for evaluation of the thesis demands careful thought. Almack (1930) and Koefod (1964) suggest that common criteria for evaluation include: significance and validity of premises, clarity and quality of reasoned arguement, validity and importance of conclusions, and quality of the contribution to learning (especially the student's learning). In addition, other criteria such as: structure of the inquiry, use of scientific methodology, power and penetration of analysis, employment of insights, quality of formal proof of hypothesis, and correctness of mechanics (grammar, style, coherence, etc.) are judged.

I find a thesis appraisal form given by Almack (1930) of considerable value. It is reproduced (with minor modification) here in Table I. The purpose is primarily for the student to use, to appraise his own work. The objective is to encourage the student to use the appraisal form as a check for his/her work and to see that each requisite is checked. It then forms the basis for the student to recognize any deficiencies and to correct them before submitting the thesis for a defense.

This form or similar forms could be used with both M.S. and Ph.D. theses although the M.S. thesis usually demands a lesser degree of originality. In the case of the M.S., original data should be the basis for the thesis if not an original theory.

ATTRIBUTES (QUALITIES) OF THESIS ADVISOR

At this point, after having alluded to the qualifications of an excellent advisor, I want, in this section, to propose specifically those attributes or qualities that mark the distinquished advisor. In this effort I blend together my own ideas with those of Goheen (1966), Krebs (1968), Pullias (1970) and Vesilind (1967) to list:

Attitude. Krebs (1968) writes, "So, above all, attitudes rather than knowledge are conveyed by the distinguished teacher". Technical skills are learned from many and, of course, are essential for research but as critical is how these skills are used. Humility may be the most important, because as a result there exists a self-critical mind and the continuous effort to learn and to improve (Krebs, 1968). Also, very important is the contagious enthusiasm mentioned by Loewi (1954) that is conveyed from advisor to student. In the spirit of contagious enthusiasm, research is not work but an exciting adventure.

Professional Competence. Vesilind (1967) points out that this is not to say that the advisor must have a greater knowledge about the specific details of the thesis subject than the student who must contribute original ideas. However, the advisor must be competent to aid in development of ideas and pinpoint fallacies. A corollary to this statement is not to advise on topics out of your major field. Another expression might be integrity or authenticity.

Imagination. Broadmindedness, imagination, and freedom of mind to encourage ideas to flow freely and an eagerness to consider many alternatives fuels the fires of creative research.

Directness. The student needs to know how he/she stands and how his/her performance is viewed. It is important that mistakes not be glossed over or successes go unrecognized.

Recognition. Individual recognition is sought by the graduate student. The advisor must treat the student as an individual in order to be sensitive to individual needs and avoid too much guidance on the one hand, and too little on the other. The advisor has to assess individual differences and adjust his role as an advisor to promote the growth of the individual student. My own experience has varied from almost daily contact with several M.S. students for a time, to a more normal weekly contact with other students. I am happy to say that in the cases of the aforementioned daily contacts, these students did develop an autonomy of research and have gone on to highly responsible employment wherein there is a significant leadership role being demonstrated.

Concern for Student. The ability of the advisor to communicate that the individual student is significant and that the student possesses great worth to be nutured is another very important attribute. The interest of the advisor is most commonly illustrated by prompt review of written materials, visits to the laboratory or office to ask questions and make suggestions, passing along pertinent information, and establishing regular meetings to discuss problems and results. A test of this concern may be as simple as, does the student feel that the door to his/her advisor's office is always open?

FINAL COMMENTS

To conclude this effort to delineate the responsibilities or role of the thesis advisor, I want to lift up four of the highlights presented by Katz and Hartnett(1976) from their studies of graduate student responses to graduate education. These are: (1) conditions crucial to the optional development of productive scholars and scientists are neglected in graduate education, (2) though graduate students are not generally regarded as individuals who are particularly vulnerable to emotional problems, there is convincing evidence to the contrary, (3) graduate students are not professional school rejects; nor is the opposite the case, and (4) though graduate departments base their choice of students on a great deal of information about students, student choices of graduate departments are often based on scant information about characteristics and features that subsequently turn out to be very important to their performance and satisfaction in graduate school.

The relationship between student and thesis advisor contains the potential for one of the most rewarding human experiences. To participate as co-discoverers on an adventure to unlock the unknown through the discovery of surprises ought to be one of the ultimate joys of life. To the extent that this paper prompts an increased appreciation for the potential of the relationship between student and thesis advisor, I consider this effort doubly worthwhile. This effort to develop a systematic presentation has proven of tremendous value to me, personally.

ACKNOWLEDGEMENT

I wish to acknowledge the wisdom and eloquence of the numerous authors and researchers from whom I have so liberally borrowed ideas and words and sensed the spirit of contagious enthusiasm.

REFERENCES

1. Allen, George R. 1973. A graduate students' guide to theses and disserations. Jossey-Bass Publ., Washington, D.C.

2. Almack, John C. 1930. Research and thesis writing. Houghton Mifflin Co., New York

3. Berelson, B. 1965. Creativity and the graduate school. In G.A. Steiner Ed. The Creative Organization. Univ. Chicago Press, Chicago, IL.

4. Causey, R. L. 1968. The importance of being surprised in scientific research. Agricultural Science Review 6(3):27-31.

5. Committee on Research and Graduate Studies of the Engineering Research Council. 1965. Problems of interaction between research and graduate studies. J. of Engr. Educ. 55(7):227-228.

6. Cox, J. E. 1964. A technique for thesis advisors. J. of Engr. Educ. 55(2):64.

7. DeNevers, N. 1965. Critical-path method for a graduate program. J. of Engr. Educ. 55(10):303-304.

8. Dubridge, L.A. 1968. Science, Technology and Society. J. of Engr. Educ. 59(2):97-100.

9. Feynman, R. 1965. The character of physical law. M.I.T. Press, Cambridge, Mass.

10. Goheen, R. F. 1966. The teacher in the university. American Scientist 54(2):221-225.

11. Katz, J. 1976. Development of the mind. Ch. 5. In Scholars In The Making. J. Katz and R. T. Hartnett. Ballinger Publ. Co, Cambridge, Mass.

12. Katz, J. and R. T. Hartnett. 1976. Scholars in the making. Ballinger Publ. Co., Cambridge, Mass.

13. Koefod, P. E. 1964. The writing requirements for graduate degrees. Prentice-Hall Inc., Englewood Cliffs, N. J.

14. Krebs, H. A. 1968. The making of a scientist. Agricultural Science Review 6(2):16-22.

15. Loewi, O. 1954. Reflections on the study of physiology. Annual Review of Physiology 16:1-10.

16. Pullias, E. V. 1970. The effective college teacher. J. of Engr. Educ. 60(7):716-717.

17. Vesilind, P.A. 1967. Attributes of a thesis adviser. J. of Engr. Educ. 58(2):151-152.

18. Weiss, P. 1945. Biological research strategy and publication policy. Science 101:101-104.

19. Whitehead, A.N. 1959. The aims of education and other essays. Macmillan, New York.

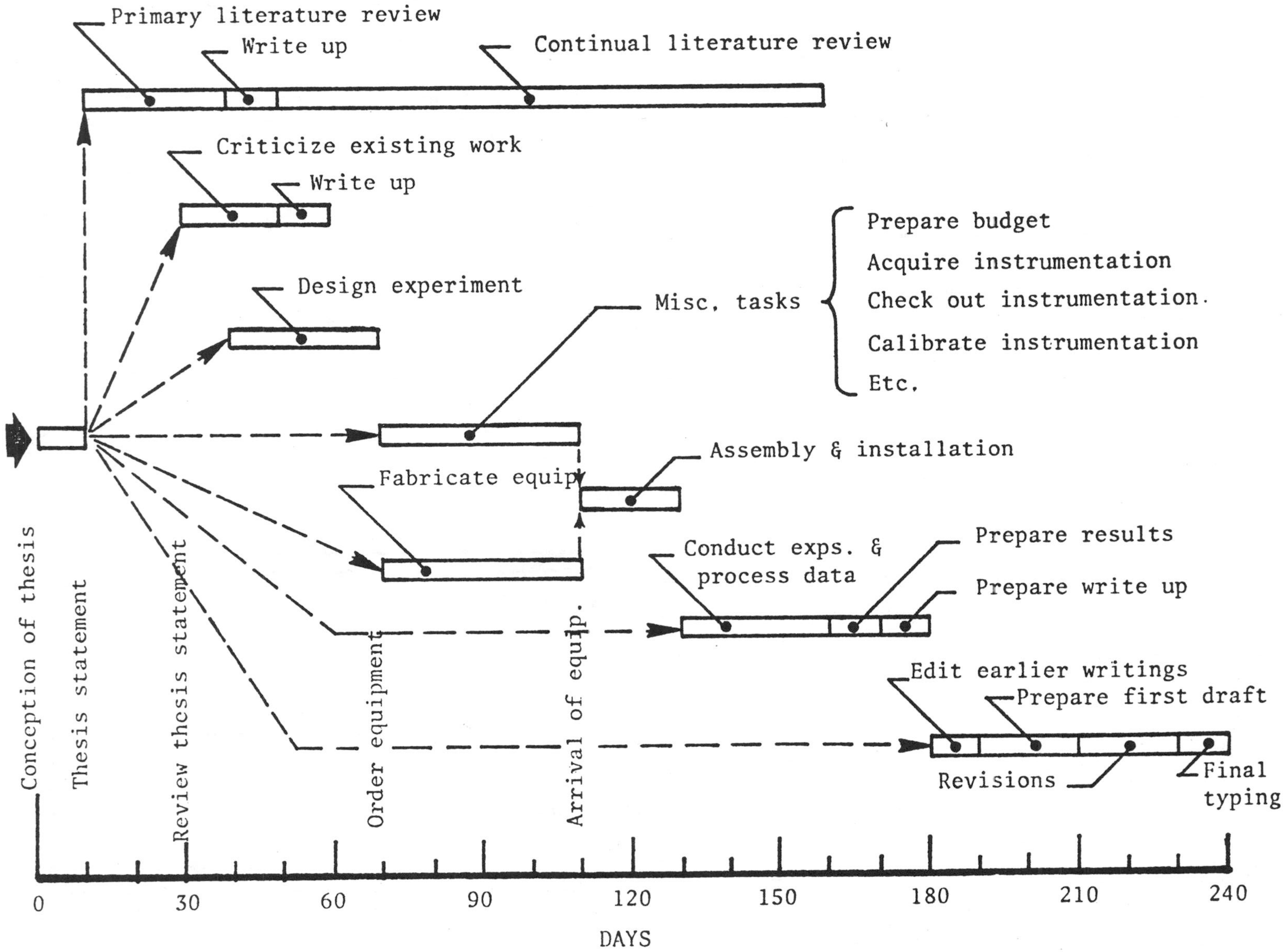

FIG. 1. Eight-month program for an experimental thesis (adapted from Cox, 1964)

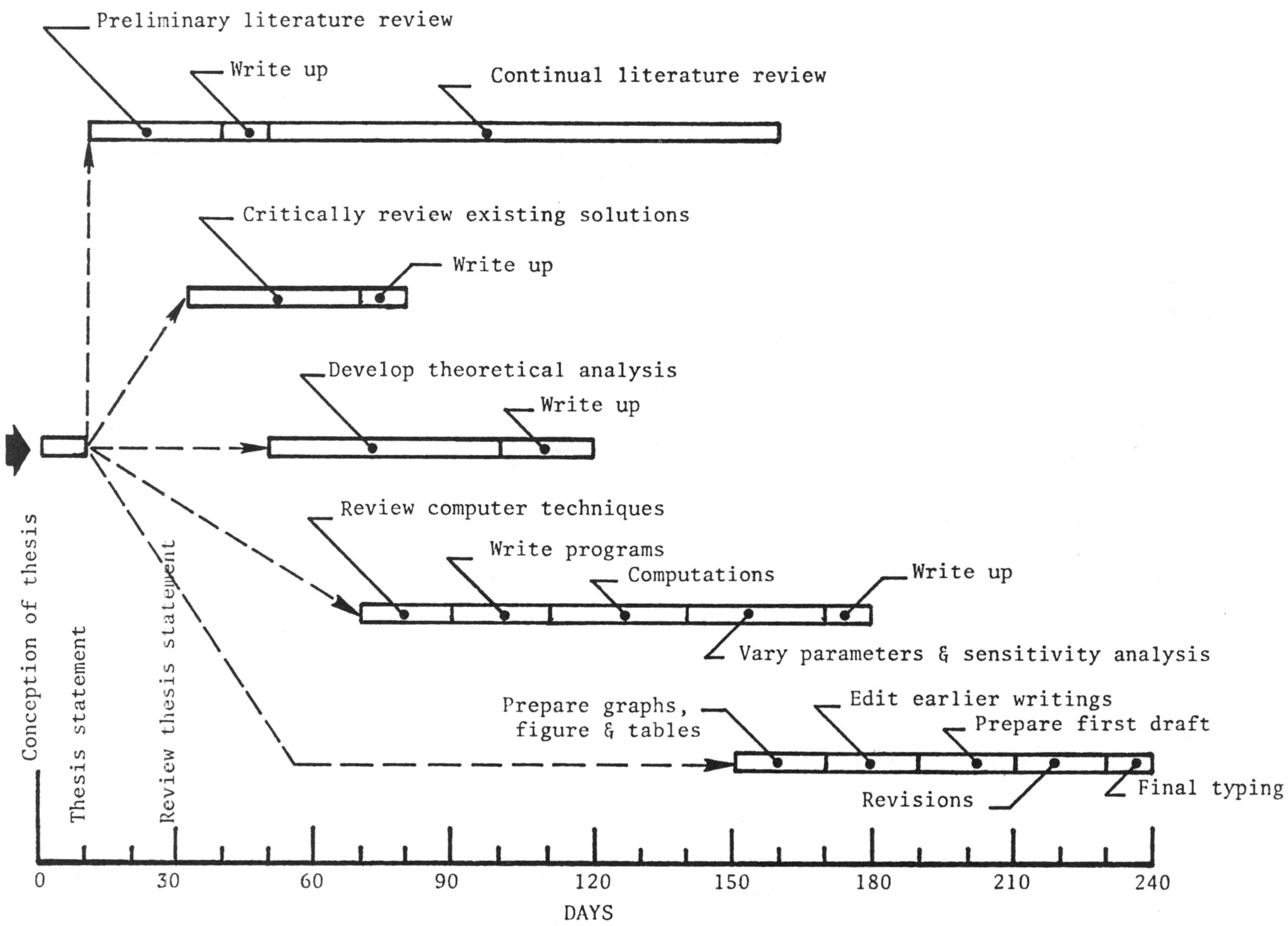

FIG. 2. Eight-month program for an analytical thesis (adapted from Cox, 1964)

Table I. Appraisal Form for Thesis (Adapted from Almack, 1930)

I.	The thesis is a contribution		
	1.	To knowledge, truth or	______
	2.	To technique or method or	______
	3.	Knowledge made available not before available	______
II.	The thesis is original		
	4.	In data and principal or	______
	5.	In technique or method	______
III.	The method is scientific		
	6.	Theoretical or	______
	7.	Experimental or	______
	8.	Combination of theoretical and experimental	______
IV.	The results are scientific		
	9.	Principle	______
	10.	Law	______
	11.	New data fitting an accepted principle	______
V.	Requirement of research process		
	12.	There is a problem	______
	13.	There is an hypothesis	______
	14.	The tests are thorough	______
	15.	The sources are valid	______
	16.	The data are reliable	______

Table I. (continued)

VI. The mechanics are correct

17.	Literature has been reviewed	______
18.	Introduction is complete	______
19.	There is a table of contents	______
20.	There are no typographical and grammatical errors	______
21.	Chart and tables are in proper form	______
22.	Conclusion is complete	______
23.	Bibliography is complete	______
24.	Form, arrangement are correct	______

Book Reviews of Scientific and Technical Books

MADHU S. GUPTA

Abstract—**Guidelines are provided for the preparation of book reviews of scientific and technical books. A comprehensive checklist of the principal issues dealt with in book reviews, developed from a survey of published book reviews, is included.**

BOOK reviews of scientific and technical books have been recognized as useful and have been published in scientific journals for some three centuries [1]. An estimated 50,000 book reviews are published annually in the various scientific fields alone [2]. Yet few guidelines are available on the preparation of a book review, and those I could find were sketchy, subjective, and superficial. Inquiries sent to the book review editors of several scientific journals that are major publishers of book reviews in different disciplines further confirmed this state of affairs. This article provides some guidelines for the preparation of book reviews and may serve as a small step toward improving the current situation.

As with any other kind of writing, there are two considerations in the preparation of a book review: the content and the presentation. First, the reviewer must identify the issues of potential value to the readers of the review, and then those issues must be brought up in a clear, concise, interesting manner. Therefore, any guidelines for book reviewers should deal with the two basic elements: *what* and *how*.

WHAT

Several years ago, when requested to review a book, I started (in keeping with my engineering training) by asking what the objectives (or "specifications" in engineers' vernacular) of a book review are. That there are no universal objectives is obvious: There are numerous examples of two or three reviews of the same book which do not even overlap in the issues they raise. How, then, can the objectives of a book review be developed in an objective way? The traditional method of obtaining such information is by surveying the opinions of book review editors and authors (and readers?). However, my conviction that actions speak better than words led me to look up the reviews that the editors and authors had published.

I read reviews of books on physical, mathematical, biological, earth, behavioral, and engineering sciences in a large number of journals. While reading each review I tried to determine what issue was addressed or question answered by each sentence and then made a note of it. As the list of issues grew, I merged similar ones and classified them in some major categories. Several years and several hundred reviews later I compiled the checklist of issues in Table I. My reading led to some interesting observations on how book reviews differ from field to field and from journal to journal. For example, some of the shortest book reviews are of books on mathematics and some of the lengthiest are about the sociology of science. Such differences were disregarded, however, in producing the table. Further refinements and extensions of the checklist are possible, but they are not likely to significantly increase its utility. In particular, there is some overlap among the items of the checklist.

I share Table I in the hope that it will be useful to writers of book reviews as a checklist of issues to raise. For a novice reviewer, this checklist can (1) serve as a prompting device, (2) supply ideas on points to consider, and (3) help guard against major lapses. The checklist may also be useful in helping authors look at their books from the point of view of a reviewer.

Clearly, only some of the issues on the checklist are important in a given book. Books differ from each other a great deal, and the important issues differ for textbooks, monographs, conference proceedings, anthologies, and encyclopedias. The checklist must be properly interpreted. For example, if a book has multiple authors, the author-related issues may apply to each author.

Rigid guidelines concerning the contents of a book review appear to be difficult to develop because many editors and reviewers have their own opinions about the suitability of various issues in a book review. For example, according to its book review editor [3], the journal *Science* edits out comments concerning the author. Similarly, a reviewer may take the position [4] that the cost of a book is not an appropriate issue to remark on in a book review; the reviewer is usually only a subject expert and the publisher is a better judge of the economics of publishing and marketing the book.

HOW

Many reviewers regard book reviews as a form of literature, evoking poetic license and freedom from rules. Nevertheless, reviews in science and technology are a subset of scientific writing and the guidelines for clarity, conciseness, and lucidity in scientific writing apply to them. Since there is a sizable literature on the "how to" of scientific and

Received February 14, 1983; revised April 5, 1983.

The author is an associate professor, Department of Electrical Engineering and Computer Science, University of Illinois at Chicago, P. O. Box 4348, Chicago, IL 60680 (312) 996-2313.

Reprinted from *IEEE Trans. Professional Commun.*, vol. PC-26, no. 3, pp. 117–120, September 1983.

technical writing, those guidelines are not repeated here but they are supplemented by the following general remarks relevant to this variety of scientific writing.

1. A book review will be read not only by those who are adequately prepared to read the book itself but also by those who are not—for example, acquisition librarians and nonexperts and novices trying to judge the suitability of the book for themselves. Some parts of a book review should therefore be accessible to an audience much larger than that of the book itself. Other parts may well be inaccessible, e.g., a critical judgment of fine points that only an expert can appreciate. Although some remarks in a review can be understood fully only after the book has been read, none should be so cryptic that it can't be deciphered without the book.

2. In the interest of balance, a review should point out both the strengths and the weaknesses of the book. Sometimes, one of these characteristics may be so overwhelming that the other seems trivial; this should be explicitly stated to put the two in proper perspective.

3. Abstract remarks should be substantiated or exemplified. For example, the statement that "the book contains imprecise results" will be more meaningful if an example is quoted and a specific page in the book is referred to [5]. Often, the characteristics of a book can be conveyed effectively by comparing or contrasting them with those of well-established books in the field [6].

4. Because it is the authors' prerogative to choose their subject matter, a book should be judged only on the basis of what it contains rather than what the reviewer would have liked to see. On the other hand, mention of omissions can be useful, e.g., when the book is intended to be a textbook. Also, a description of the state of the art in the field may be appreciated.

Just as books span the range from shoddy workmanship to life time labors of love, book reviews range from superficial to scholarly. Some reviewers produce a mini-introduction to the subject of the book, while others simply fall for the blurb on the dust jacket. Some reviews read like the table of contents or a summary, while others are a critical analysis of a small part of the book with no comment on the remainder. Some are factual and some are unabashed opinion. Obviously, book reviewing is a very human activity. This article is not intended to reduce this form of expression to a set of rules but to help in unlocking the flow of that expression.

REFERENCES

1. Farr, A. D. "Book Reviews in Scientific Journals." *Medical Laboratory Science*. March 1981; 38(2): 75-76.
2. Garfield, E. "A Swan Song for IBRS." *Current Contents*. November 30, 1981; 12(48): 5-8.
3. Livingston, K. Personal communication, Washington, DC., December 20, 1982.
4. Ellis, H. "The Art (or Science) of Book Reviewing." *British Medical Journal*. 1980; 280(6221): 1079-1080.
5. Gupta, M. S. Review of *Noise in Measurements* by A. van der Ziel. *Journal of Vacuum Science and Technology*. March/April 1977; 14(2): 753-754.
6. Gupta, M. S. Review of *Noise and Fluctuations in Electronic Devices and Circuits* by F. N. H. Robinson. *Journal of the Franklin Institute*. February 1977; 303(2): 220-221.

TABLE I
ISSUES TO CONSIDER IN REVIEWING SCIENTIFIC AND TECHNICAL BOOKS

1. The Field or Subject Matter

Origin, age, history
Scope, domain, central theme
Background, its main concepts, definitions of important terms
Importance or significance
Applications or reasons for interest
Nature of problems of interest
Significant or open questions, or frontiers of the field
Major hypotheses
Recent breakthroughs, advances, turns, or developments
Current or future directions
Needs for further work
Measures of activity (e.g., number of annual conferences or published papers)
Field's ripeness for a book
History of books in the field
Need for still other books in the field

2. The Book's Position in the Field

Status of publishing on the subject (e.g., recent spate or long neglect)
Timeliness
Contribution
Other fields in which the results of the book are of interest
Other sources of similar or related material
Complementary books
Competing books (dates)
Similarity to other books
Points of departure
Details of any companion volume or series
Differences from other volumes of the series
Reviewer's suggestions or advice for future books or subsequent editions

3. Basis of the Book

Origin (e.g., class notes, symposium lectures, committee report, translation from another language)
Earlier work updated by author
For new editions: added and deleted topics; correction of errors in previous edition; changes in organization and ordering of topics, or level of treatment
Sponsoring or funding agency
Sources from which data or material is drawn (e.g., interviews, author's laboratory, congressional testimony)

Reliability and limitations of data
Any reviewing (by colleagues, students, or a committee) from which the book has benefited
Origin of results, interpretations, examples, or material (e.g., standard or author's own)
Simultaneous publication of same or similar material elsewhere (e.g., in a journal)

4. The Author and the Author's Effort

Past and present affiliations
Qualifications and experience
Status in the field (e.g., authority or outsider)
Credentials and achievements
Expertise, experience, and involvement in the field
Contributions by the author or the author's research group or institution
Other books by the author
School of thought
Philosophy or viewpoint
Inclination, bias, or prejudice
Flaws in approach or reasoning
Labor and care in preparation (e.g.,compilation of widely scattered material)
Exercise of judgement (e.g., critically evaluated and well-digested material rather than uncritically assembled or unassimilated reproduction)
In multiple-authored books, commonalities or disparity among authors

5. The Anticipated Readership

Characteristics (e.g., specialists vs. generalists)
Prerequisites assumed (particularly if the readership includes students or new comers to a field)
Success in writing for the intended audience
Readers for whom the book is likely to be useful
Readers' likely reaction
Effort required by the reader (e.g., must fill-in details, tortuous, easy)
Sources recommended by the author, or by the reviewer, to develop adequate background

6. Goals and Success of the Book

Rationale behind the book
Goals as stated by the author and as viewed by the reviewer
Success in meeting goals
Purposes served and not served

7. Title and Theme of the Book

Central theme or highlights
Appropriateness of title in view of content
Definitions of the words appearing in the title
A more suitable, descriptive title, or qualifiers needed
An overview of the contents

8. Book's Results and Conclusions

Major claims or conclusions
Novelty of principal results (e.g., revolutionary or well-known)
Significance of principal results
Validity of principal results, as judged by the reviewer (e.g., speculative or firmly established)
Evidence or arguments conflicting with the principal conclusions
Recent breakthroughs or results in the field confirming or invalidating the conclusions

9. Book's Methodology or Approach to the Subject

Approach (experimental or theoretical; empirical or analytical; inductive or deductive)
Freshness or unusualness
Methodology or techniques used (e.g., factor analysis, spectroscopic)
Novelty or prior use of approach
Validity or limitations
Emphasized aspect, phase, topic, viewpoint, framework
Suitability or utility of the methodology either in the field or for practical problem solving
Appropriateness in view of goals
Mention of alternative methods or approaches
Intentional avoidance of some technique or method (e.g., due to dubious value or unfamiliarity to readership)
A unified or integrated approach throughout or a diversity of approaches

10. Topic Selection

General plan or outline of the book
Major questions or issues addressed
List of topics (e.g., chapter headings)
Topics or aspects emphasized
Utility/appropriateness of inclusion of individual topics
Novelty of topics or selection
Topics unavailable elsewhere
Length of individual parts (e.g., chapters or topics) in number or portion of pages
Balance between the lengths of various topics; completeness, well-roundedness
Unrelated or superfluous topics
Topics or issues omitted, slighted, or treated too briefly
Overlap between topics, particularly in multiple-authored books
Range of topics (e.g., broad or selective)

11. Topic Arrangement

Subdivision of material among chapters
Basis of organization or topic sequence (e.g., chronological, geographical, by classification, or by technique)
Order of topics (e.g., systematic, haphazard, continuity of thought from topic to topic)
Stand-alone or interdependent topics or chapters
Novelty of arrangement
Alternative reading strategies or chapter order

12. Topic Treatment

Specific topics treated at length, in depth, or superficially
Breadth of treatment (e.g., comprehensive or encyclopedic vs. narrow or specialized)
Kind of treatment (e.g., qualitative vs. quantitative, rigorous vs. intuitive, abstract vs. concrete)
Level of treatment (e.g., detailed vs. shallow, introductory vs. advanced, simplistic vs. sophisticated)
Uniformity of the level of treatment
Completeness
Compactness or brevity
Novelty
Balance (e.g., one or all sides of controversial issues)

13. Accuracy and Bias

Reliability of facts and figures
Unwarranted extrapolations, overstatements; questionable interpretations or inadequately supported conclusions

Inaccuracies, factual errors, contradictions, or omissions
Speculative or factual basis
Misleading or pretentious writing or topics
Typographic errors
Inappropriate or inadequate credit for borrowed material; incorrect reference citations
Neutrality, fairness, or evenhandedness on controversial issues
Clear and complete statement of assumptions, axioms, or hypotheses
Mention of alternate axioms

14. Writing Style and Readability

Style (lucid vs. tortuous, informal vs. pedantic, clear vs. obscure)
Compactness (terse vs. verbose; efficient vs. repetitive)
Use of jargon or nonstandard notation or terminology
Coherence
Adequate definitions of terms
Adequate cross-referencing, particularly in multiple-authored books
Needless detail, distracting asides
Feeling provoked (e.g., stimulating, thought provoking, pessimistic, optimistic, entertaining, enthusiastic, idealistic)
Clear statement of objectives; identification of what is important or incidental

15. References and Bibliography

Size of bibliography; thoroughness
Up-to-dateness as indicated by the latest cited work or the ratio of recent to old citations
Selectivity (e.g., comprehensive or items restricted to one viewpoint)
Purpose (e.g., for further pursuing a topic or for acknowledging original literature)
Type of material cited (e.g., references to monographs and reviews or to journal articles)
Distribution of references (e.g., throughout or confined to certain parts)
Arrangement (e.g., by author or topic, as footnotes, or at the end of each chapter)
Presence and succinctness of annotations or remarks
Biases (e.g., omission of, or emphasis on, the literature from some laboratory, language, or country)
Major omissions

16. Usability and Usefulness

Strengths, limitations, and shortcomings
Usefulness (e.g., as textbook or handbook or roadmap)
Introductory remarks to establish goals or place material in perspective
Summaries or concluding remarks to aid in focusing
Material not found elsewhere in book form
Condensation of material into tables, charts, and graphs
Up-to-dateness of material
Expected longevity
Utility or futility of individual parts
Arrangement of bibliography from the point of view of utility
Exhaustiveness of author, subject, or other index

17. Physical Attributes

Length
Size of the page; portability
Quality of paper, printing, and binding
Particularly convenient, inconvenient, or unusual feature of arrangement or style of the printed page (e.g., pagination, running headings, footnotes, margins)
Features enhancing readability or scannability (e.g., variety of type fonts or gray shading)
Number, quality, and variety of illustrations
Special features (e.g., use of color, thumb tabs)
Price and affordability by intended readership
Materials accompanying the book (e.g., instructor's manual, plastic overlays, audio cassettes)

18. Other Noteworthy Features

Innovative ideas
Critical review of the literature
Suggestions for future work, or still-open questions
Biographic/historical notes

19. For Student Texts

Suitability for classroom use or self-study
Exercises and problems; their number, difficulty, novelty, and adequacy
Illustrative examples
Level and preparation required of students
Duration of classroom use (e.g., one-semester course)
Student reactions (if post-use review)
Clarity of educational objectives

20. The Reviewer

Name, affiliation
Involvement in the field of the book
Philosophy, beliefs
Qualifications
Authorship of books on related subjects

Writing a College Textbook

An engineering professor/author/publisher looks at the task of writing a textbook. Although time-consuming, and sometimes tedious, it can also be a satisfying and rewarding project.

Donald G. Newnan
San Jose State University

One of the most interesting and challenging activities an established college professor may undertake is the preparation of a college textbook. Probably every engineering professor has at one time or another wondered if he or she should write a textbook. Some of the steps in the process are described here to give prospective authors a better understanding of the problems and pleasures of writing such a book.

Why Write a Textbook?

At the outset a college professor must recognize that a textbook writing project is a substantial one that probably will adversely affect other professional writing. And since textbooks are, by their very nature, something less than state-of-the-art expositions, the effort in their preparation may not be fully appreciated by one's peers. Young faculty members, in particular, are encouraged to do original research and publish papers, rather than books.

For the seasoned professor, however, there is little so fulfilling and rewarding as writing a textbook. Few technical papers ever bring their author the same broad recognition that is associated with a widely adopted textbook. Few professors write a textbook to produce another publication for their resume or to seek acclaim. Rather, they are dissatisfied with existing textbooks and feel that they can write a better one. If you hold this view, then you have the nearly universal reason for beginning to work on a manuscript.

Getting Started

Most textbook writing projects are associated with courses that professors are actively teaching, and this is a wise restriction for an author's first text. Once you have selected your textbook topic, you must prepare an outline. At this point you find out if you really have anything new and innovative to offer. If the outline is a rehash of an existing book, with topics to be presented in the same order, in about the same way, then maybe the writing project should be abandoned. But if you have outlined some new approaches, sifted the content for current and future instructional needs, and have new things to say and better ways to say them, then you have the basis for a textbook. Having said this, one caution is in order. A new textbook gains adoption because faculty members think it is superior to existing texts, or they have tired of the one they are using and want to try another one. As a practical matter, the content of course work (and hence the text adopted) is often an evolutionary process. The author of a first textbook therefore should not attempt to revolutionize the content of a particular course unless he or she has strong evidence indicating the desirability of such a step. Many textbooks fail to achieve commercial success because they embody substantial shifts in course content that are not generally viewed as desirable.

With the outline in hand, you will begin accumulating materials for the book. Although I am sure some would disagree, I have found writing a large set of homework and examination problems a major task. To make the job easier, I prepare original questions for all assigned homework and for examinations, and then test them in the classroom. After doing any needed rewriting, the questions are ready for inclusion in the manuscript. At the same time one must make notes of ideas, gather il-

Reprinted with permission from *Engineering Education,* vol. 71, no. 4, pp. 279–282, January 1981.

> "You will realize that the rewards of being a successful author will not be achieved quickly or easily. You must ask yourself if this is how you want to spend your evenings and weekends for the next year or two."

lustrations, prepare numerical examples and do the many things that form a manuscript. Experience indicates that only about half of the material assembled will actually be used.

At this point one is excited at the prospect of writing the book, but the acid test is still to come. The test of a textbook author comes in sitting down and writing a chapter. The first chapter written should not be chapter one; it should be a more representative chapter intended for the middle of the book. Preferably it will be material that can be handed out to a class as part of the instruction.

After writing two or three chapters for classroom testing, it is a good idea to reexamine the textbook writing project. At this point you will realize that the rewards of being a successful author will not be achieved either quickly or easily. You must ask yourself if this is how you want to spend your evenings and weekends for the next year or two. If not, maybe you should shift your attention to other scholarly writing.

The Manuscript

An author looks at a manuscript as the finished product. But from the publisher's viewpoint, it is the raw material in the publishing process. To produce a suitable manuscript, the author must understand how a manuscript is used. A complete manuscript has three components: 1) text, 2) line drawings and figures, and 3) photographs. Each of these is treated differently in the publishing process.

The text (words, mathematics and tables) makes up the bulk of the manuscript. It will be copy edited for grammar, style and consistency. (Different publishers have different rules, of course, but an excellent guide is *A Manual of Style,* from the University of Chicago Press.) The copy editor will revise the text, write questions to the author, and give typesetting instructions. Since much of this must be put in the margins of the manuscript, the margins should be generous on all four sides of the pages. Double or triple spacing should be used throughout. The typesetting is done directly from the original manuscript pages.

Line drawings and figures do not go to the typesetter, but are sent to the illustrator. He or she will convert the author's rough sketches into finished professional figures. Because the typesetting and illustration work is done separately (maybe 3,000 miles apart), publishers want the line drawings and illustrations prepared on separate pages, rather than put into the manuscript text.

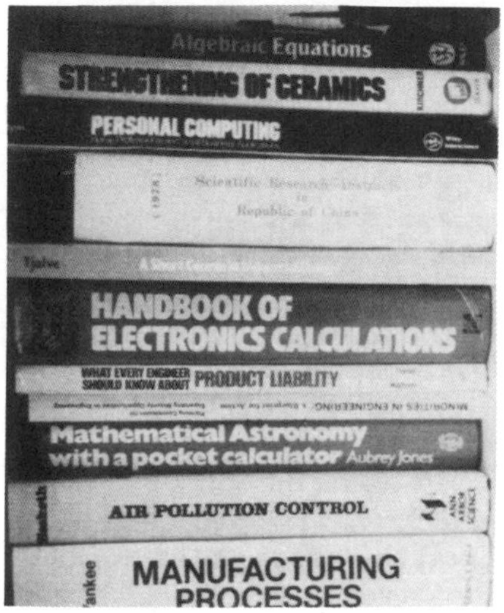

For testing a manuscript in the classroom, however, authors usually want to put their sketches in the manuscript in their proper position. The solution is to keep the original sketches separate to satisfy the publisher, and put photocopies of the sketches in the text to facilitate classroom use.

Simple photographs, without lettering or other features added, are not part of the text or sketches. Instead, they are held for the necessary camera work to prepare them for insertion into the typeset text as the final page-by-page layout of the book is prepared.

Selecting a Publisher

Frequently, publishers decide what they wish to publish and then seek an author. Assuming it is the author who has the idea, how do publishers select manuscripts and authors? A first-time author needs to assemble a proposal consisting of an outline, one or two sample chapters, a statement of viewpoint, and an analysis of the relationship of the proposed book to competing books.

An author may seek a publisher early in the process with a proposal, or wait until the manuscript is well along. Publishers would like an opportunity to suggest ways to improve a manuscript, but they also want manuscripts they can publish in the not-too-distant future. One possible approach is to seek a publisher as soon as one feels the need for assistance and guidance.

When a manuscript or manuscript proposal is sent to a publisher, it will be given to the acquisition editor working in the subject area. If the editor thinks the project has potential, it will be sent out for review. The first reviewer will be, in many cases, a consulting editor who is a college professor-author. For a small royalty, these consultants help the publisher develop books in a particular field.

A manuscript or proposal may be rejected if neither the acquisition editor nor the consulting editor have a strong favorable opinion of the project. If they like the material, it may be sent to one or more professors for review, a process that may take several months. The long time lag in getting an answer from a publisher suggests that an author may find it desirable to contact several publishers if the manuscript has been completed or is nearly finished.

The Book Contract

With about 17 U.S. publishers of engineering textbooks, there inevitably will be some variations in their book contracts. Probably all, however, will include the nine items listed in the contract shown on this page. It is not customary to pay advance royalties in engineering textbook agreements, but if needed, a publisher will advance a few hundred dollars for manuscript preparation costs. Contrary to the contract shown here, many publishers limit the royalty on mail order sales to 5 percent.

Any change an author makes to the wording of a book after the original typesetting has been completed is called an author's alteration (AA). To restrain an author from making excessive changes, most contracts (e.g., paragraph 9 of the contract shown) will provide an allowance of from 10 percent to 20 percent of initial composition cost for AA's. AA's in excess of this amount could be charged against royalties, although that is rarely, if ever, done. Finally, it should be noted that all parts of a publishing contract are subject to negotiation. Publishers, if they really want a particular author or manuscript, may be induced to make some contract concessions.

The Final Tasks

The book contract places on the author the sole responsibility for obtaining permission to use copyrighted material in the manuscript. This is a time-consuming task that should be worked on concurrently with the manuscript. The convenient way to seek permission is to write the copyright owner and carefully describe the material being sought, and its intended use. At the bottom of the letter type a permission form granting use of the material in all editions and revisions of the book. If all goes well, the permission form will be signed and returned to you, and the procedure will have been completed. Frequently, permission will not be granted so easily. The copyright owner may require payment ranging from a token amount (like $20) to hundreds of dollars. Sometimes the copyright holder may be unwilling to allow any use of the material, or to allow as much material as the author would like, or to grant permission for its continued use in all editions of the book. In any of these situations there may be protracted correspondence over a period of months. For this reason work on permissions must be begun early to avoid delaying publication of the manuscript.

Submitting a completed manuscript to the publisher is far from being the culmination of the project to the author. The publishing process may take from six to 12 months (smaller publishers tend to proceed more quickly than larger ones). Ini-

Memorandum of Agreement

____

hereinafter called the Author(s), hereby agrees to prepare and supply a manuscript of about pages on or about __________ to Engineering Press, Inc., hereinafter called the Publisher, a Work tentatively titled ____________________

____

in form and content satisfactory to the Publisher.

1. PUBLICATION RIGHTS. The Author hereby grants to the Publisher during the full term of the copyright and all renewals, the exclusive right to print, publish, and sell the Work in all forms and languages throughout the World, and to license or contract with others to print, publish or sell the Work. The Publisher at its expense will copyright in its own name the Work in the United States.

2. AUTHOR'S WARRANTY. The Author warrants that the work is original except for such excerpts from copyrighted works as may be included with the permission of the copyright owners thereof and excerpts from works in the public domain, that it contains no libelous statements and does not infringe upon any copyright, trademark, patent, proprietary, or statutory right of others, and that he will indemnify the Publisher against any costs, expenses, and damages arising from any breach of the foregoing.

3. ROYALTY. The Publisher agrees to pay the Author a royalty on net sums received from the sale of all copies of its edition of the Work (except as hereinafter set forth) as follows:

Fifteen percent (15%) of Publishers net receipts on all US and Canada sales.

Ten percent of Publishers net receipts from all sales by the Publisher of inexpensive editions (editions published in less expensive form than the original) and from all sales by the Publisher outside of the United States and Canada. Fifty percent of Publisher's receipts from any rights licensed to others. Payments to the Author shall be in February and August for the preceding calendar half years, provided however that the Publisher may defer any accounting until the total royalty due reaches fifty dollars.

4. PUBLICATION. The Publisher shall proceed at its own expense to publish and promote the Work in editions and prices it deems appropriate. Author shall receive 10 copies of the Work without charge.

5. RELATED WORK. The Author shall not, without the written consent of the Publisher, publish, prepare or assist others to publish any other edition of the Work, revised or otherwise, or any work of a character that might interfere with or injure the sale of the Work.

6. REVISED EDITIONS. The Author agrees to revise the Work when it is determined by the Publisher that a revision is desirable. Should the Author be unable or unwilling to perform such revision or be deceased at the time at which it is determined that such a revision is desirable, the Publisher shall have the right to arrange for the preparation of a revision. In such case, the reviser shall be compensated by a share of the royalty on the sale of the revision or by a fee paid by the Publisher and charged against the royalty and other income accruing to the Author on the sale of the revision. The Publisher shall have the sole right to determine the reviser's fee or royalty percentage.

7. OUT OF PRINT. When the Publisher determines that the demand for the Work is not sufficient to warrant its continued sale, the Publisher may allow the Work to go out of print. If the Work shall be out of print, the Author may request Publisher to reissue it. If the Publisher fails within eight months to reprint the Work, all rights herein granted to the Publisher shall revert to the Author, subject however to any agreements made by the Publisher to others.

8. ALTERATIONS. If Author's alterations to proof are necessary, the cost thereof, up to 15% of initial composition shall be borne by the Publisher, and the excess cost, if any, may be charged against royalties payable to the Author.

9. HEIRS AND ASSIGNS. This agreement shall enure to the benefit of the heirs, executors, administrators and assigns of the Author and the successors and assigns of the Publisher.

____________________ Engineering Press, Inc.
Author

__________ ____________________ ____________________
Citizenship Social Security Number

____
Date

A brief, but otherwise typical, book contract.

tially the author must provide guidance to insure that the copy editing does not alter the technical accuracy of the manuscript. Later, as typesetting proceeds, the author must read and correct the first proofs (galley proofs) and proofs of the artwork. When the galley proofs and the artwork are corrected, they are made up into the finished pages of the book. Page proofs, sent to the author, are the last opportunity to make changes (AA's) or to correct errors. While reading the page proofs, the author will be preparing the book's index. During these months it always seems the author has something to do and is expected to do it promptly. After the index is completed, the author can only wait patiently for the two or three months needed to print and bind the book.

The process of writing a textbook manuscript, testing it in the classroom, arranging for its publication, and finally reading galley and page proofs is both long and, at times, tedious. But once it is over, most authors would agree that it is a satisfying experience. Further, of course, it may be personally, financially and professionally rewarding.

Annotated Bibliography

Brinker, Russell C. "Should I Write A Textbook?" *Journal of Engineering Education*, April 1962. This engineering professor-author-consulting editor describes the world of engineering textbook writing and publishing in 1962.

Dessauer, John P. *Book Publishing: What It Is, What It Does*. New York: R. R. Bowker Company, 1976. A highly readable description of the business side of publishing, including college textbooks.

Lee, Marshall. *Bookmaking: The Illustrated Guide to Design and Production*. 2d ed., New York: R. R. Bowker Company, 1979. Detailed information on book production.

Skillin, Marjorie E. *Words Into Type*. 3d ed., Englewood Cliffs: Prentice-Hall, Inc., 1974. Similar to A Manual of Style. Topics include the manuscript, copy and proof, copy-editing style, typographic style, grammar, use of words, typography and illustration.

The University of Chicago Press. *A Manual of Style*. 12th ed., Chicago: U. of Chicago Press, 1969. (13th edition due spring 1981). Authoritative volume on bookmaking, style, production and printing. An indispensible guide for authors.

Getting started in consulting

Thinking of striking out on your own as an independent consultant? Many engineers enter consulting every year, but only a few continue in it. Before you make your move, find out exactly what you will be getting into.

Arthur S. Diamond, *Diamond Research Corp.*

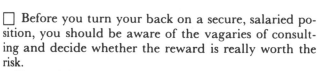

☐ Before you turn your back on a secure, salaried position, you should be aware of the vagaries of consulting and decide whether the reward is really worth the risk.

You might begin by examining your reasons for seeking this new life-style and by learning what other engineers have encountered as independent consultants. This will give you a fair idea of what is required in time and savings to build the sort of business that will meet your professional and financial goals.

You will find yourself asking a lot of questions about the future. Can I make a career of consulting? Is it only a temporary situation for most people, or does it provide permanent employment? Is it something one does between salaried positions, or does it come after one's career is over? Do people ever start out consulting, or do they only end up that way?

How these questions are answered depends upon various factors, such as the individual, his background, and the type of consulting he wants to do. For example, consulting is a natural choice for the engineer or scientist who has had a distinguished academic or industrial career. Such a person may enter private practice as a form of semiretirement. But the appeal and the opportunities are certainly not limited to individuals in such special circumstances.

The lure of private practice

The services provided by consultants need to be cataloged, for they cover a broad spectrum of activities. From these, you can identify those pursuits that best fit your interests, personal traits and intellectual talents.

What follows, then, is an attempt to describe these services. Before setting off on your entrepreneurial odyssey, you might take this moment to review some of the pros and cons of private practice.

The major impetus to consulting is the desire to work independently. It is this longing to apply one's creative faculties in an atmosphere unpolluted by organizational politics and policies that lures many engineers into consulting. And it is the opportunity to reach for the greatest recognition and reward that one's abilities will permit that gives consulting its luster.

Of course, some may enter consulting as a result of a poor job situation, a layoff, a merger, or a business failure. Others may be unwilling to relocate from a particularly desirable living area when their employer requires a move. The reasons for getting started in consulting are numerous; but of particular interest here is why many have stayed with it and how they have dealt with the problems.

Your finest credential

What is a consultant? We can probably define him, or her, as an individual having expertise in a particular field and whose income is principally earned by providing professional advice, reports, testimony, analyses, surveys, personal contacts, representation, or other specialized services to one or more clients.

The word "professional" should also be defined. This has been a controversial topic for years, as evidenced by the endless chain of letters written to the editors of

Reprinted with permission from *Chemical Engineering,* vol. 82, no. 19, pp. 139–140, 142, 144, September 1975.

technical publications. And yet, the numerous definitions seem to agree upon what essentially characterizes an individual or group as professional [1]. These include:

1. A high degree of individual responsibility, and pride in one's work.
2. A type of activity that deals with problems on a distinctly intellectual plane.
3. A strong motivation to make a contribution to the state of the art.
4. A respect for the public interest.
5. Possession of a common body of knowledge that is continually being extended through united effort.
6. A standard of conduct, or a code of ethics, that governs relations with clients, colleagues and the public.

The problem is not whether an engineer or scientist is, by virtue of his education, training and dedication, of professional caliber. Of this, there can be no doubt. It is whether his job enables him to function with sufficient independence to preserve his professional integrity and individualism. Too often, he is engaged as a professional but employed as a technician.

This is not to say that consulting alone allows an engineer or scientist to function as a professional; rather, that it is only consulting that assures him total independence.

Such considerations were involved in a situation that developed about five years ago, when several of the nation's top consulting firms changed their ownership structure [2]. Some went public, others were merged into larger companies, and at least one of the leading management-consulting firms was acquired by a major bank. This upheaval rocked the consulting fraternity. At the center of the controversy was the question: "Will outside ownership affect the ability of these (consulting) firms to function objectively as true professionals?"

Many spokesmen for these firms claimed to foresee no conflict of interest and no change in the quality of their work. Others, however, disagreed. Marvin Bower, a former director of McKinsey and Co., said that independence is the heart of professionalism, and that a consultant who works for a publicly held corporation, or for a division of a company or a bank, does not have that independence. In such a situation, he explained, the consultant is accountable for current profit to stockholders or "to his superiors in the corporate office, who are neither consultants nor professional men."

Whatever the outcome of this debate, one fact remains: The consultant in private practice is a professional and, as such, personal integrity is his finest credential.

A long, hungry road

Becoming an independent consultant is no simple task. "Getting established is a long, hungry road," observed Denzel Dyer, an independent microbiologist and past president of the Los Angeles-based Consulting Chemists Assn.

It can take years to get established. One needs time, temperament and money to weather the storm. And it can be an expensive venture, especially if equipment and a laboratory are required. The consensus of a dozen independent chemical consultants interviewed in 1972 was that anyone getting started in his own consulting practice should be prepared to finance the new business for about two years [3].

There are certain liabilities—personal, professional, and financial—to be assumed. To a family man, consulting is a "marital joint-venture," which means risking more than just dollars on the success or failure of the business.

Insurance is a major expense: fire, theft, health, disability, workmen's compensation (if there are other employees) and other forms of coverage may be required. There are taxes and license fees that must be paid, and countless forms that any business must fill out to satisfy local, county, state and federal agencies.

Legal, bookkeeping and accounting fees take their

toll. Membership in professional societies, and technical-journal subscriptions, all come right off the top of earnings. There is no one to pick up the tab for transportation and other costs incurred in attending local and national association meetings, yet such participation is vital if one is to keep abreast of what is happening in his field.

One factor often overlooked is that the average engineer or scientist has not been trained to be a businessman. He lacks the mind of a shopkeeper, whose business acumen is almost intuitive. Yet, in many ways, the engineer in private practice must think like a shopkeeper in order to make ends meet.

To these problems are added difficulties in dealing with clients. The independent engineer is an outsider and, as such, must expect to be carefully scrutinized by any research director or other member of a company's management.

The consultant may find himself caught up in an interdepartmental political struggle; he may embarrass some individual unwittingly by uncovering an error in approach or by coming up with an "instant" solution where others have failed. He may soon become an unwelcome visitor; he may be a thorn in the side of the inside research group. Also, there are complex problems of trade secrecy and intellectual property, which make it unwise in many instances for a company to even listen to the ideas of an outsider. This last is perhaps the greatest barrier the creative consultant will face during his career.

Services rendered

How do you go about generating an income on the basis of your technical education and experience? Here is an annotated list of various activities that fall under the broad definition of consulting:

1. *Consultation*—This may be a per-hour or per-diem arrangement. It might involve the simple discussion of a problem or the evaluation of a design or a proposal. It could concern a technical matter, a management problem, an economic study, etc. The point is that it represents a limited engagement with a client, whether in person or via telephone or letter.

2. *Retainer arrangement*—This represents a longer association—for six months, a year, or more—and could involve continuing consultation in any one of a number of areas: research, production, quality control and marketing, to name a few.

3. *Compliance*—The operation of certain industrial facilities—such as food, drug and beverage manufacturing—is closely regulated by state and federal agencies. Some consultants specialize in providing comprehensive, up-to-the-minute knowledge of these regulations. They offer continuing inspection and consulting service to ensure that their clients stay in compliance. New environmental restrictions have recently expanded opportunities for this type of consulting service.

4. *Expert testimony*—This may require the consultant to appear in court as an expert witness or to serve as an affiant in a legal or patent matter.

5. *Single-client study*—Often, the need develops for a technical, economic or marketing study. Management

usually favors the independent consultant for this type of survey because of his anonymity. Occasionally, rights to publication of the study may revert to the consultant after some time has elapsed, thereby permitting him to offer it for general sale.

6. *Mulpiple-client survey*—This type of service forms a major part of the projects undertaken by many leading industrial-research and -consulting firms. Subscription contracts are signed before the survey is conducted so that payment is assured. Such surveys usually involve a new product or technology for which published market data are lacking.

7. *Technical surveys or reports*—A study offered for general sale commands a lower price than a multi-client survey. As a finished report, it does not offer subscribers the opportunity to modify the scope or emphasis of the field work. However, such a publication is usually sold in greater numbers, so that the money earned is comparable to that from a multi-client study.

8. *Free-lance writing*—This includes articles, regular columns in newsletters or trade publications, and textbook preparation. A word of caution here: writing is tough work and time-consuming. Any journal article you might publish will take at least two weeks of work, for which you are likely to receive one day's pay (at your usual rate). Also, publication deadlines impose an enormous strain and can cut deeply into your consulting schedule. However, publishing is important because there is no better way of putting your ideas, yourself and your firm before thousands of prospective clients.

9. *Seminars*—Plant seminars can be quite lucrative, especially if two or more consultants can assemble a one- or two-day program covering a topic of major interest. The limited-attendance, executive seminar takes a more established organization, a secluded meeting facility, and a major investment in direct-mail advertising.

10. *Research contracts*—Many consultants follow the

Commerce Business Daily and bid on research contracts in their particular field. It's not difficult to bring together a few associates who would be interested in being part of your proposal. If you're fortunate enough to be awarded a contract, you may have the beginnings of a new company. Alternately, you may have an idea for a product that can be demonstrated to a prospective manufacturer; if you can prove feasibility and market potential, it could be the basis for a long-term development contract.

11. *Research grants*—A number of private and public agencies will consider your application for a research grant. Completing the forms and preparing the detailed proposal may be an arduous task, and the gestation period could seem like forever, but many independent consultants depend entirely upon this type of funding.

12. *Representation*—There are a number of opportunities in this category, which includes manufacturer's representatives, technical sales representatives and, possibly, commission agents. Perhaps this activity is too close to sales to be rightfully included in the realm of technical consulting, yet it seems to meet the definition given earlier. Income may be from commissions, retainer fees, or a per-diem arrangement. With the right product, such activity can provide a sound and continuing income.

13. *Finder's fees*—This is a highly speculative area, usually associated with acquisitions, business contracts, sourcing rare materials, or finding used equipment. In many cases, it will prove to be a waste of time, unless you're wise enough to get a retainer fee. Even so, the legal pitfalls can be treacherous and the out-of-pocket expenses formidable.

Some consultants concentrate on just one or two of these different activities. In this way, they are able to hone their approach to its sharpest economic advantage. In my experience, these people represent the skilled professionals; their fields may be narrow but within these realms they are extremely effective.

In consulting, success breeds success; recognition is swift and publicity is long-lasting. A job well done will open up other opportunities and put you in touch with many prospective clients.

What it takes

"The primary skill you need in consulting," says Ben S. Cole, president of a Boston consulting firm, "is the ability to listen and a reputation as a person who can be trusted" [*4*].

"The successful consultant likes to solve problems, especially new ones. And he develops the ability to solve problems without needing to become deeply enmeshed in the countless details particular to a business." So says professional engineer W. H. Weiss [*5*].

Photographic consultant Howard W. Hoadley believes writing is important. "If it is done assiduously, it will force you to keep abreast of your field," he claims. "It will help you to write better reports for your clients, and it will discipline you in other forms of communication."

Wilbur C. Myers, one of the foremost consultants in micrographics and microimaging technology, has some further thoughts about writing. "Having a regular column is an excellent way of getting known in the industry," he says. Although it entails giving away information—your stock-in-trade—it does guarantee your receiving all pertinent press releases and announcements in advance of their publication elsewhere. This gives you can important advantage in your consulting activities.

If I had to compile a list of operating instructions based upon my own experience, it would include these items:

- Maintain a high standard of ethics and integrity.
- Know exactly what you are selling and how to avoid giving it away.
- Know how to trim your expenses, especially when times get tough.
- Scrupulously avoid taking an equity position with your clients.
- Maintain as many contacts throughout the industry as possible, with friends, former associates, and clients.
- Learn to become a good business man.
- Be willing to turn down assignments that either lie outside your area of expertise, or are unprofitable, regardless of their appeal.
- Be discreet about any personnel placement opportunities—be willing to help out a friend or a client as a favor, but not to engage in recruiting for profit.
- Communicate as high up in a client's organization as possible to avoid spinning your wheels.

As you take on assignments, you will recognize your strengths and weaknesses. In many ways, consulting is a continuing aptitude test during which your bookkeeper tells you where your high scores are being made. If you are smart enough to listen and to concentrate on those high-performance areas, the chances are you will soon become well established in private practice.

References

1. Professional Practices in Management Consulting, Assn. of Consulting Management Engineers, Inc., New York, New York, 1959, pp. 4-5.
2. Consultants Clash Over Ownership, *Business Week*, Nov. 27, 1971, pp. 66-69.
3. Consulting Profession Attracts More Chemists, *Chem. & Eng. News*, Vol. 50, July 3, 1972, pp. 8-9.
4. Higdon, H., "The Business Healers," Random House, New York, 1969, p. 75.
5. Weiss, W. H. Should You Employ Consultants? *Chem. Eng.*, April 17, 1972, pp. 114-118.

Research in Engineering Education: An Overview

William K. LeBold
Purdue University

Research in engineering education has often centered around national studies conducted in response to demands for examining engineering education in terms of contemporary practices and anticipated changes and social demands.* Among the major investigations were the Mann Study of 1908, which emphasized the need for balance between general and practical technical education.[3] The Wickenden investigation, 1928-32,[4] studied institutions, students, faculty, curricula and industry needs, and stimulated the organization of the Engineers' Council for Professional Development (ECPD). The Hammond Reports of 1940[5] and 1944[6] attempted to define the objectives and content of engineering education with emphasis on both the "technical-scientific" and the "humanistic-social" stems. The Grinter Report stimulated the development of stronger mathematics, science and engineering science programs in engineering, and proposed a bifurcated engineering program that would stress both professional-scientific and professional-practical programs.[7] In spite of initial rejection by major engineering colleges and universities, this bifurcation has become a reality today in the form of engineering science-oriented and engineering technology-oriented programs.

The Goals of Engineering Education study (1968),[8] in its controversial preliminary report, strongly supported graduate programs to at least the M.S. degree. In spite of strong opposition, especially from engineers in industry and government, it gave impetus to the growth and development of master's programs in research development and design. The Engineering Technology Education Study[9] provided further impetus for the parallel development of engineering technology programs, which now offer a broad span of degrees.

The Goals study also focused on the increasing responsibilities of engineers and on the impact of technology on society. The Olmsted study, Liberal Learning for the Engineer,[10] called on engineering educators to consider communications, the humanities and the social sciences not just as part of the liberal or general education of engineers, but as an integral part of their "professional" education. Interest in interdisciplinary studies is growing, especially in those related to such problems as energy, environment, transportation, communication and productivity.

The Goals and the Liberal Learning studies relied heavily on a variety of educational research methods. The Goals study collected extensive information from engineers in industry and government about employment, career development, education and training, and attitudes towards working conditions, professional development and engineering education.[11] Self-study reports from institutions, professional societies and industries were content-analyzed, and manpower needs were estimated using inputs from key personnel in industry and government.[12] The Olmsted Report did special analyses of existing studies and sent field representatives to a carefully selected cross-section of U.S. engineering institutions. Almost all the studies mentioned have involved the collection and analysis of data on students, faculty, curricula and the supply and demand for engineers. These data constitute some of the most important research and historical information available on U.S. engineering education.

*For a look at issues of importance in the history of engineering education in this country and of engineering in western civilization, see references 1 & 2.

Focus of Current Research

Recent engineering education research focuses on four major areas: students, faculty, curricula and instruction, and systems. There are parallel national and institutional efforts to collect, analyze and synthesize information in each area.

Reprinted with permission from *Engineering Education,* vol. 70, no. 5, pp. 406–409, 422, February 1980.

Students

At the student level, there is a continuing effort to assess the characteristics of incoming engineering freshmen, to assess their abilities and academic performance and the factors influencing their retention or attrition.[13] There are also significant efforts to assess the types of employment, salaries, and graduate and continuing education of B.S. and recent engineering graduates. Since the profession of engineering in the U.S. has been dominated by white males, attention has focused recently on attracting and retaining more women[14] and minorities.[15]

Data on enrollment and degree trends indicate that between 1973 and 1978, there was a five-fold increase in the number of women freshmen, undergraduates and B.S. recipients. More modest increases have been reported for American Indians, Hispanics and foreign nationals.[16]

Connally and Porter found that engineering institutions that admitted a high proportion of women into their undergraduate programs were larger and more selective institutions, and were more likely to have personalized, comprehensive support programs.[17] There is evidence that men and women engineering students are more alike than different,[18,19] but some of the differences require sensitivity to women's needs, such as having dual careers and dealing with social pressures.[20,21]

The National Academy of Engineering, the Sloan Foundation and NSF have launched major efforts at recruiting and retaining minorities in engineering. Conferences at MIT[22] and IIT[23] and a National Research Council Task Force[24] recently reported on a wide range of research on minority retention. Davis and her colleagues have examined factors associated with retaining men and women in engineering and indicated that academic success, interest and commitment to an engineering career were the most important factors.[25]

Characteristics of engineering students and graduates have been widely studied, including research on cognitive and non-cognitive variables.[26] McCampbell and DeLauretis have conducted comprehensive surveys, stressing interest measures.[27,28]

Faculty

At the faculty level, research has focused on providing opportunities and resources for continued growth and development,[29] especially of younger faculty, in teaching, research and public service. Although research and publications still dominate promotion and tenure decisions, engineering institutions have been giving more attention and recognition to quality and innovations in engineering instruction. The role of faculty members in defining and solving national problems has grown, especially in energy, transportation, environment and communications. These efforts have resulted in increasing cooperation and interdisciplinary work.[30]

Among the outcomes of these interdisciplinary efforts are the identification of effective intervention strategies, including problem identification, method analysis and the importance of breaking down artificial barriers between disciplines and of encouraging close student and faculty interaction.[31]

The increasing complexity of technical problems and their impact on society has resulted in rapid increases and proliferation of interdisciplinary courses and research efforts. At the undergraduate level, society problems have been emphasized,[32] usually with team-taught courses using faculty from engineering management, social sciences, humanities and the fine arts. Evaluations have generally been positive;[33] problems cited usually concern organizational process and integration.[34] At the graduate level, basic and applied interdisciplinary research has involved engineers, physicists, chemists and mathematics.[35]

Curricula and Instruction

Research on engineering curricula has reflected the rapid changes taking place in science and technology and increased demand for using systems engineering in solving engineering problems. The computer has revolutionized engineering course content and instructional methods. As a result, IEEE and the Association of Computing Machinery have cooperated in several studies on course and curriculum content[36-38] that have focused on the importance of elementary, advanced, and assembling languages and software engineering education.[39]

The computer's impact on instructional methods[40] is evident in CAI (Computer Aided Instruction),[41] CMI (Computer Managed Instruction) and Computer Aided Design. Particular instructional methods have been studied, such as aptitude treatment interaction[42] and the Keller Plan.[43] Student evaluation of instruction and testing[44] and grading methods[45] have also been researched.

Systems

System research studies are being conducted nationally through the collection of data on engineering enrollment, degrees granted, employment and professional activities. A wide range of studies are being conducted on attitudes and personalities of students, faculty, and practicing engineers. At the institutional level, increased resources and capabilities are being devoted to examining the changing characteristics of incoming students, their academic progress (including grades), retention and follow-up studies. The latter focus on the graduates' employment, education, professional growth and personal development. These data provide a basis for the development of policies and practices that will improve and enhance the quality of engineering education.

For example, in a comparative study of UCLA and Berkeley graduates, Case and others found few differences among post-college activities of B.S. graduates whether they had graduated from a unified or traditional engineering program.[47] Perrucci and LeBold also found only a few differences in the post-college activities of B.S., M.S. and Ph.D. graduates in engineering and science, with a degree level being the most significant factor cited.[48] Pace studied eight types of higher education institutions and found that graduates of colleges of engineering and science ranked near the top in vocational training and science-related activities and in critical thinking, knowledge and independence. In personal and social development, civic and political affairs and in benefits and outcome related to liberal education, however, they ranked near the bottom.[49]

To provide an institutional focus on the role of research in engineering education, current and recent studies at Purdue University (see below) are illustrative.

Conclusion

Current problems and issues in engineering education include the role of interdisciplinary studies, the supply and demand for engineering personnel, engineering vs. technology programs, and faculty obsolescence.

Engineering educators everywhere can use educational research to plan and build for change. Every institution, department and faculty member has an opportunity to conduct some "shirt sleeves" research. Retain what is good from the past, but be responsive to the present; be innovative, but evaluate and re-evaluate; use evaluative feedback to keep what is effective and discard what is not.

To provide information on areas of concern and to evaluate the effectiveness of traditional and innovative teaching methods is in the best tradition of both engineering and education.

References

1. Finch, J.K., *Engineering and Western Civilization,* McGraw-Hill, 1951.

2. Grayson, L.P., "A Brief History of Engineering Education in the United States," *Engineering Education* (Annals), vol. 68, no. 3, Dec. 1977, pp. 246-264.

3. Mann, C.R., *A Study of Engineering Education,* Bull. no. 11, Carnegie Foundation for the Advancement of Teaching, 1918.

4. Wickenden, W.E. (chairman), *Report of the Investigation of Engineering Education,* Society for the Promotion of Engineering Education, 1930.

5. Hammond, H.P. (chairman), "Report of the Committee on Aims and Goals of Engineering Curricula," *Journal of Engineering Education,* vol. 31, 1940, pp. 279-357.

6. Hammond, H.P., *et al.,* "Engineering Education, After the War," *Journal of Engineering Education,* vol. 14, 1944, pp. 589-614.

7. Grinter, L.E. (chairman), "Report on Evaluation of Engineering Education (1952-55)," *Journal of Engineering Education,* vol. 46, 1955, pp. 25-63.

8. Walker, E.A. (chairman), "Final Report: Goals of Engineering Education," *Engineering Education,* vol. 58, no. 5, Jan. 1968.

9. Defore, J.J. and L.E. Grinter, "Engineering Technology Education Study," *Engineering Education,* vol. 63, no. 4, Jan. 1972.

10. Olmsted, S.P. (director, Humanistic/Social Research Project), "Liberal Learning for the Engineer," *Engineering Education,* vol. 59, no. 4, 1968, pp. 303-342.

11. Perrucci, R., W.K. LeBold and W.E. Howland, "The Engineer in In-

Research on Engineering Education at Purdue

Purdue's Engineering Education Research Studies, part of the Freshmen Engineering Department, has a small staff that tries to keep a balance between projects funded from internal and outside sources. During the past few years, the group has presented many papers that reflect a wide spectrum of study and that are consistent with our primary purpose, namely, to provide information to evaluate and improve the quality of engineering education—primarily at Purdue, but nationally and internationally as well.

• As part of the national effort to increase the number of women and minorities in engineering, Purdue developed a program that enables eighth graders to try engineering for a week. Sponsored by a Sloan Foundation grant, the program brings about 100 young women and men on campus during the summer, and gives them a chance to learn about the opportunities for them in engineering and the importance of pursuing a strong mathematics and science program in high school. The students are surveyed before and immediately following the workshop, and once a year thereafter. Our results indicate that students learn and tend to retain their knowledge of how mathematics and science relate to engineering, and they tend to plan and actually take more mathematics and science courses and encourage other students to do so as well.[49]

• Another recent project focused on providing more equitable educational opportunities for women entering engineering. With a grant from the Women's Educational Equity Act, an experimental course was developed to help beginning students understand how engineering deals with problems in the environment, energy and transportation. Experimental and control groups of men and women were pre- and post-tested and also surveyed before and after the course. The results were generally very positive; significantly higher gains in knowledge and confidence were observed among those men and women who participated compared with the control groups. The results are being disseminated regionally, nationally and internationally.[50]

• Other recent studies of the Education Research Studies group have been concerned with the development of the Purdue Interest Questionnaire (PIQ)[51] to assist students in selecting engineering specialities; evaluation of a counselor tutorial program to help students who begin engineering with deficiencies in mathematics and science;[52] the optimal placement of engineering freshman in beginning mathematics, chemistry and physics courses;[53] declining college admission scores and grade inflation and their relationship to engineering attrition, retention and graduation rates.[54]

Under a grant from NSF's Comprehensive Aid to Undergraduate Education (CAUSE), Purdue has developed the Center for Instructional Development in engineering.[55] The Center provides staff and facilities to assist faculty members interested in developing and evaluating improved and innovative teaching-learning procedures, including the use of transparencies, slides, audio-video tapes, and computers. The Center has sponsored faculty development seminars and workshops on teaching large classes, guided design and simulation. A small evaluation group has just completed two surveys, one an appraisal by graduating engineering seniors of the nature and quality of their undergraduate experiences, the other a survey of engineering faculty to determine the types of facilities, services and programs they would like to see developed by the Center, and an appraisal of current practices and degree of satisfaction with Purdue's programs and policies.

dustry and Government," in Goals of Engineering Education information document no. 7, ASEE, Jan. 1968.

12. LeBold, W.K., *et al.*, "Reactions to the Preliminary Report of the Goals Study," *Journal of Engineering Education*, vol. 57, no. 6, Feb. 1967, pp. 434-437.

13. Gatewood, R.D., "An Investigation of the General Achievement of Engineering, Social Science and Humanities Seniors at Purdue University," master's thesis, Purdue, 1967.

14. Frohreich, D.S., "How Colleges Try to Attract More Women Students," *IEEE Transactions on Education*, vol. E-18, no. 1, Feb. 1975.

15. Padulo, L. (chairman), *Minorities in Engineering—A Blueprint for Action*, Alfred P. Sloan Foundation, N.Y., 1974.

16. Doigan, Paul, "Engineering Manpower Planning Information," *Proceedings, Frontiers in Education Conference*, Niagara Falls, Canada, ASEE/IEEE, 1979, pp. 374-381.

17. Porter, A.L. and Terry Connolly, "Women in Engineering: Policy Recommendations for Recruitment and Retention," Georgia Institute of Technology, 1977; and "The Process of Entry of Women into Engineering: Two Hypotheses," *Proceedings, Frontiers in Education Conference*, Niagara Falls, ASEE/IEEE, 1979, pp. 367-373.

18. Ott, Mary O., "Female Engineering Students: Attitudes, Characteristics, Responses to Engineering Education," Cornell University, College of Engineering, June 1978.

19. Durchholtz, P., "The Hidden Career: How Students Choose Engineering," *Engineering Education*, vol. 69, no. 7, April 1979, pp. 718-722; see also, "Women in a Man's World: The Female Engineers," *Engineering Education*, vol. 67, no. 4, Jan. 1977, pp. 292-99.

20. Ott, Mary, "Experiences, Aspirations and Attitudes of Male and Female Freshmen," *Engineering Education*, vol. 68, no. 4, Jan. 1978, pp. 326-333, 338.

21. Davis, S., "Beyond Tokenism," *New Engineer*, vol. 7, 1978, pp. 22-29.

22. *Proceedings, Workshop on Retention of Minority Undergraduate Students in Engineering*, MIT and the Committee on Minorities in Engineering, 1977.

23. *Proceedings, Midwest Conference on Minorities in Engineering*, Illinois Institute of Technology, May 31-June 1, 1979.

24. Committee on Minorities in Engineering, *Retention of Minority Students in Engineering*, a report of the Retention Task Force, NAS, Wash., D.C. 1977.

25. LeBold, W.K., "Engineering Education," *Encyclopedia of Educational Research*, 4th ed., R.L. Eble (ed.), Macmillan Co., 1969, pp. 435-443.

26. McCampbell, M.K., "Differentiation of Engineers' Interests," doctoral dissertation, U. of Kansas, 1966; (*Dissertation Abstracts*, 1966, vol. 27, 1662A-1663A, and University Microfilms no. 66-13051).

27. Foecke, H.A., "Engineering Faculty Development and Utilization," *Journal of Engineering Education*, vol. 50, no. 9, May 1960, pp. 757-828. See also: Goodwin, W.M. and W.K. LeBold, "Interdisciplinarity and Team Teaching," *Engineering Education* (Annals), vol. 66, no. 3, Dec. 1975, pp. 247-254.

28. Walsh, W. Bruce, G.L. Smith and M. London, "Developing an Interface Between Engineering and the Social Sciences," *American Psychologist*, vol. 11, 1975, pp. 1067-1071; and *ERM Magazine*, ASEE, vol. 8, no. 3, Spring 1976, pp. 58-62.

29. Demetry, J.S., J.M. Boyd and T.H. Keil, "Societal Interactions in Engineering Education: Five Years of Experience at WPI," *Proceedings, Frontiers in Education Conference*, Lake Buena Vista, Fla., 1978, pp. 17-23.

30. Wales, Charles, "There Is a 'How to Teach' Frontier," *Proceedings, FIE*, Lake Buena Vista, Fla., 1978, pp. 2-5.

31. Ratledge, E.T., "The Technological-Cultural Interface," *Proceedings, FIE*, Lake Buena Vista, Fla., 1978, pp. 33-37.

32. Ballantyne, J.M., "Impact of the National Research & Resource Facility for Submicron Structures on Education Graduate Research," *Proceedings, FIE*, Lake Buena Vista, Fla., 1978, pp. 6-13.

33. Curriculum Committee on Computer Science (C^3S), "Curriculum '68, Recommendations for Academic Programs in Computer Science," *Communications of the ACM*, vol. 11, no. 3, March 1968, pp. 151-197.

34. Austing, R.H., *et al.*, "Curriculum Recommendations for the Undergraduate Program in Computer Science: A Working Report on the ACM Committee on Curriculum in Computer Science," *SIGCSE Bulletin 9*, no. 2, June 1977.

35. Engel, G.L. and O.N. Garcia, "Curricula Development in Computer Science and Engineering," *IEEE Transactions on Education*, vol. E-21, no. 4, Nov. 1978.

36. Doelling, Arvin, "Computer Programming Languages and Expertise Needed by Practicing Engineers," 87th Annual Conference, ASEE, Louisiana State U., June 1979.

37. Levien, R.E., *The Emerging Technology: Instructional Use of the Computer in Higher Education*, a Carnegie Commission on Higher Education and Rand Study, McGraw-Hill, 1972.

38. Sugarman, R., "2nd Chance for Computer-Aided-Instruction," *IEEE Spectrum*, vol. 15, no. 8, 1978, pp. 29-37.

39. Kaplan, M.N., "An Aptitude-Treatment Interaction Study of Student Choice and Completion in a PSI Course," *Engineering Education* (Annals), vol. 69, no. 3, Dec. 1978, pp. 273-284.

40. Hereford, S.M., "The Keller Plan Within a Conventional Academic Environment: An Empirical 'Meta-Analytic' Study," *Engineering Education* (Annals), vol. 70, no. 3, Dec. 1979, pp. 250-260.

41. Boyle, T.A. and G.L. Wright, "Computer-Assisted Evaluation of Student Achievement," *Engineering Education* (Annals), vol. 68, no. 3, Dec. 1977, pp. 241-245.

42. Work, C.E., "A Nationwide Study of the Variability of Test Scoring by Different Instructors," *Engineering Education* (Annals), vol. 67, no. 3, Dec. 1976, pp. 241-248.

43. *Engineering Manpower Bulletin*, no. 39, Engineering Manpower Commission of EJC, Dec. 1977.

44. Harrington, T.F., "The Interrelation of Personality Variables and College Experiences of Engineering Students Over a Four-Year Span," doctoral dissertation, Purdue, 1965.

45. Case, H.W., W.K. LeBold and W.D. Diemer, "A Comparative Study of University of California Engineering Graduates from Berkeley and Los Angeles, a report of the UCLA Engineering Educational Development Program, EDP report no. 4-67, Sept. 1967.

46. Perrucci, C.C. and W.K. LeBold, "The Engineer and Scientist: Student, Professional, Citizen," *Engineering Bulletin of Purdue University*, Engineering Extension Series, Jan. 1967.

47. Pace, R.C., *The Demise of Diversity?* (A Comparative Profile of Eight Types of Institutions), the Carnegie Commission on Higher Education, 1974, pp. 114-115.

48. Hancock, J.C., "The REETS Recommendations: A Progress Report," *Engineering Education*, vol. 70, no. 2, Nov. 1979, pp. 163-168.

49. Blalock, M.W. *et al.*, "Engineering for Minority 8th Graders," *Proceedings, Frontiers in Education Conference*, ASEE/IEEE, Buena Vista, Fla., 1978, pp. 343-353.

50. Butler, B.R., *et al.*, "An Action Research Program to Provide Educational Equity Opportunities for Women in Engineering," *Proceedings, FIE,* Champaign, Ill., 1977, pp. 131-139.

51. Shell, K.D. and W.K. LeBold, "A Guidance Tool for Engineering Students: The Purdue Interest Questionnaire," *Engineering Education* (Annals), vol. 69, no. 3, Dec. 1978, pp. 243-249.

52. Deputy, G.O., "The Counselor-Tutorial Program: A Second Chance," *Engineering Education,* vol. 68, no. 4, January 1978, pp. 313-317.

53. Hannan, R.J. *et al.*, "Discrimination Analysis for the Placement of Freshman Engineering Students in Physics, Mathematics and Chemistry Courses," *Proceedings, Frontiers in Education Conference,* Buena Vista, Fla., 1978, pp. 354-366.

54. LeBold, W.K. *et al.*, "Grade Inflation: Trends and Relationships in Grades, Admission Scores and Engineering Retention," *Proceedings, FIE,* Buena Vista, Fla., 1978, pp. 419-426.

55. Lindenlaub, J.C. and F.S. Oreovicz, "Activities of the Center for Instructional Development in Engineering," Purdue University, January 1978.

Part X
The Teaching Profession and Career Development

Scope and Purpose

WHILE a discussion of the career development of a teacher may seem out of place in a volume on teaching and learning, it is the belief of the editor that the two issues cannot be separated, particularly for beginning teachers. Career related issues are often uppermost in the minds of ambitious, hard-driven professionals and if one ignores what is on the teacher's mind, one cannot expect to make a real impact on his teaching. Those who organize teaching workshops, or run teaching clinics, often find that a teacher who is seriously concerned about his career is not amenable to, or is even cynical of, efforts aimed at improving teaching; is too preoccupied mentally to benefit from such programs; and may even undermine the value of such programs for other participants. An open and early discussion of career related issues is desirable, if for no other reason than because it brings the real concern of the teachers out in the open.

The Issues

A number of issues may be of interest to the beginning teacher:

• *The Nature of Academia.* It might appear that an introduction to the university is redundant, since all those who begin college teaching have themselves spent a number of years as a student in a similar environment. The perception of that environment from the viewpoint of a teacher is sufficiently different to make the discussion useful. Every university or college has its own organization, administrative structure, policies and procedures, and operating style. But there are some underlying principles and philosophies which are nearly universal. These principles may not always be explicitly stated, and their status may vary at different institutions, from the gentlemen's agreement or tradition to legally enforceable statutes. Nevertheless, a new teacher might find an understanding of these useful in some circumstances (e.g., if involved in a grievance proceeding). These principles relate to academic freedom, university governance, code of behavior, tenure, sabbatical leave, etc.

• *The Teaching Profession.* This includes such issues as what is academic life like, what are the joys and frustrations of being an educator, what are the responsibilities of an educator, what should an aspiring teacher be prepared for, what might a teacher reasonably expect from his work environment, what are the economic and political realities of the profession, what keeps some teachers going, and what makes others quit the profession. Although the answers to some of these questions can only be discussed somewhat subjectively, and depend on the individual teacher's circumstances and institution, there are enough commonalities in the experiences of most faculty to make this discussion relevant.

• *Performance Evaluation.* It is an academic cliche that the performance of the faculty is measured in terms of three areas—teaching, research, and service, which are the declared goals of a university. Beyond such generalities, the description of faculty duties becomes elusive. Even definitions of these three terms, and the relative emphasis on them in reward determination, will vary from institution to institution. It is often difficult for a new teacher to distinguish between the verbalized and the real value attached to these performance factors, because the lip service, euphemisms, and personal opinions of individuals can be misleading. Perhaps the most hotly debated issue in faculty evaluation is the teaching vs. research controversy. There is also concern about the proper recognition for service activities without which an educational institution could not survive.

• *The Reward Structure.* The reward system in academia is sufficiently different from that of most other professions and industry that an individual who has picked up his value system form the society at large may find surprises. In short, a new teacher needs to understand two aspects of the academic reward system. First, the rewards in academia (retention, tenure, promotion, salary raises, support of professional needs like research equipment, conference travel, and technician time, and even such subtle things as physical facilities, and the size and location of one's office and laboratory space) are a little more complex, because they cannot be measured by a unidimensional variable like salary. Second, the measures on which these rewards are based are also more complex than in industry, because they are multidimensional, and cannot be measured by a few numbers like project size or number of team members supervised.

• *Economic Matters.* Faculty salaries have long been an actively discussed topic in the literature. In particular, the apparent independence of the academic salaries from the laws of supply and demand and the "market rate for professionals," the substantial difference between academic salaries and net earnings of the faculty due to their other activities, and the proper accounting of the faculty work load with respect to which the salaries must be normalized, have all attracted debate in the literature. The recent trend toward collective bargaining in academic settings has introduced a new subject in this debate.

Only some of these issues are brought up in the papers reprinted here. Others may be found in the literature cited.

The Reprinted Papers

The following five papers are reprinted in this part.

1. "Perspectives on a Career in College Teaching," L. W.

Potts, *Analytical Chemistry,* vol. 53, no. 14, pp. 1603A–1608A, Dec. 1981.

2. "The World of an Educator," J. B. Angell, J. D. Meindl, and H. H. Skilling, *IEEE Student Journal,* vol. 6, no. 6, pp. 37–39, Nov. 1968.

3. "The Professor as Researcher," N. L. Gonzalez, *National Forum (The Phi Kappa Phi Journal),* vol. 67, no. 1, pp. 7–10, Winter 1987.

4. "Service: The Neglected Person of the Academic Trinity," W. V. Hohenstein, *National Forum (The Phi Kappa Phi Journal),* vol. 60, no. 2, pp. 18–19, Spring 1980.

5. "Promotion and Tenure Policies in U. S. Engineering Schools," E. E. Cook, C. Dodd, and S. Sami, *Engineering Education,* vol. 69, no. 7, pp. 750–751, Apr. 1979.

The first paper reprinted in Part X, by Potts, discusses the liabilities and assets of a teaching career. Although the author is an analytical chemist, and appears to be addressing the potential teachers of analytical chemistry, his comments apply equally well to engineering, due to the similarities of the two fields in some important regards (such as the alternative of industrial employment).

The second reprinted paper, by three electrical engineering educators, brings out the significant responsibilities of a teacher toward his students, his institution, and himself. These two papers together will help the readers who are asking themselves whether teaching is the right career for them.

The next paper on the research-teaching controversy attempts to examine the rationale, role, and utility of research on campus, and the desirability, justification, implications, and consequences of the emphasis on research in the academic reward structure.

Interestingly enough, in academic jargon, the phrase "service activity of the faculty" does not imply teaching, which is presumably the principal service to the university. Instead, it includes a number of other things, for instance, committee work, administrative work, faculty statesmanship, and public service. Hohenstein's short paper on the service activities of faculty points out that service to the university should be viewed as "institutional maintenance," without which the university, as we know it, cannot exist. This article is also candid about the value of service activities in one's career development, especially during the early years.

The last paper by Cook, Dodd, and Sami reports the results of a survey of the promotion and tenure policies in the U. S. engineering schools. Based on the data from 37 universities, the results are representative only of the norm and may not apply to the reader's own institution. However, there is a fair amount of uniformity in the criteria for tenure throughout the U. S., with the difference being only in the relative emphasis among various factors.

The Reading Resources

A. The Nature of Academia

[1] Perhaps the best method of familiarizing oneself with the nature and current concerns of academia is by browsing through the recent issues of some of the following periodicals:

(a) Academe. Published by the American Association of University Professors. AAUP champions academic freedom, and many articles deal with issues related to this cause, but there are also articles on subjects of current interest or controversy.

(b) Change. Subtitled "A Magazine of Higher Education," it carries essays on academic life and social issues on campus.

(c) The Chronicle of Higher Education. A weekly newspaper with emphasis on administrative issues, and position open advertisements for largely administrative vacancies.

(d) Engineering Education News. A monthly newspaper, published by the American Association for Engineering Education, with news and commentary on issues of current importance in engineering education, along with mostly senior administrative position open advertisements.

[2] D. R. Goddard and L. C. Koons, "Intellectual Freedom and the University," *Science,* vol. 173, no. 3997, pp. 607–610, Aug. 1971. Presents one view of the meaning of intellectual freedom, and the threats to it from within academia and outside.

[3] R. Hofstadter and W. P. Metzger, *The Development of Academic Freedom in the United States.* Columbia University Press, New York, 1955. A scholarly study of the history and basic tenets of the academic freedom tradition in the U. S. universities.

[4] G. Highet, *The Immortal Profession: The Joys of Learning and Teaching.* Weybright and Talley, 1976. An adaptation appears as G. Highet, "How To Teach College Students," *Today's Education,* vol. 67, no. 1, pp. 39–42, Feb.-Mar. 1978. Observations on the nature and demands of the profession, the duties of a college teacher, and some advice to new teachers on a miscellany of topics.

[5] J. Newman, "Academic Freedom and the Power of the Guild," *Improving College and University Teaching,* vol. 30, no. 1, pp. 8–11, Winter 1982. Points out that academic freedom also leads to the concentration of power in the hands of the faculty collectively, who determines what fields may be explored and doctrines embraced through the process of faculty selection and retention.

[6] T. Walden, "Tenure: A Review of the Issues," *Educational Forum,* vol. 44, no. 3, pp. 363–372, Mar. 1980. A very brief summary of the history of tenure; arguments in favor and against tenure; and quotations from major writings on the subject.

[7] *Academe* (Amer. Assoc. Univ. Profs.), vol. 68, no. 1, Jan.-Feb. 1982. A selection of five opinions on faculty participation in directing the institutions of higher learning, with different perspectives (university president, trustee, researcher, etc.), some of them based on actual experience at various institutions.

[8] L. G. Geiger, "Academic Codes and Governance," *Intellect,* vol. 101, pp. 225–230, Jan. 1973. An introduction to the academic tradition of faculty governance and code of conduct; historical viewpoint; and comments on the AAUP statement.

B. The Teaching Career

[9] E. Eisenberg and A. V. Galanti, "Engineer and Academia: An Analysis of the Changing Relationship," *Engineering Education,* vol. 73, no. 3, pp. 232–234, Dec. 1982. Carries out a literature survey to determine the motivation of engineers in academia, relationship between it and the nature of academic environment, and the importance of non-salary issues.

[10] R. G. Baldwin and R. T. Blackburn, "The Academic Career as a Developmental Process: Implications for Higher Education," *Jour. Higher Education,* vol. 52, no. 6, pp. 598–614, Nov.-Dec. 1981. Based on a survey of liberal arts faculty, the academic career of a professor is divided into five stages (from new assistant professor to professor nearing retirement), and characteristics of each stage are described.

[11] M. S. Gupta, "EE Hunt in Academe," *IEEE Spectrum,* vol. 20, no. 7, p. 16, July 1983. A tongue-in-cheek description of the ways of academia.

[12] D. G. Ullman, "Maintaining a First Rate Faculty: The Base Data," *Engineering Education,* vol. 71, no. 7, pp. 700–708, Apr. 1981. Examines the various factors that affect the supply and demand of engineering faculty, supported by data from the past two decades; recommends some actions to exert deliberate influence on these factors.

[13] R. L. Wolke, "Faculty Development Explained," *Jour. Chemical Education,* vol. 57, no. 12, pp. 838–840, Dec. 1980. An overview of the need and types of programs, typically for older faculty, providing opportunities for instructional development and professional renewal. This issue of the journal has a number of other articles on faculty development.

[14] M. S. Baratz, "Academic Tenure and Its Alternatives," *National Forum (The Phi Kappa Phi Journal),* vol. 60, no. 2, pp. 5–9, Spring 1980. An exceptionally clear, succinct, readable, and candid statement of the rationale behind the academic tenure system, and the implications of a few of its suggested variants.

[15] J. O'Toole, "Tenure: A Conscientious Objection," *Change,* vol. 10, no. 6, pp. 24–31, June-July 1978. Discussion in *Change,* vol. 10, no. 9, pp. 4–5, 44–47, Oct. 1978. States the case against the traditional tenure policy, based on such factors as accountability, adequacy of existing protection against arbitrary personnel actions through court actions, and the reduction of faculty mobility, salary, academic quality, and public confidence in higher education due to tenure.

[16] J. O'Toole, "Academic Tenure," *Center Magazine,* vol. 12, pp. 53–60, Spring 1979. Further arguments against the system of academic tenure.

[17] J. C. Nitzsche, "How to Save Your Own Career," *Change,* vol. 10, no. 2, pp. 40–43, Feb. 1978. A faculty member's advice to other beginning teachers, based on personal experience; the author is not in the engineering field.

C. Performance Evaluation

[18] R. D. Evans, "Teaching Loads in Engineering: A Move Towards Obsolescence," *Engineering Education,* vol. 68, no. 7, pp. 735–738, Apr. 1978. Relates contact hours (or classroom hours) to actual work load in hours; estimates time devoted to other faculty activities; compares faculty work load with those in other professional fields; and warns of the dangers of overload in the form of professional obsolescence.

[19] R. H. Long, Jr., R. C. Jordan, N. Y. Wessell, and H. P. Eichenberger, "Research and Education—Conflict or Harmony," *Mechanical Engineering,* vol. 90, no. 5, pp. 18–21, May 1968. Several authors point out the beneficial aspects of faculty research, and its justifications: an increasing percentage of graduates are doing research work upon employment, and quality education is positively related to creativity.

[20] J. J. Sharp, "Reflections on Teaching, Research, and Publications," *Engineering Education,* vol. 64, no. 2, pp. 141–142, Nov. 1973. Points out that undergraduate teaching benefits from research only if the research is in a related area; makes a strong case for encouraging faculty publications; and demolishes some of the arguments for not publishing.

[21] A. L. Gosman, "The Coupling of Teaching with Research: The Administrator's Role," *Engineering Education,* vol. 63, no. 4, pp. 243–246, Jan. 1973. A statement of some of the benefits of research to the individual faculty, and to the university, and of the responsibilities of the administration in recognizing, facilitating, and clarifying the role of research on campus. Several other articles in this issue touch on the research-teaching interface.

[22] E. Garfield, "How to Use Faculty Citation Analysis for Faculty Evaluations, and When Is It Relevant," Parts 1 and 2, *Current Contents,* vol. 14, no. 44, Oct. 1983, and no. 45, Nov. 1983, pp. 5–14. An examination, by one of the pioneers in citation analysis, of one of the recent methods of quantitatively determining the impact of a faculty member's research and publications.

[23] C. A. Coberly, "Faculty Development—A Department Responsibility," *Engineering Education,* vol. 62, no. 2, pp. 101–103, Nov. 1971. Takes the position that faculty development is a group responsibility at the department level; the responsibility includes establishing goals, assessment of the status quo, careful faculty selection, and establishing the right environment.

[24] R. E. Fadum, "The Role of Dean in Faculty Development," *Engineering Education,* vol. 62, no. 2, pp. 104–106, Nov. 1971. Summarizes a number of administrative mechanisms through which a college dean can help the development of the faculty.

[25] E. Chesson, "The Future Shortage of Faculty: A Crisis in Engineering," *Engineering Education,* vol. 70, no. 7, pp. 731–738, Apr. 1980. Data comparing engineering faculty salaries with other engineers, Ph. D. engineers, and other educators; also brief remarks on current factors reducing the attractiveness of teaching as a career.

[26] D. S. Peters and J. R. Mayfield, "Are There Any Rewards for Teaching," *Improving College and University Teaching,* vol. 30, no. 3, pp. 105–110, Summer 1982. Reports on the results of a survey, not confined to engineering disciplines.

[27] J. V. Jucker and R. E. Lave, Jr., "Engineering Education Lost Among the Steeples: Let's Reward Good Teaching," *Engineering Education,* vol. 61, no. 6, pp. 537–540, Mar. 1961. A scathing attack on the current academic

reward structure which emphasizes research at the expense of quality teaching; presents some good counter-arguments refuting the claimed research-teaching synergy, and makes a plea and some suggestions for alternative reward structure and reform.

D. Economic Matters

[28] J. H. Wilson and R. S. Wilson, "Teaching-Research Controversy," *Education Digest,* vol. 38, no. 2, pp. 56–59, Feb. 1973. A brief statement of reasons for the desirability of research orientation among faculty, and a suggestion for reducing the tension between researchers and teachers.

[29] K. E. Dillon and R. H. Linnell, "How and For What are Professors Paid," *National Forum, (Phi Kappa Phi Journal),* vol. 60, no. 2, pp. 21–23, Spring 1980. Examines the data on the difference between academic salaries and actual income of the faculty members; deduces the relative importance of the various income producing activities of the faculty; and comments on some policy matters.

[30] P. Nevaldine, "Collective Bargaining at a State University," *Engineering Education,* vol. 63, no. 1, pp. 33–36, Oct. 1972. Views on, and experience with, collective bargaining at one university where unionization occurred in the early 1970's, along with some details of the problems and conflicts posed by collective bargaining.

[31] K. R. Kleckner, "If Collective Bargaining Comes to Your Campus...," *Engineering Education,* vol. 64, no. 8, pp. 600–602, May 1974. An account of the events, strains, and readjustments following unionization of faculty, based on actual experience.

[32] J. Weatherford, "Collective Bargaining in Academe: Professional Associations and Bargaining Agents," *Library Jour.,* vol. 100, pp. 99–101, Jan. 1975. Contains bibliographies on the subject.

Lawrence W. Potts[1]
Department of Chemistry
Gustavus Adolphus College
St. Peter, Minn. 56082

Perspectives on a Career in College Teaching

Those of us who teach analytical chemistry in colleges and universities see more clearly than ever a developing crisis. At the present time there is a two-to-one salary differential between an industrial position in analytical chemistry and a position as an assistant professor in most four-year colleges. We are rapidly approaching the point where sharp, aggressive young PhDs in analytical chemistry will not even consider making the financial sacrifice to teach. Furthermore, many who are already teaching are finding it harder to rationalize continuing to do so when they are losing spending power at an increasing rate. The people who know analytical chemistry will soon not be teaching it; industry is indeed beginning to "eat its seed corn," an analogy to acts of desperation in Native American culture drawn recently by D. Allen Bromley (*1*). Who will teach the next generation of analytical chemists?

It is my purpose here to encourage new PhDs to carefully consider college teaching as a career by pointing out some assets and liabilities of the profession, and discussing these within the context of rewards. I also wish to discuss what I believe are serious problems that young college teachers may encounter that relate to the increasingly bleak salary picture.

Table I lists the major categories of rewards that most people seek in their employment. Certainly we all respond to somewhat different rewards, but this list covers what I think are the general categories. I have arranged the list in order of increasing abstractness: Financial security and recognition are much more tangible than is the feeling that one's work will be of lasting value. Traditionally, teaching has attracted people who appreciate the more abstract rewards and have less interest in the concrete rewards. This may come from the fact that most American private colleges and many universities were originally sponsored by churches and religious orders, and teachers were thought of as missionaries. Regardless of the cause, teachers have never been very assertive about asking for large salaries for their efforts. It is also quite clear that the values of most people in American society are skewed toward concrete rewards. A conflict is inevitable. We are encouraging college students to value more abstract rewards long after they have learned to be accomplished materialists. When there is a salary differential of two-to-one, most students will not be very willing to listen.

Table I. Rewards

- Financial security
- Recognition
- Approval of colleagues and friends
- Service
- Feeling that the work is of value

The two categories of rewards that graduate students really ought to hear more about are service and the feeling of the worthiness of the effort. These are values that liberal arts colleges aspire to instill in students. Unfortunately, they are also values that most of us feel quite awkward about discussing. They are the rewards that are important to the clergy and those who work in social services. Although premed students can be counted on to badly overstate it, service is doubtless an important factor in motivating people to study medicine. Thinking that one's work is worthwhile really involves faith. Most teachers count on the worth of their work being shown in the future—their productivity is evident in the minds that they discipline. The results may not be tangible for years.

The relative importance of various rewards will determine the extent to which a person sees aspects of a career in college teaching as assets or liabilities. In Table II I have made a list of what I believe are the important assets and liabilities of college teaching, arranged like an accountant's balance sheet. Rather than offer my own view of the balance, I would like to discuss the items on the sheet in detail.

Table II. A College Teaching Balance Sheet

Liabilities	Assets
Financial insecurity	Academic freedom
Nomadic life of the untenured	Job security (tenure)
Working with young people	Working with young people
	Campus community
	Faculty role in governance

[1] 1981–82 mailing address: Department of Chemistry, Oberlin College, Oberlin, Ohio 44074

Reprinted with permission from *Analytical Chemistry,* vol. 53, no. 14, pp. 1603A–1608A, December 1981.

Academic Freedom

The principal asset of college teaching is *academic freedom*, the freedom to discuss controversial viewpoints in the classroom and direct one's own research without fear of recrimination. The irony of academic freedom, like that of freedom of any kind, is the responsibility carried by those who have it to treat ideas with intellectual honesty and discipline. Academic freedom is often misinterpreted as leisure time. The formal classroom and laboratory contact time required of most faculty members is about 12 to 16 hours per week, 30 weeks per year. Even with an additional 20 hours of preparation time per week, office hours, and committee work, there is time available to pursue one's own interests. But this does not make life any easier. On the contrary, in colleges of quality, tenure and promotion come only to those who are both effective in the classroom and engaged in scholarship.

There is really no counterpart to academic freedom in industry. Rarely will a research chemist be able to work on a project simply because it is interesting. The final arbiter of what is done in the private sector is market pressure, or perhaps worse, a marketing specialist's interpretation of market pressure.

Financial Insecurity

The main liability of college teaching is poor financial reward. Current starting salaries at the assistant professor rank for PhD-holding chemistry teachers are in the range of about \$15 000 (colleges) to about \$19 000 (universities) (*2*). Starting salaries for PhD industrial chemists range from about \$25 000 to about \$32 000. According to the 1981 ACS salary survey, the median salary of all PhD chemists in colleges and universities was \$26 200, while the median salary for their counterparts in industry was \$39 000.

It seems to have been a popular notion at one time that highly experienced teachers command salaries comparable to those paid to industry people, but the facts show that such a situation must be quite rare. In the 1980–81 academic year the *95*th percentile salary for a full professor in liberal arts colleges was \$30 580. The ACS March 1981 survey showed that the *10*th percentile salary for PhD industrial chemists with about 25 years experience (after the BS degree) was \$32 400. This means that at these levels of experience all but 10% of PhD industrial chemists have higher salaries than 95% of PhD academic chemists in liberal arts colleges. A further disturbing note is that PhD chemists in neither area have beaten inflation in the last decade. When 1980–81 median salaries are normalized to 1971 dollars using consumer price index figures, industrial chemists have lost about 12% of their salary, while academic chemists have lost about 17%.

The bleak message in all of these figures is that academic chemists can look forward to a career-long financial pinch if they must live on a single nine-month salary. There are several ways out: marry someone who is employed, teach summer school every summer, or look for work "on the outside." The trouble with the latter two strategies is that they consume valuable research and writing time, and make it almost impossible to show the kind of growth needed to obtain a tenured position or promotion. Of course the best strategy is to find a way to get paid for doing one's own research, as many of the most successful academic people have managed to do.

A fifth, far less desirable alternative, is to become a "temporary associate dean." These two- or three-year rotating positions seem to attract new associate professors who are exploring administrative work and are lured by higher salaries. During these few years of exploration one is almost surely going to lose touch with developments in the field, and the return to teaching is apt to be painful. This kind of professional drifting can severely damage, even ruin, a teaching career.

Another ramification of the financial pinch involves *morale*. In our society a person's value is usually confused with his or her wealth. Someone who has "really made it" has lots of money and its symbols. It is difficult for a teacher to learn how to brush all of this aside and pretend to be above it. The result can be bitterness and cynicism that can poison relationships with students.

Job Security

Job security for tenured faculty is a great asset of the teaching profession. In fact, many argue that job security compensates for the difference in salary between teaching and industry. A tenured teacher has a permanent contract until retirement. Dismissal can occur only after adequate cause has been shown in hearings before faculty committees. Traditionally there have been three causes: moral turpitude, incompetence, or the imminent bankruptcy of the institution. There is really no alternative to tenure if faculty are to be free to contribute to the governance of a college and academic freedom is to exist.

Unfortunately, a largely tenured faculty tends to result in inflexibility in staffing and program development and makes it difficult for a college to adapt to the increasingly volatile student market. Some colleges have responded by establishing tenure quotas and adopting policies limiting the number of tenure track positions available. For example, one or two positions in a six-person chemistry department may be three-year terminal positions, without the possibility of tenure, regardless of the qualifications of the individuals who hold them. There are teachers who have become academic nomads, wandering from one temporary job to another, perpetual instructors or assistant professors, spending years at the bottom of the pay scale. While this situation seems to pose little threat to analytical chemists, whose services are in great demand, it is a very serious problem in many disciplines in the humanities.

Students

The students themselves are at once an asset and a liability. The fictional archetypal teacher, Mr. Chips, observed that the faculty and administration of a school get older, but the students are always young. Youth is the characteristic of students that can make working with them so stimulating one minute and so infuriating the next. Students come out of high school at the age of 18 or 19, still in the emotional upheaval of their teens. Many of them were never challenged by their high school curriculum and are largely unprepared to deal with the pressures of a college education. Even though many are bright and eager, few have had to develop the discipline needed to sit and concentrate in order to really learn something. Few, as college freshmen, are able to carry an argument to its logical conclusion or express themselves clearly. The best of them try to be assertive and independent, but falter, unsure of their abilities. All of a sudden they must work with a *professor*, someone who has been portrayed in television and the movies either as a mixture of brilliance and preoccupation, or as a fool. Many actually resist being taught and fight every effort the teacher makes. It may take a half dozen approaches to a concept before such a student will take hold of it. Even a class of college seniors can be told how to prepare for an exam, and most will ignore the advice. These are all the ingredients of the *challenge* in teaching and are some of the things that make teaching so incredibly interesting.

In fact, most of the important but intangible assets of college teaching involve the students. Watching students transform from somewhat disinterested spectators in a large general chemistry class to enthusiastic, aggressive class participants in their senior courses is a real source of satisfaction. Watching heads nod in agreement and faces light up as ideas are grasped (the "aha!" phenomenon) keeps many teachers going. The news that a former student is doing well in a highly competitive graduate program is one of the greatest sources of pride to a college teacher.

Most of the readers of this JOURNAL are familiar with the assets of working in a college campus community. Simply put, colleges are interesting and exciting places to be. Except during the summer months, there is more happening on the typical college campus than one could ever hope to get involved in: drama, musical performances, seminars and guest lectures, athletic events, and perhaps most important, a rich, often informal, intellectual life.

Governance

An important asset of college teaching, although it is not universally available, is the chance for even the youngest members of a faculty to play an important part in directing the course of a college. For genuine academic freedom to exist, faculty must be involved in matters relating to personnel and curriculum. At least on campuses where faculty are not members of a union, faculty committees recommend candidates for tenure and promotion, hear grievances, and recommend changes in curriculum, majors, and even programs. The realities of the competitive business world make this kind of cooperative governance rare (probably nonexistent) in industry.

Committee involvement can, however, become a trap for young faculty who are too eager to get involved in campus affairs. It can rob energy and time that might be better spent on scholarship. Particularly annoying are the distractions of ad hoc committees, which usually seem to generate only directives for standing committees. In the absence of leadership, monthly faculty meetings can degenerate into forums for the longwinded with righteous causes, and hours can be spent bickering over issues of no greater moment than campus parking. Involvement is a matter of balancing the important functions of faculty governance against the demands of one's career.

Summary

Whether the assets of college teaching are sufficient to outweigh the liabilities depends entirely on an individual's values, and these are very difficult to sort out. If a graduate student sees the balance as favorable, then I would hope that he or she would at least try to find a suitable teaching position, being aware from the beginning of the problems that might arise. Graduate advisors have a tremendous influence on the opinions of their students. Students may hear grumbling about poor salaries and long hours, but may not hear about the more positive aspects of teaching. Advisors should start talking about what they like about their work with as many students as will listen.

The chemical industry must realize its stake in the education of analytical chemists. Industry cannot be faulted at all for trying to hire the best qualified students available. However, it could make a tremendous contribution to the morale of those who choose to teach by stepping into the vacuum being left by the diminishing support of the federal government in science education. With or without federal help, colleges still need to replace worn-out equipment and purchase modern instrumentation. Increased private support of undergraduate and graduate research in analytical chemistry could go a long way in helping to replant some of the "seed corn."

People are the principal product of an educational institution. Just as the success of an industrial organization is measured by the profit it makes, so the success of a college or university is measured by the changes it makes in people. This basic distinction in institutional motivation gives rise to most of the unique obligations and opportunities with which a person is confronted as he prepares to join the world of an educator.

What would this mean in your life? Here are the obligations, and the opportunities, in the world of an educator: 1. To continue your education. 2. To communicate with students. 3. To advance knowledge. 4. To gain professional stature. 5. To aid in administration. 6. To seek visibility.

The world of an educator

For those to whom a continuation of the academic life seems attractive, cheerful acceptance of the six-point summary of obligations and opportunities is a necessary requirement, beyond the feeling of being called and proving qualifications.

by James B. Angell, James D. Meindl and Hugh H. Skilling

Reprinted from *IEEE Student Journal,* vol. 6, no. 6, pp. 37–39, November 1968.

Continuing education

The one universal obligation of an educator in a fast-moving technical field such as electrical engineering is that he continue to advance his own education. This requirement is, of course, self-evident to the young person who has just earned his B.S. or M.S. degree and hopes eventually to join a college faculty, where a Ph.D. degree is almost universally required. However, the need for continued education is equally true for Ph.D.'s. It is estimated that the half-life of the usefulness of factual information that is taught in electrical engineering curriculums is now less than six years, and that this time constant is continually decreasing. Thus, any educator must work hard to maintain the timeliness of his own knowledge through a concerted effort to learn through his own work, the work of his colleagues, the work of his students, the work of other educational institutions and the technical output of industry.

Communication with students

The most obvious activity of a college educator is that of communication with his students. However, the actual process of such communication consumes a comparatively small fraction of his working time. He may teach one or more lecture or laboratory classes, with perhaps three or four hours per week per course actually spent in class. He may work with one to five graduate research students, normally spending up to one or two hours per week with each student discussing the research work, reviewing the results of the student's research or planning a new program. An educator may also spend time counseling students on their programs, their future plans or special problems that they may wish to discuss.

These times with students in class, in research discussions and in private consultation, are the high points of professional activity for many educators. Such occasions provide the most direct opportunity for satisfying the fundamental human urge to be of service and help others who need your help.

Advancing knowledge

For most educators, the opportunity to advance the state of knowledge in their profession goes hand in hand with the requirement that they disseminate knowledge. For most young professors, advancing knowledge comes about through participation with colleagues and with students in research work that forms the basis for subsequent technical papers. However, for some educators this activity is more in the form of arranging and distilling the results of the research of others into a clear, concise, communicative curriculum or textbook. Many educators are helped both in disseminating and advancing their knowledge through the process of consulting with industrial or government organizations whose work closely parallels their own technical interests.

The *freedom of action* in a university environment normally exceeds that of almost any other professional climate. This freedom comes about from the fact that very few respected educators or educational administrators would knowingly wish to restrict the intellectual pursuits of their colleagues into any valid inquiry. However, this freedom is a mixed blessing, because it entails additional responsibilities. Each professor has the responsibility to select the research area he will pursue. Most research pursuits in a technical field require some form of financial support—for facilities, for graduate student and staff salaries and for materials. The professor also has the responsibility for seeking this support; he is the best salesman for his ideas. In recent years the majority of this financial support has come from the Federal Government. Now that many colleges and universities are establishing new graduate programs leading to advanced degrees with research dissertations, the competition for the limited funds available to finance this academic research is increasing. Thus, many a researcher finds that the freedom of action granted by an educational institution will be restricted by his ability to sell to a funding agency the desirability of performing such work as he wishes. This hurdle is almost universally an annoyance to the educator, even though it sometimes helps to ensure against irrelevance in his work.

Professional stature

In no job is an individual's professional stature more important than in an educator's. A professor finds that his students, his colleagues, the university administrators that govern his academic future, his financial sponsors and his casual visitors will all be more willing to accept his ideas and programs if it is known that his peers throughout the professional world likewise accept and respect his work. The most clearcut evidence of this professional acceptance is normally acquired through publications in reviewed journals, such as the journals and transactions of the various IEEE technical groups. Textbooks likewise aid in establishing the educator's professional ability. While the "publish or perish" dictum has sometimes been deprecated as having excessive influence on an educator's future, nevertheless in a fast-moving, highly technical field such as electrical engineering there is ordinarily no better indication of a man's technical excellence than that obtained from the judgment of his peers throughout the profession. No small group of administrators or faculty members could be expected to maintain the breadth of background needed to assess an educator's contributions without a good calibration from specialists throughout the entire professional world.

This article describes an educator's world that is characteristic of a research-oriented university. Some of the obligations mentioned here may not be as common in other graduate schools.

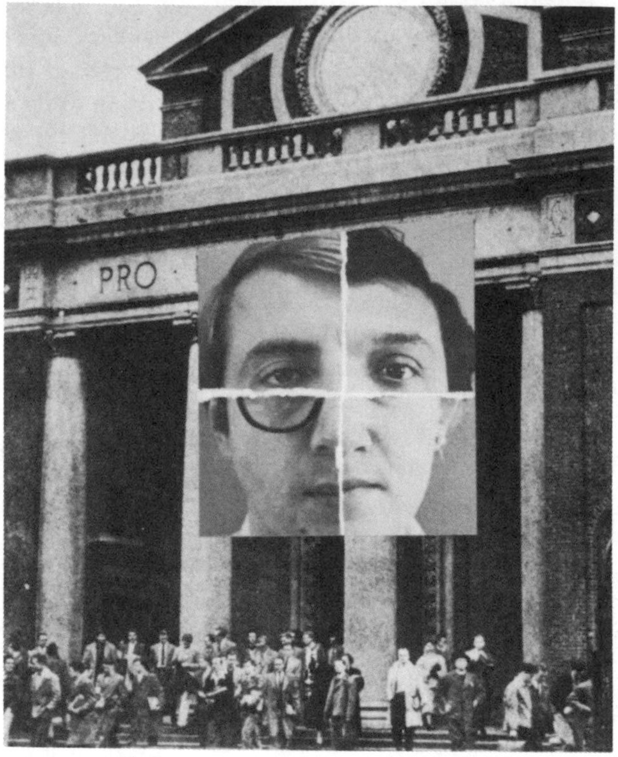

Administration

Normally every educator, from acting instructor through professor, is called upon to provide assistance in certain of the administrative activities of his organization, such as admission of students, administration of examinations, planning curriculums and obtaining equipment. Many such people consider this work a necessary evil—necessary because they do not wish to have someone less qualified then they making decisions that will affect their environment, and evil because it takes them away from tasks they would much rather be doing. To be envied, they would say, is the educator who spends less than 10 percent of his time in such work, and unfortunate is the man who spends more than a quarter of his time in such activities.

Visibility

The young man embarking on a career which would eventually take him into senior educational circles will find it necessary to "display his wares" in a manner that will call attention to his capabilities in competition with his young colleagues in a similar educational environment and in competition with energetic young technical people in industry who may eventually hope to return to an academic environment. University administrators have certain highlights that they constantly look for in their assessment of prospective members of their faculty. They will normally be seeking an exceptionally able man for whom a job will be found after he has been employed, rather than accepting just any man to fill a specific short-term job they have in hand. In order to help to convince these administrators, a young prospect should

a) try to earn a good record in a high quality graduate school program,
b) be willing to devote the required time and effort to see that his work is visible, normally by publication, throughout the professional community,
c) welcome the challenge that he will have the opportunity and the obligation to continue to learn just as rapidly as he did during his student days throughout the remainder of his professional career.

Nancie L. Gonzalez

The Professor as Researcher

In 1969, as a recently hired full professor, I was called upon to help review the credentials of a colleague up for promotion at the University of Iowa. Turning for guidance to that university's faculty handbook, I was intrigued to read that the "bottom line" for professors, so to speak, was that they should have "something to profess." Presumably, as learned individuals, they should have something of value to relate to the scholarly and mundane worlds—or at least to those who care to listen. And the further implication of the statement is that they should do more than merely "teach"—worthy though that activity may be in and of itself. "To profess" is to present one's *own* data, synthesis, or way of viewing the world. "To profess" is to exercise intellectual leadership; "to profess" is to maintain a continuous and independent search for new knowledge and to teach others both what one has learned and how to do the research itself.

Reasonable as this might appear at first, the thoughtful reader may soon turn to the question of how those who wish to become educated are to evaluate whether the "professor" indeed has anything worthwhile to profess. If a student is looking for concrete answers to specific problems, it is relatively simple to evaluate how much a teacher knows and how well he or she gets it across. But teaching alone does not a professor make.

On the other hand, if one is looking for insight, inspiration, a model of clear thinking, ways to deal with the data of experience and thereby increase, enhance, or otherwise control one's experiences for personal or social benefit, then that person would do well to seek a true professor—one who relates and interprets, as well as transfers, knowledge. But it is crucial to choose preceptors carefully, for it is not always easy to distinguish the truly profound from the merely fashionable and/or glib. In this effort, most of us rely upon advice from those with more experience than our own—whether these are students who have already taken classes from a particular professor, or recognized scholars in the same field. This is, in essence, the basis for the credentialing process that we call "tenure and promotion review," which determines which neophytes will be encouraged to stay and which will have to seek other employment. In the process, younger professors are evaluated by their more experienced disciplinary peers, and a decision is reached based upon the candidate's past performance and upon whether he or she is likely to continue to contribute significantly to the college or university community and to the scholarly field in question.

The assumption is, when tenure is granted, that the new associate professor will eventually mature to the full professorial level, at which point would-be consumers (students, purchasers of books, government or private sector contractors) may expect interesting, original, and useful intellectual products. The tenure itself and the title that goes with it (usually associate professor) constitute a certain assurance that the individual has been examined and found not wanting by his peers.

Now, as is well known, tenure and promotion decisions are based primarily on research performance, even though teaching and service (to the institution, to the disciplinary associations, and to various local community organizations and government) are often worked into the profile in various ways. Increasingly, a really poor teaching record is unacceptable, especially for advancement to full professorial status. But in the long run, an objection to tenuring on the basis of poor teaching becomes difficult to uphold if the professor wins acclaim through research. At all levels of university rank, a merely mediocre teaching record is more than compensated by outstanding research; the reverse, although it occurs, is rare. The rationale, which contains more than a grain of truth, is that excellence in teaching is really more difficult to define and assess than is excellence in research. In judging teaching, students tend to be swayed by personal appearance and style, and even the best and brightest may not be able to evaluate substance—after all, they are enrolled in the course presumably because they do not know as much as the professor. In-class visits by the professor's peers, even when socially acceptable, are inadequate because their very presence may influence their colleague's behavior for better or for worse; and although they can

Reprinted with permission from *National Forum,* vol. 67, no. 1, pp. 7–10, Winter 1987.

evaluate substance, their assessments of style and its effects may be highly subjective.

Service is even more difficult to evaluate than is teaching, since no one knows just what it may or should entail. Yeoman service probably never compensates for either poor teaching or poor research when tenure decisions are being made. It is not clear what, if any, relationship there may be between service and either of the other two components. In marginal cases, service *might* tip the scale, but by itself it is not highly valued, nor will its absence necessarily even be noted where teaching and research are satisfactory. Thus, the service element is, for the most part, redundant and more a matter of window-dressing for a critical public than a meaningful category for evaluation. If the service has been exemplary in the public sector, governors, legislators, influential alumni(ae), and trustees may express opinions on a candidate's suitability for advancement, but such endorsements are not necessarily heeded and may even work to the detriment of the candidate in some instances.

"There are a number of ways to document research excellence, and counting the number of publications is only the beginning."

Research, on the other hand, is evaluated by its product, and may even be done anonymously, as in the editorial reviews favored by many major journals. Acceptance of an article in one of the latter is evidence that the ideas and data presented have been taken seriously by the intellectual community. When others begin to cite the publications, one's reputation grows and eventually the researcher is invited to lecture in other settings, to write articles and books, and to conduct collaborative research. There are a number of ways to document research excellence, and counting the number of publications is only the beginning.

But, it might be argued, does anyone outside the academic world itself actually know or care about this kind of credentialing? Does tenure fit a person for anything other than a lifetime in the ivory tower—does it, perhaps, even work otherwise? Do professors, once granted tenure, simply relax and tend their rose gardens, abandoning the search for knowledge that they had seemed so intent upon in their youth? Do students, as well as outside contractors, find that the tenured professor is less interesting, less capable of teaching, less able to formulate questions and answers regarding the real world, than others who may have been turned down for tenure? These charges are often made—unfairly, in my opinion. They derive from mistaken notions of what a university is and, therefore, of what "professing" is, or ought to be. The academy is not like a government bureaucracy or an industrial corporation. And the research enterprise upon which the first depends and thrives is not the same as the information or fact finding upon which an executive in either of the latter kinds of institutions bases decisions.

It is true that practical decision making may indeed require prior fundamental research, and in such cases executives may find it useful to call in as a consultant a professor who has done research on a specific topic or even to ask that new research be undertaken. Sometimes there are even cadres of in-house researchers whose work differs little from that which takes place in the academic world. These are most often found in the biological or physical sciences, as in the famed Office of Naval Research or Bell Laboratories. But one may also find political scientists, sociologists, and historians doing basic research in places such as the CIA or the Library of Congress.

Still, in spite of the fact that much respectable research has been done outside the academy, most scientists and humanists seeking a research-oriented career would probably agree that, other things being equal, they would prefer to be employed in a university. In part, this preference has to do with the collegial atmosphere generated within an academic institution—the sense that one is among peers who understand what one is trying to do and who respect the researcher's need for privacy, balanced by an equally strong need for intellectual exchange. The freedom to pursue topics merely because no one knows the answer—never mind whether anyone else cares at that moment—is also an important reason to remain within the academic world. Basic research usually does not thrive when faced with deadlines, pressure to answer specific questions, or too much publicity.

This is not to say that applied research should never be done by academics. There are always specific questions about the universe, or society, or behavior, that some company or agency or institution needs to have answered quickly, and many academic researchers—particularly in the sciences—spend much of their time responding to, rather than thinking up, research projects. But even applied research thrives best when control over conditions and procedures is with the researcher—and the academic environment ensures that control. In part for this reason many corporations and agencies of the government provide funding to allow professors to conduct part-time research on both specific problems and general themes of potential interest to the contractors.

The presence of graduate students in the universities who learn by doing and later become available for consultation or employment themselves should be seen as another plus, for their presence ensures that the nation's research effort will have continuity over time—a circumstance which might occur less frequently or with less brilliant results if all research were conducted outside the academy.

Although I know of no "hard data" on the subject, there is some anecdotal evidence on what happens to those assistant professors who do not receive tenure in the American system. What we know suggests that there is life outside the academy for the highly educated, knowledgeable, articulate, and personable former assistant professor whose only failing may have been that his/her research was not sufficiently impressive. But the opportunities beyond academe are not likely to be in basic (original) research. It may have been that such individuals, like many lay-persons, misunderstood the nature of research—confusing it with information or data gathering—or the nature of college teaching, thinking it merely a matter of passing on information.

Outside the academy, information gathering, as opposed to original or what is usually called basic research, may be all that is required, and most professors learn to do that while in graduate school, if not before. Indeed, most undergraduate term papers are a matter of finding and reporting on information originally produced by someone else. Success in producing papers on a variety of subjects using secondary data may be the best predictor of one's ability to succeed as a purveyor of knowledge or a facilitator of action in the business or governmental world. In such settings, insisting on doing *original* research before coming up with a suggested solution to an urgent problem may be inimical to effective action on intervention, especially when an eager or agitated client or public is waiting. Of course, there are times when there is really little or nothing in the way of solid information available, and the most effective administrators will know how to identify such situations. They will probably then call in an academically oriented person—a professor—to conduct the research. Although both basic and applied research are done within the academy, the former is usually the primary measure of the liveliness of a professor's mind and the test of how up-to-date one is.

Although it is frequently attempted, it cannot be fairly argued, I believe, that only some institutions of higher learning should demand original, creative research of their professors. I take the stand that all professors—whether they are at private or state universities, liberal arts colleges, even two-year institutions—should be involved in ongoing original research as opposed to mere information gathering on the subjects they teach, keeping abreast of the journals, and rehashing the ideas of others. The quality of the resulting product can only be assessed when the professor writes and publishes results, which can then be read and criticized by peers throughout the world. Different institutions may well maintain different standards of quality for their researchers—not all of us will win Nobels or Pulitzers—and that is a matter for each academic community or its leaders to decide. The size of the teaching loads, the quality of libraries, and the availability of laboratories, equipment, and research assistants clearly affect the quantity and the standards from institution to institution.

But it is the *doing* of the research that is important in defining the special kind of activity we call "professing," or, if one prefers, "professoring." Ironically, we often refer to what we do as "teaching," in spite of both the ideal and the practice being to the contrary. "What do you do?" "I teach at a university." No wonder the public thinks we do not earn our keep—especially when they learn that we "teach" only three, six, eight, perhaps twelve hours a week! Even the eighteen hours expected at junior or community colleges is thought to be a soft job by those who do not understand the system. We unconsciously contribute to the misunderstanding when we refer to ourselves as teachers, knowing full well that good performance in the classroom demands research—and I do not mean just fact finding, although a certain amount of that is inevitable, as well.

It always astonishes me when I encounter the argument, sometimes even among academics themselves, that research is overemphasized in tenure decisions. It is pointed out that one need not generate new knowledge in order to understand and teach what others have done. That is patently true, and most of us must teach some courses in subjects on which we have not conducted research. And for these, we must depend upon our fellows' research and writing for the information we convey. Indeed, the professors who teach only one or two courses per year are few and far between, and they tend to be those who have reached the apex of the research ladder. The more and better the research a professor does, the fewer courses she teaches, but the workload may well be greater and more intense, because most of the students will be graduates. But research-oriented professors are important at the lower academic levels as well.

To my mind, however, research is and must remain paramount for professors for several reasons. Quite apart from the value of the final product of research—whether it be a new scientific law or mathematical formula, the description of a genetic component or a previously unknown pattern of animal behavior, improved understanding of the sociopolitical situation in a less-developed country, or a work of art or philosophical treatise—is the effect of the research process itself. The active researcher is simply a different kind of human being from the one who only teaches. Not only does research keep one abreast of new developments and new performers in one's immediate subject-matter, but it hones and refines the critical faculties and thus makes one better able to judge the quality of related work, even in areas of nonspecialization. There is no substitute in the classroom for firsthand professorial commentary on books read, ideas presented by other researchers, and views expressed by students. For a teacher merely to parrot or paraphrase someone else's work, even when several views are presented, is less interesting and less stimulating than if she speaks from personal analysis of the materials themselves. And the habit of making such analyses is part and parcel of the research process.

The teaching effort is also improved when the professor is able to use examples spontaneously to illustrate points or to assist in answering questions or criticisms from the floor. Those who carry on active research have a wealth of anecdotal material upon which they may draw, thus adding substantive information to the discussion, as well as letting students have a glimpse of what their professor "really" does for a living. Advanced students, and sometimes even those at a lower level, may respond in ways that assist the research, as well. Most professors acknowledge that the feedback generated by such teaching is invaluable. It also is the stuff of which dreams and future careers are created. Role models must project something of the excitement of whatever it is they do, and the professor who teaches what he or she researches is more likely to be emulated.

There are still other ways in which ongoing research makes for a better teacher. The discipline and rigor required to keep at the problem or the creative effort, the disappointment and frustration when things do not fall into place as and when expected, the excitement when they do, the sense of being on the frontiers of knowledge, the prestige garnered when publications appear or when prizes are won—all these, when experienced by their teachers, contribute to the education of college students, at whatever level and in whatever kind of institution. Not all of one's professors can be expected to produce great and lasting works, but so long as they produce *something*, their students will understand a bit more of what it means to be creative. In this sense, the creative process itself generates creativity when others are permitted to watch, to eavesdrop, perhaps actually to participate. Cognitive understanding of what is taking place is ideal, but as I have suggested, even without that, there are benefits to the student in having a professor who is committed to and engaged in research.

To the extent that colleges and universities accept research as part of their institution's mission, they should provide a hospitable environment for it. This includes, at a minimum, space for laboratories, workrooms, library and computer facilities, support staff, time, relative peace and quiet, and recognition of achievement. Equipment and supplies, assistance in preparing manuscripts and graphic displays for publication, and sometimes even funds for publication costs themselves are often provided by the more affluent and committed institutions. Finally, providing travel funds for professors to attend conferences and other professional meetings, whether they make formal presentations or merely keep abreast of what is happening through informal discussion, is essential to the continuing pursuit of knowledge. Researchers need to discuss their ideas and problems with others in order to avoid unnecessary duplication, including pitfalls which others may have experienced. At the same time, discussions among peers may trigger creative thrusts in completely new directions.

However, there is a hierarchy within the academy, ranging from institutions that seemingly value research more than the teaching function, to those that provide no research support to their professors at all. Indeed, sometimes professors must manage all of the above facilities, including time, as best they can. Obviously, to do so is difficult, and many are unable to cope. An elitist stance is that the system sorts and culls efficiently and that only the least talented drop by the wayside when they are forced to accept an academic position in an institution with less than adequate research support. Unfortunately, there is no way to know what such individuals might have produced had they managed to land different jobs, and the fact that there are too few academic positions in the highest-quality institutions for all our talented graduates means that some untold number will likely not achieve their potential. Thus, it is even more important than ever to make tenure decisions carefully and wisely.

But my point here is that even when research itself is not conceived as part of the institution's mission, it should still be considered important for the teaching function. It is sometimes difficult to convince legislators of the value of research to the state (even though special interest sectors, such as agriculture, may be perfectly well aware of the potential benefits), and it may be nearly impossible to make them understand that all institutions of higher education should seek professors with some interest and competence in research. But some consciousness-raising is in order here. Legislators should be reminded of the more "practical" benefits. Demanding that a professor have something "to profess" and providing a suitable milieu for the continuing *creation*, as well as the transmission, of knowledge will provide a better learning environment for students and will help reduce professorial dissatisfaction, burnout, and turnover.

In the best of all possible worlds, everyone would value the pursuit of knowledge for its own sake as well as for its social benefits. There have been relatively few societies in which research has been institutionalized at all and still fewer where it has been linked with the educational enterprise and supported by both governmental and private agencies. We enjoy such a system in the United States today. It may not be perfect, but it has produced a society that can boast artistic, philosophical, and technological breakthroughs that are envied world-wide. But there is no guarantee that things will continue to operate as they have operated. Those of us who understand the delicate balance we have achieved must all work together to preserve the peculiarly effective combination of professional activities to be found in the American professoriate today. Let us continue to insist that a professor must have something "to profess." NF

NANCIE L. GONZALEZ is professor of anthropology at the University of Maryland, College Park.

Service: The Neglected Person of the Academic Trinity

Walter V. Hohenstein

The "theology" of the present American educational system asserts that public universities perform three functions: teaching, research, and service. Verbal obeisance to this trinity is rendered by almost all elements of the university community—but in reality few give equal honor to the three, supposedly equal persons.

Although teaching is first listed, it ranks far behind research as an element of reverence. Lagging even farther is service. Service is a recognized obligation of a few specific units connected with some universities. The Cooperative Extension Service is probably the best example. PENNTAP in Pennsylvania and the Technological Extension Service in Maryland are additional examples. These, however, represent functions far from the heart of the university systems.

Why "theology" and "reality" differ so substantially is of vital significance to all universities and their faculties, as the service component will be an important factor in institutional survival in the "dark days" ahead. Faculty dedicated to performing service functions must also know whether their efforts will ever be rewarded.

There are six factors that can be advanced as reasons why service does not hold a place of equality with the other university functions.

1. LACK OF TRADITION. Public service is a uniquely American graft onto the tradition of higher education. The research function developed, and was almost deified, in the German university system. The teaching function, of course, has existed from the very beginning. It should be no surprise, then, that a relative newcomer must struggle for a role in so complex a social institution as the university.

2. NOT STRESSED IN TRAINING. The service function receives little attention in the training of our fledgling faculty members. Research, on the other hand, is the heart of the Ph.D. program. Most doctoral candidates also receive some training in teaching as a by-product of their efforts at economic survival through holding graduate teaching assistantships. The few fortunate may actually receive some teacher training through more formal programs—such as the Ford Foundation Teaching Internships in which I was privileged to participate.

   In contrast, almost no formal or informal plans have been introduced to prepare faculty to perform public service functions. A few institutions have major programs to prepare students to participate in one of the few university units dedicated to service—programs in agricultural and extension education being excellent examples. A few students on their own have taken leave from their studies to participate in Vista or Peace Corps projects. Such formal and informal activities, however, remain the exception.

3. COMPLEXITIES AND BREADTH OF THE SERVICE FUNCTION. Service is so complex a topic that it can actually stagger the mind. The basic bifurcation separates those activities performed for the institution (internal service) from those performed for all other entities (public service).

   Internal services could be performed for departments, colleges, divisions, or the entire institution. They could also be undertaken to support officials within any of these units.

   Public services could pertain to businesses, cities, counties, states, nations, and even international entities. The persons supported could range from an Appalachian mother needing information concerning nutrition to the governor of the state needing health statistics on an area underserved by present facilities.

   What is done as service also varies widely—from taxation studies for legislative committees to advising a student; from writing the preamble to the UNESCO Charter to analyzing the causes of death of ten chickens in a distant village.

4. HARD TO MEASURE AND IMPOSSIBLE TO COMPARE. Research and publication can be measured in simplistic ways. Pages in a published article may be counted—or the number of articles published in a year may be calculated—or even the number of times one's articles are cited may be tallied.

   There is no simple way, however, to measure service. How does one quantify advising a farmer on the best plowing methods—or serving on a faculty senate committee on curriculum development—or reorganizing the record keeping system for a small city?

   Even if one could quantify each, how, in a period of limited resources, do we compare one to

WALTER V. HOHENSTEIN is Director of Articulation for the Central Administration, University of Maryland. He presently serves as Regent for the National Honor Society of Phi Kappa Phi.

Reprinted with permission from *National Forum,* vol. 60, no. 2, pp. 18–19, Spring 1980.

the other in order to decide equitably to whom scarce faculty raises should be given?

5. PRESENT PEER SYSTEM OF EVALUATION. Faculty evaluation—at least that relating to promotions—is initially carried on by peers. Those peers who have gained their positions through research appear to have little willingness to consider service as a factor of importance. Moreover, because of the wide diversity in service functions even within one department, many faculty just don't understand the service activities undertaken by their colleagues.

   This problem is compounded by college and university faculty review committees. Faculties in political science have little understanding of the public service activities of those in chemistry, as faculty in physics would have similar problems evaluating service by their colleagues in physical education.

   The critical peer review is in the department and, at this level, faculty often tend to understand little concerning the overall institutional goals. Their judgments are more often in terms of departmental prestige or status within subject-area professional organizations—both determined predominantly by quality and quantity of research and publication.

   The addition of students to these peer review committees in only a few ways may strengthen the role of service—and those would relate directly to student services. Students otherwise tend to use personal criteria, and not those pertaining to the broad institutional needs for service, in evaluating faculty.

6. CONFUSION BETWEEN TIME SPENT IN SERVICE AND THE RESULTS OF THAT SERVICE. Some who emphasize the importance of their service activities stress only the time they've devoted, not the results. They are encouraged in this by the trend toward time reports that are being foisted upon many public institutions.

   In an academic community, little value is attributed to the length of time it takes to accomplish a task. The significant factor is the positive achievement itself.

   Those advocating the need to recognize service are in a sound position only if they stress the results of service, not just the simple measure of the time devoted to it.

At the same time that little recognition is given to the public service function, it appears possible that service may be the key to survival of some institutions. Service can, if used correctly, stimulate essential public and political support that could mean the difference between continuation and disaster.

The service role is also vital in providing the faculty with a reality test for their ideas. A commentator wrote that universities should not be places where one intellect feeds incestuously on other intellects. Faculty need to try out their ideas in the outside world, and service provides an ideal vehicle.

Service also provides desirable faculty contacts with their counterparts in industry, government and outside research facilities. If faculty lose such contacts, they tend not to be abreast of the latest developments, not to be in a position to best prepare their students, and not to be so successful in placing their graduate students.

Universities absolutely could not exist without the internal service provided by faculty. Members of search committees, senate chairpersons, and directors of departmental advising are examples of the very basic machinery by which a university operates. Persons holding these responsibilities, of course, will not have as much free time to carry on their normal research efforts.

The recognition of service is also most vital to organizations such as The Honor Society of Phi Kappa Phi. If a university gives no recognition for work in perpetuating and advancing these activities, such groups will inevitably wither on the vine.

Since the traditional procedures do not reward faculty for their vital service activities, the proposal of alternate procedures appears not only appropriate, but essential. A possible approach if an institution actually wants to reward service would be the establishment of a Plan for Annual Results (PAR). Important decisions would first have to be made by the institution and the faculty:

1. Of the multitude of results of service, which is the institution willing to reward?
2. How are the results to be measured?
3. What efforts are needed to apply the standards as uniformly as possible?
4. Who will measure the results?

Within the framework of these decisions, each faculty member would meet once a year with the department head to develop what might be thought of as that individual's annual service plan. (Since I'd like to focus on results of that service, I refer to it as a "results" plan.)

"Universities absolutely could not exist without the internal service provided by faculty."

This plan would state specifically what the results of the faculty member's service are expected to be for the next year and what the response of the institution would be for such successful results. This plan would then be reviewed by the departmental faculty and the dean. Some amendments may have to be made to this plan during the year in order to respond to unanticipated requests for service, but generally one would expect minimal changes. A year later the actual results would be reviewed and decisions made concerning institutional responses to such results in terms of rank and salary.

Without some change, service will continue to be the neglected person of the academic trinity. Until PAR or a similar plan is initiated, it will continue to be difficult to encourage faculty service activities with any hope of reward in this world. ■

Promotion and Tenure Policies In U.S. Engineering Schools

Echol E. Cook, *Professor,* Curtis Dodd, *Associate Professor, and* Sedat Sami, *Professor, School of Engineering and Technology, Southern Illinois University, Carbondale*

The members of committees charged with developing promotion and tenure guidelines for engineering departments at their schools must often wonder how such issues are handled at other institutions. At Southern Illinois University, our committee decided to answer this question before formulating our guidelines.

A questionnaire was designed and sent to SIU's engineering faculty for comments and additions. It was then mailed to the engineering deans of all public institutions in the United States.[1] The data yielded by the questionnaire is presented here.

The Population

The schools selected to receive the questionnaire were separated into four groups:

1) Engineering schools having a Ph.D. program and more than $2 million of research funding.
2) Engineering schools having a Ph.D. program and between $500,000 and $2 million in research funding.
3) Engineering schools having a Ph.D. program and less than $500,000 in research funding.
4) Engineering schools having a master's program and less than $500,000 in research funding.

These categories were established to determine whether there are any differences in policy between schools having Ph.D. programs and a large amount of research funding and schools having a master's program and little research funding. Little or no difference in policy was found, however, among these groups.

Of 73 questionnaires mailed, 37 were returned. Since not all questions were answered on the returned questionnaires, there are fewer than 37 responses given to some questions.

Results & Discussion

Following are the questions on tenure and promotion policy and the responses received. Also included are some pertinent statements from the engineering deans. The number of replies received appears in parenthesis after each question.

1) Does your college have guidelines for promotion and tenure, and were they established by the engineering faculty? (n = 37)

Most respondees (89%) indicated they did have guidelines; 36 percent indicated they were established by the engineering faculty.

2) Are tenure and promotion based on faculty assignment or are all faculty expected to meet the same criteria? (n = 36)

Of those responding, 36 percent said promotion and tenure was based on each individual's assignment, whereas 64 percent indicated that all faculty members are expected to meet uniform criteria regardless of an individual's assignment.

One dean commented that "a high minimum standard is required for all faculty members; however, consideration of each candidate certainly takes into account the relation between the assignment and performance of a given faculty member."

3) In order to receive tenure or be promoted, must an individual's performance be outstanding in all three of the areas of teaching, research and service? (n = 36)

Of those responding, 18 percent indicated that to receive tenure an individual must excel in all three areas, 17 percent said these standards apply for promotion from assistant to associate professor and 28 percent said excellence in all three areas was required for promotion from associate professor to professor.

Two deans commented that excellence must be exhibited in two of the three areas. A third dean indicated that promotion should be based on merit and should recognize a sustained period of distinctive performance and achievement in one or more of the areas. Another commented, "I know of no group in the country which can insist upon outstanding performance in teaching, research *and* service. Of necessity the standard of performance is relative to the local inhabitants beyond the minimum established."

4) For tenure or promotion, how would you weigh the relative importance of performance in the areas of teaching, research and service? (n = 26)

Reprinted with permission from *Engineering Education,* vol. 69, no. 7, pp. 750–751, April 1979.

This question was designed so that the respondent could assign different percentage weights to each of the areas, depending on whether the individual was being considered for tenure or promotion to associate professor or professor. The percentage weights assigned to the areas of teaching, research and service, however, did not vary appreciably between the categories of tenure or promotion. The weighted relative importance of teaching, research and service was respectively 47 percent, 39 percent and 14 percent.

Some of the comments: "If a faculty member were truly outstanding in any one area we would probably promote him." "Cannot answer in a meaningful way—so much of what we do is subjective." "We have a high minimum standard for all areas and then expect (by our standards) excellence in either teaching or research and a very good performance in the other (research or teaching). No weighting factors are used or implied. Clear evidence of exceptional quality is required along with continued promise for the future. A set of high marks in all three areas without excellence in one is not good enough."

5) In evaluating teaching, how much significance is attached to each of the following: (n = 28):

- *a. Student evaluation in classes*
- *b. Alumni evaluation*
- *c. Peer evaluation*
- *d. Other*

Student evaluations in class comprised 40 percent of the total evaluation, alumni evaluation 10 percent, peer evaluation 30 percent and "other" 20 percent.

For "other," four deans indicated that the department chairman's or the dean's evaluation should constitute from 25 to 50 percent of the teacher's evaluation, and two said that some personal evaluation is appropriate.

One dean commented, "Student evaluation of teaching is a required component of promotion papers. Alumni and peer evaluation is taken seriously when available. Comparisons are made with other instructors in the same courses, courses at the same level, and in all courses. Curriculum and course development also enter into consideration, as does the quality of thesis guidance and undergraduate advising."

6) In evaluating teaching performance (for the purpose of granting tenure or promotion) do you utilize a teaching rating formula or an overall minimum rating (e.g., 3.0/4, 3.5/4 or any other ratio? (n = 36)

Only 14 percent of those responding answered yes to this question.

7) In evaluating "research productivity," do you require a minimum number of publications for tenure or promotion? (n = 36)

Only two of those responding (6%) indicated any requirement for some minimum number of publications to receive tenure or promotion to any rank.

Various comments were: "We try to evaluate each candidate's total performance. While some bean counting is inevitable, too much of it is deadly and actually counterproductive." "No numerical value is assigned to refereed publications, since these are expected to vary from field to field, and it is recognized that experimentalists are likely to have fewer publications than theorists." "Quality, not quantity, governs the assessment of research and publication. Refereed journals are the key to outside evaluation for junior people, invited papers and presentations are measures of peer admiration for more senior people. Funded research also is a measure of outside approval, provided the research is of suitable quality. The difficulty of funding in some areas must, of course, be taken into account. We have promoted people with as few as two papers (of superb quality) and turned down others with dozens in refereed journals (but of mediocre quality)."

8) What is the relative weight of the listed scholarly activities in relation to a refereed journal article? (Example: 1 textbook = 2 journal articles)

- *a. Textbook (undergraduate) (n = 10)*
- *b. Research-related book (graduate level) (n = 10)*
- *c. Funded research project (n = 13)*
- *d. Patent (n = 8)*

On the question of books, the general consensus was that both a graduate and an undergraduate textbook should count as the equivalent of 2.7 refereed journal articles. As to the funded research project, those responding to this question felt a research project was equivalent in weight to one refereed journal article. Those responding to the patent thought 1.2 refereed journal articles were equivalent to one patent.

One pertinent comment was offered on this question: "Trying to relate texts to papers or patents to papers will involve you in endless useless debate."

9) Is there a minimum time requirement between ranks for promotion? (n = 36)

Of those responding, 70 percent indicated there was a minimum time requirement for promotion from assistant to associate professor, and the time averaged 4.35 years. For promotion to professor, 68 percent indicated there was a minimum time requirement, which averaged five years.

10) Is registration as a professional engineer required for tenure or promotion? (n = 37)

Only 10 percent indicated that registration was required for tenure; 12 percent indicated it was required for promotion.

A few of the deans mentioned that registration is certainly encouraged and is considered during promotion and tenure decisions.

11) In your school (college), what is considered the average teaching load per semester for each professor? Example: If a professor were teaching 9 hours consisting of one 3-semester-hour graduate course and two sections of a 3-semester-hour undergraduate course, he would be teaching one graduate course, one undergraduate course and have 2 preparations.

For semester hours taught, 35 deans responded, indicating an average assignment of 9.3 semester-hours. On the number of graduate and undergraduate courses, 23 deans indicated an average of 1.4 and 1.9 courses respectively per semester. As to the number of class preparations, 15 deans indicated that the number of class (course) preparations was 2.9.

Conclusion

It was the purpose of this study to provide information for those facing the task of establish promotion and tenure guidelines in their schools. For those already beyond that hurdle, it may validate existing guidelines or provide ideas for making them more relevant. It is hoped that publication of these selected policies will prompt and encourage others to offer their views and present additional data which might prove most valuable in the formulation or modification of promotion and tenure policies in engineering schools.

Acknowledgements

The authors are thankful to all the deans and associate deans for their participation in this study. The response was heartening, and we are most appreciative of their willingness to share some personal thoughts and comments to help those of us in need.

References

1. As listed in *Engineering Education,* vol. 68, no. 6., March 1977 (Research & Graduate Study issue).

Epilogue

As might be expected in any finite-sized volume on an open-ended subject, there are many other topics related to teaching that are of potential interest to beginning teachers of engineering, and that could not be accommodated in this volume. A small number of these topics, frequently arising in discussions on pedagogy, are mentioned here, along with some suggested readings where more information on them may be found.

A. Individual Instruction

What are some of the ways in which a teacher can use an individualized system of instruction? What are the advantages and shortcomings of such systems? Why aren't such systems more commonplace?

[1] J. G. Sherman, Ed., *Personalized System of Instruction: 41 Seminal Papers.* W. A. Benjamin, Menlo Park, CA, 1974. A collection of papers on the well-known Keller Plan.

[2] S. N. Postlethwait, J. Novak, and H. T. Murray, Jr., *The Audio-Tutorial Approach to Learning,* 3rd ed. Burgess Publishing Co., Minneapolis, MN, 1972. Describes a system based on audio tapes, visual aids, and modular course material, that can be used for self-paced instruction.

B. Educational Technology

How might the various technological aids be used for increasing the efficiency or effectiveness of instruction? What has been learned from the past efforts at introducing educational technology in the classroom? What developments might be expected in the future?

[1] National Academy of Engineering, *Educational Technology in Engineering.* National Academy Press, Washington, DC, 1984. An assessment of the role of educational technology in engineering education as viewed by a committee of this august body.

[2] J. Canelos and B. W. Carney, "How Computer-Based Instruction Affects Learning," *Engineering Education,* vol. 76, no. 5, pp. 298–301, Feb. 1986. A status report of the current practice in computer-assisted and computer-managed instruction at college level, based on a literature search.

C. Teaching Problem Solving and Creativity

How can an instructor increase the problem solving skills of his students? What is creativity, and how does one teach for enhancing the creative potential of the students?

[1] H. A. Simon, "Studying Human Intelligence by Creating Artificial Intelligence," *American Scientist,* vol. 69, no. 3, pp. 300–309, May-June 1981. A survey of the known results on human problem solving, and the distinction between problem solving strategies of novices and experts.

[2] N. Whitman, "Teaching Problem-Solving and Creativity in College Courses," *Bulletin of the American Assoc. for Higher Education,* vol. 35, no. 6, pp. 9–13, Feb. 1983. A literature survey of the numerous attempts at teaching creativity in various fields at the college level.

[3] B. Skinner, "The Myth of Teaching for Critical Thinking," *The Clearing House,* vol. 45, no. 6, pp. 372–376, Feb. 1971. Summarizes the definitions of critical thinking, the characteristics of the problems that are suitable as a vehicle for teaching critical thinking, and the process and evaluation necessary for success in that endeavor.

D. Teaching Effectiveness

What makes a teacher effective? Are carefully prepared lectures sufficient to increase the learning of the students? What human factors limit learning after the classroom presentations have been improved to the extent where they are no longer the principal bottleneck to learning?

[1] R. G. Kraft, "Bike Riding and the Art of Learning," *Change,* vol. 10, no. 6, pp. 36, 40–42, June-July 1978. Reminder that teaching cannot be effective unless it actively involves the learner and meets his goals.

[2] P. R. Halmos, E. E. Moise, and G. Piranian, "The Problem of Learning to Teach," *American Mathematical Monthly,* vol. 85, no. 2, pp. 466–476, May 1975. Addresses by three highly regarded teachers of mathematics on the importance of the active involvement of students, the students' belief that the instructor is also actively involved, the interaction among the students, and the use and pacing of homework.

E. Great Teachers

What makes former students remember a professor decades later when nearly everything else learned in the college has been forgotten? Will strict adherence to the advice appearing in this volume and elsewhere on syllabus design, lecturing, and test construction make a teacher a superstar? (Answer: No, but it can prevent the teacher from being a poor one). What distinguishes a truly outstanding teacher from one who is merely good?

[1] G. H. Flammer and P. M. Flammer, "The Master Teacher," *Engineering Education,* vol. 71, no. 5, pp. 359–362, Feb. 1981. Lists characteristics of master teachers, and the challenges they face.

[2] J. Epstein, Ed., *Masters : Portraits of Great Teachers.* Basic Books, New York, NY, 1981. Reprints memoirs on distinguished teachers, written by their former students.

Appendix A
Further Search In Educational Literature

THE annotated bibliographies at the end of each part of this volume have been kept short on purpose. There is a law of diminishing returns governing the length of bibliographies: it states that the longer a bibliography, the smaller the likelihood that it will actually be used. The literature on engineering education, and on higher education in general, is extensive, and this Appendix is intended to suggest some of the sources that beginning teachers of engineering may find useful.

Those who have a specific information need should perhaps begin by seeking the assistance of a reference librarian. The following brief listing is meant only to assist independent searchers in getting started.

A. Journals

Those readers who like to browse may occasionally find an interesting article in some of the journals listed below; these include journals in higher education, journals in science and engineering teaching, and some professional research journals.

Academe (formerly *Bulletin of the Amer. Assoc. Univ. Profs.*)
American Journal of Physics
Canadian Journal of Higher Education
Change
Chemical Engineering Education
The Chronicle of Higher Education
Civil Engineering Education
College Student Journal
College Teaching (formerly *Improving College and University Teaching*)
Engineering Education
ERM (Education Research and Methods; ERM is a division of ASEE)
European Journal of Engineering Education
Higher Education
Higher Education Review
IEEE Transactions on Education
Instructional Science
International Journal of Electrical Engineering Education
Journal of Chemical Education
Journal of College Science Teaching
Journal of Educational Psychology
Journal of Educational Measurement
Journal of Educational Research
Journal of Higher Education
Journal of Professional Issues in Engineering (Amer. Soc. Civil Engrs.)
National Forum (Phi Kappa Phi Journal)
New Directions in Higher Education
Phi Delta Kappan
Physics Education
The Physics Teacher
Planning for Higher Education
Research in Higher Education
Review of Educational Research
Review of Higher Education
Teaching of Psychology.

B. Abstracts

The principal abstracting and indexing journals for educational literature in the English language are the following:

1. Current Index to Journals in Education (from the Educational Resources Information Center, or ERIC).
2. Resources in Education (devoted to non-serial items such as reports; also from ERIC)
3. The Education Index (from H. W. Wilson Co.)

C. Conferences

The two major annual conferences in the field of engineering education are:

1. Frontiers in Education Conference (held in the Fall; sponsored by ASEE and IEEE; proceedings available from IEEE, New York, NY).
2. Annual Conference of the American Society for Engineering Education (held in summer; sponsored by ASEE; proceedings available from ASEE, Washington, DC).

D. Data Compilations

There are several sources of data and information concerning the engineering colleges and programs in the United States. These include the following:

1. Annual (March) issue of *Engineering Education,* on Engineering College Research and Graduate Study.
2. Annual and occasional reports from the National Science Foundation on degrees awarded and utilization of scientific and technical manpower.
3. *Peterson's Annual Guide to Graduate Study: Engineering and Applied Science,* and *Peterson's Guide to Undergraduate Engineering* (Peterson's Guides, Princeton, NJ).
4. Annual Report of the Accreditation Board for Engineering and Technology (ABET, New York, NY).

E. Books

There are a number of other books for the guidance of college teachers. A small sample of these is listed below:

1. T. H. Buxton and K.W. Pritchard, Eds., *Excellence in*

University Teaching. University of South Carolina Press, Columbia, SC, 1975.

2. S. M. Cahn, Ed., *Scholars Who Teach: The Art of College Teaching*. Nelson-Hall, Chicago, 1978.

3. B. Cronkhite, *A Handbook for College Teachers, an Informal Guide*. Harvard University Press, Cambridge, MA, 1950.

4. K. E. Ebel, *The Craft of Teaching: A Guide to Mastering the Professor's Art*. Jossey-Bass, San Francisco, CA, 1976.

5. H. A. Estrin, Ed., *Higher Education in Engineering and Science*. McGraw-Hill, New York, 1963.

6. H. A. Estrin and D. M. Goode, *College and University Teaching*. W. C. Brown Co., Dubuque, IA, 1964.

7. B. S. Fuhrmann and A. F. Grasha, *A Practical Handbook for College Teachers*. Little, Brown, and Co., Boston, MA, 1983.

8. N. L. Gage, *The Scientific Basis of the Art of Teaching*. Teachers College Press, Columbia University, New York, NY, 1978.

9. M. M. Gullette, Ed., *The Art and Craft of Teaching*. Harvard Danforth Center for Teaching and Learning, Cambridge, MA, 1982.

10. The Harvard Educational Review, *Breakthrough to Better Teaching*. Harvard University Graduate School of Education, Cambridge, MA, 1966.

11. J. Justman and W. H. Mais, *College Teaching: Its Practice and Its Potential*. Harper, New York, 1956.

12. R. D. Mann, *et al, The College Classroom*. John Wiley, New York, NY, 1970.

13. W. J. McKeachie, *Teaching Tips: A Guidebook for the Beginning College Teacher,* 7th ed. D.C. Heath, Lexington, MA, 1978.

14. H. R. Mills, *Teaching and Training. A Handbook for Instructors,* 2nd Ed. MacMillan, London, 1972.

15. F. C. Morris, *Effective Teaching. A Manual for Engineering Instructors*. Sponsored by ASEE Committee on Teaching Manual, McGraw-Hill, New York, 1950.

16. H. H. Skilling, *Do You Teach? Views On College Teaching*. Holt, Rinehart, and Winston, New York, 1969.

Author Index

Subject Index

U

V

W

Editor's Biography

Madhu Sudan Gupta (S'68–M'72–SM'78) was born in Lucknow, India, on June 13, 1945. He received the B.Sc. degree in physics from Lucknow University, Lucknow, India, in 1963, and the first degree in electronic and radio engineering from Allahabad University, Allahabad, India, in 1966. Thereafter, he continued his graduate study in the United States, receiving the M.S. degree in Engineering Science from Florida State University, Tallahassee, FL, in 1967, the M.A. degree in Applied Mathematics from the University of Michigan, Ann Arbor, MI, in 1968, and the Ph.D. degree in Electrical Engineering, also from the University of Michigan, in 1972.

During his graduate study, he held the Department of Atomic Energy Fellowship at Allahabad University, a Teaching Fellowship at Florida State University, and a Teaching Fellowship and a Research Assistantship at the University of Michigan. During the years 1966–68, he also worked in the areas of ionospheric sounding, noise measurements, radar antenna analysis, and system analysis and software development for an information retrieval system. His doctoral research in the area of microwave semiconductor devices was carried out from 1968 to 1972 at the Electron Physics Laboratory of the Department of Electrical and Computer Engineering at the University of Michigan, and was on the noise in avalanche transit-time microwave diodes and oscillators. He was awarded an outstanding teaching fellow award at the University of Michigan in 1970.

From 1972 to 1973, he served as an Assistant Professor of Electrical Engineering at Queen's University, Kingston, Ontario, Canada. From 1973 to 1979, he was at Massachusetts Institute of Technology, Cambridge, MA, first as an Assistant Professor, and later as an Associate Professor of Electrical Engineering and Computer Science. As a member of the Research Laboratory of Electronics at M.I.T., he carried out research on microwave semiconductor devices and their characteristics, materials, circuits, and noise, and on thermal fluctuations in nonlinear systems. During 1974–75, he was awarded a Lilly Fellowship in the Division for the Study and Research in Education at M.I.T., in a program of teaching improvement.

Since 1979, he has been at the University of Illinois at Chicago, initially as an Associate Professor, and subsequently as Professor, in the Department of Electrical Engineering and Computer Science. His current research interests are in the area of electron devices that are active, nonlinear, noisy, or very small, and particularly in the thermodynamic limitations and fluctuation phenomena in semiconductor devices. He also has significant interest in the history of electronics (particularly the origin of early solid-state devices and of electrical theorems and laws), in the literature of electrical engineering (specifically its classification, retrieval, and effectiveness), and in engineering education (specifically in the evaluation of students, teachers, curricula, and new innovative programs, and in the preparation of new engineering educators).

Dr. Gupta is a Senior Member of IEEE, having also served as the Chairman of the IEEE Microwave Theory and Techniques Society, Boston Chapter (1977–78), and of the Chicago Chapter of the same society (1986–87) and was the editor of *Electrical Noise: Fundamentals and Sources* (IEEE PRESS, 1977). He has also been a member of Eta Kappa Nu, Sigma Xi, Phi Kappa Phi, and the American Association for the Advancement of Science, and is a Professional Engineer, registered in the Province of Ontario, Canada. He is listed in several reference works, including American Men and Women of Science, Who's Who in the Midwest, Who's Who in Technology Today, and International Who's Who in Engineering.